The Third Planet

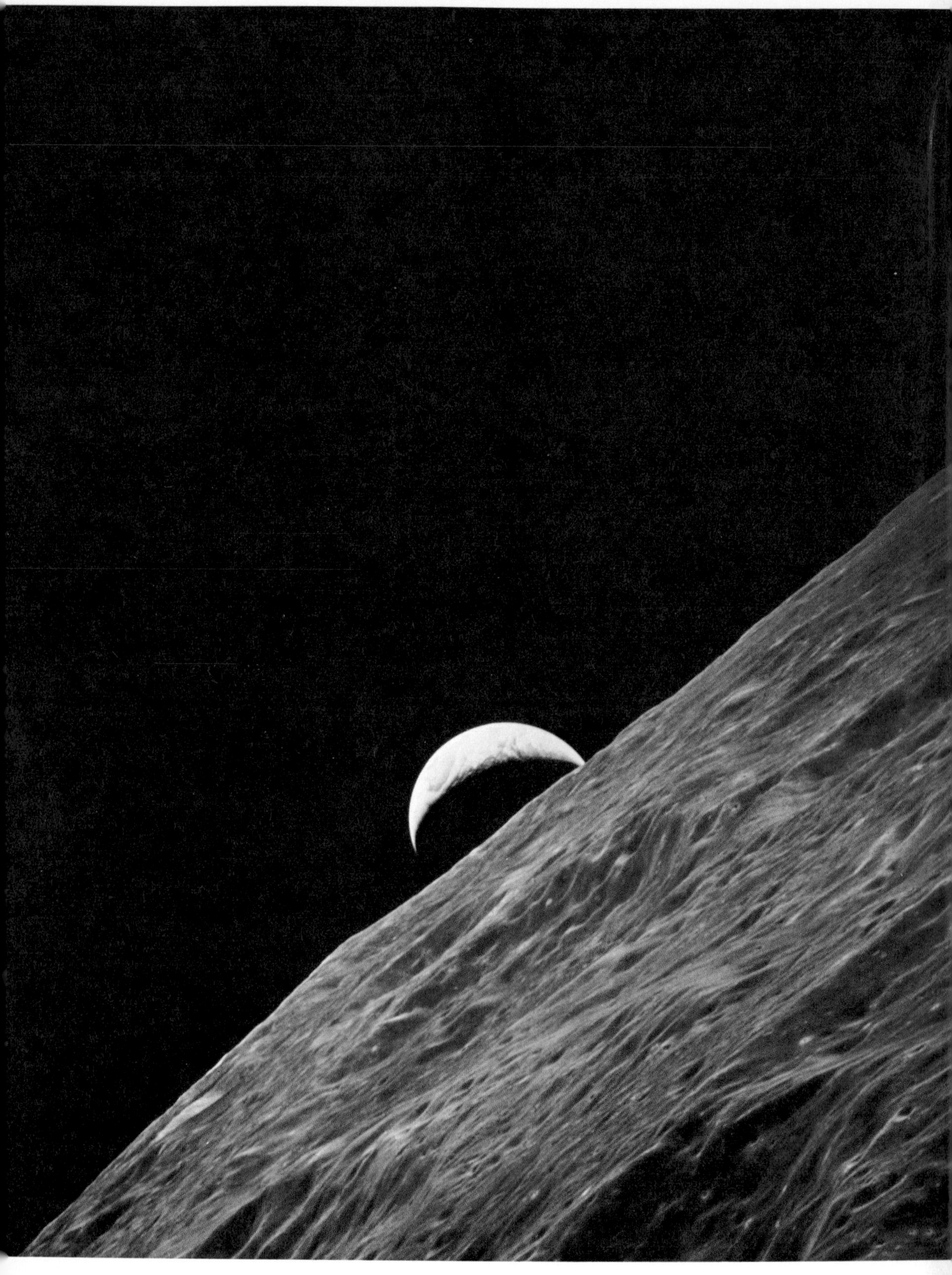

The Third Planet

AN INVITATION TO GEOLOGY

Konrad B. Krauskopf

STANFORD UNIVERSITY

FREEMAN, COOPER & COMPANY

1736 Stockton Street, San Francisco, California 94133

Cover Photograph: David Muench

Front End Paper: Map of earthquake belts from National Earthquake Information Center, U.S. Geological Survey
Map of world's volcanic belts from the Seismographic Station, Department of Geology and Geophysics, University of California, Berkeley

Rear End Paper: Maps of drifting continents from the paper by J.D. Phillips and D. Forsyth, published in the *Bulletin* of the Geological Society of America, June 1972

Title Page Photograph: *Earthrise* from NASA

Printed in the United States of America
Library of Congress Catalogue Card Number 74-77823
ISBN 0-87735-359-X
0-87735-360-3

Contents

COLOR PLATES

Preface

Geology is presented in this book as a personal enterprise of the reader. From the beginning he is asked to pose questions to himself about things he has seen. He takes immediately the role of explorer and detective, as should any beginner in science. In this role he can feel the excitement of observing things more carefully than he has done before, of piecing together fragments of evidence to check hypotheses, of discovering relationships that he had not suspected, of using his discoveries to reach out beyond immediate experience and bring order to the rocks and landscapes of wide regions. In the questions he is asked to formulate he must push his vision not only to the farthest reaches of the earth as it exists today, but backward into the long history revealed by rocks and forward to the state of the planet in a distant future. As the structure and history of the planet take shape, he develops a background for considering the uses that he and his fellows have made of their planetary environment, and the adequacy of that environment for their continued survival. Emphasis is kept on the conclusions the reader can draw for himself, not on his ability to learn the conclusions reached by others.

Of all the sciences geology is best suited to such an approach, for many of its principles are evident from simple application of common sense to everyday observations. The deeper insights that come from more specialized knowledge grow naturally from the broad principles, and the steps can be followed, at least in outline, by any student who brings a modicum of curiosity and enthusiasm to his work. The curiosity and enthusiasm should be self-generating, because geology seeks answers to basic questions that everyone has asked in childhood, questions that range from simple queries about how things work to the ultimate enigma of man's place in the universe.

To present geology in this way requires sacrifice of any hope of complete coverage. For many readers this will not be important. Students already committed to geology and seeking the shortest route to detailed knowledge will need additional material, of a sort that is readily available in many excellent paperbacks on particular geological subjects. For either kind of reader, specialist or nonspecialist, the book will have served its purpose if it helps to widen his vision of time and space, if it increases his awareness of the strange and beautiful planet on which we live, and if it helps him to formulate the penetrating questions without which there can be no progress in science.

ACKNOWLEDGMENTS

To the many who helped to make this book possible, colleagues and students and secretaries, I would express my gratitude, and I wish the expression could be more personal to each one. Three must have special mention: my colleague Charles F. Park, Jr., who read parts of the manuscript and gave help and advice with the illustrations; Margaret Freeman, whose unerring eye for the niceties of syntax, word meanings, and punctuation caught many gaucheries and improved the manuscript immensely; and my wife, whose ability to sense over-detailed, over-abstruse, and over-pompous explanations was a constant and much-needed restraint in the writing.

Sources of illustrative material are acknowledged in the captions, but special thanks for help are due the following organizations and individuals:

Alabama Geological Survey, Aluminum Corporation of America, California Division of Mines and Geology, Geological Society of America, Illinois Geological Survey, Los Angeles County Museum of Natural History, Luray Caverns, Michigan Department of Conservation, National Geographic Society, National Park Service, NASA, New York State Museum and Science Service, Peabody Museum, Smithsonian Institution (U.S. National Museum), Thermal Power Company, U.S. Bureau of Mines, U.S. Department of Agriculture, U.S. Geological Survey, Utah Division of Travel Development, Washington Division of Mines and Geology.

Paul Bailly, Bruce Bolt, W.C. Bradley, Ernest Carter, Charles Chesterman, Robert R. Compton, John Crowell, Stanley N. Davis, William R. Dickinson, Carl Dutton, Gerald M. Friedman, Edward L. Hamilton, Bruce Heezen, Mary Hill, Patrick M. Hurley, Arvid Johnson, Stanley Kling, Kurt Lowe, Gordon Macdonald, Perfecto Mary, Ely Mencher, Carleton Moore, John Ostrom, Ben M. Page, Charles F. Park, Jr., Austin Post, Enrique Prado Ruiz, George I. Smith, Jon K. Solnes, Marie Tharp, Kenneth M. Towe, T.R. Walker.

Konrad B. Krauskopf

The Third Planet

AN INVITATION TO GEOLOGY

Fig. 1-1. Storm waves. See also Plate 1. (David Muench)

1

Where Land Meets Sea

As a student of geology, your laboratory is the world outside your door. You will bring pieces of that world inside to examine them more closely, but your center of interest is the world in its natural state. Like any other scientist, a geologist seeks to understand his world, to make predictions about it, and to gain some measure of control over it. Unlike others, he seeks such understanding and control primarily by sharpening his ability to observe things as they are rather than by analyzing their behavior in the artificial setting of an indoor laboratory.

So we begin with observation, and the easiest part of nature to observe is a landscape. It doesn't matter much what landscape we choose, so long as it is a place not greatly modified by man and his machines. Rolling hills, a flat plain, rugged mountains, a swamp or a desert—for any of these we could ask the same questions, and in framing replies get a feel for the way a geologist approaches his subject.

Take any landscape that is easily accessible and look at it *carefully.* The important thing is the care: don't just glance at the trees and sky and pass on, but sit down, relax, keep your mind open to all the details, and let your imagination have free play. Has this little fragment of the earth's surface always been as you see it now? What evidence do you note of processes that might be changing details here and there? Suppose that these processes go on unhindered for a very long time: could they ultimately change the entire scene before you? Suppose you had a movie camera focused on the landscape and timed so that a single frame would be exposed every decade; if you could take the film representing exposures for ten thousand years and run it through a projector at normal speeds, what sort of an animated picture would you get?

Fig. 1-2. A landscape in the high mountains of Colorado, with Long's Peak in the background. Is there any evidence in the picture that the landscape may be slowly changing? (National Park Service)

Ask such questions of any familiar landscape, and you will be starting to think like a geologist.

Meditation on a sea cliff

We need a particular landscape to concentrate on, and we pick a place where geologic activity is evident even to the most skeptical. We sit at the edge of a sea cliff, where the waves are pounding on the rocks beneath. The day is clear and breezy, with a few clouds in a bright blue sky. When we have grown used to the splendor of the scene, to the noise of the waves, to the feel of salt spray, we start looking at details. For each detail we ask the eternal scientific question "Why?", and the answers require a generous use of imagination (Plate 1).

Beneath us is a cliff of solid rock, vertical or nearly so. The cliff surface is colored in splotches and streaks of dull gray and brown, more yellow brown and red brown above and gray to black near the base. The cliff is cut by cracks and grooves in an irregular pattern; between the cracks the surface in places projects out a few feet and in places is deeply recessed. Plants cling to parts of the cliff, more numerous near the top and along some of the cracks. At the base of the cliff is a beach of sand and gravel, wide where the cliffs are recessed and absent where rocks project farthest into the sea. The material of the beach is constantly moved by the waves, shoreward as each wave breaks and runs up the beach, seaward as the water retreats. Coarse gravel and large loose blocks mixed with driftwood at the upper edge of the beach are mute evidence of the larger waves raised by occasional storms. Out to sea away from the shore are rocky islets where shorebirds sit in rows and seals crawl up to sun themselves. Patches of white water betray other rocky projections from the sea floor, now only reefs under the surface but probably exposed when the tide is lower. Among the reefs and islets float brown sea plants, probably attached to the bottom but with upper ends free to move with the waves. Shallow pools of quiet water nearer shore are lined with other forms of sea life, often in brilliant colors.

Now what meaning can we attach to this scene of jagged rock and moving water? Put in that form, the question admits of many answers. A seascape would have one sort of meaning for an artist, a different meaning for a philosopher, still a different one for a fisherman. Here we seek the geological meaning, a meaning that looks behind appearances to explain how details of the scenery have been formed.

Fig. 1-3. A desert landscape in southeastern California—looking westward across Death Valley to the Panamint Mountains. Would you expect changes in this landscape to be faster or slower than in the one of Fig. 1-2? (Mary Hill, California Division of Mines and Geology)

The cliff beneath us, for example: why should it stand as a vertical wall of rock, bare of vegetation except for a few plants in crevices? There is no mystery about the answer, if we imagine the gravel and angular blocks on the upper beach being lifted and hurled against the base of the cliff by storm waves. Hard as the rock of the cliff may seem, it must be chipped and abraded by this violent attack. Slowly, in successive storms, the base of the cliff is worn back. The upper part of the cliff is undermined, and once in a while a huge block must break off to be added to the other blocks on the beach. Erosion at the base, gravity pulling on loosened material above: surely these processes are sufficient to maintain the bare, vertical cliff.

The irregularity of the cliff, the parts that project and the parts that are deeply indented: how do we account for such details? Imagination suggests that the rock is not all of the same hardness, that the soft parts are eaten away more rapidly by waves, by slumping, and by rivulets during rains. We could check this guess by examining the rock in detail, but even without a check it seems a pretty obvious explanation.

What about the islets and reefs offshore? Again we appeal to imagination: if the cliff at present is being worn away, it must once have stood farther out to sea; as it was eroded to its present position, remnants of its harder materials have been left behind. The intricate crags and grooves and caves on the islets show how discriminating waves can be in picking softer materials to erode.

Then look at the beach: what determines its gentle seaward slope? A complete answer would not be easy, but the slope must somehow be determined by the movement of sand shifted back and forth by advancing waves and retreating currents. It must be affected by the nature of the sand and the average speed of water movement, although the precise relationship is not obvious. What about the slope offshore, between the rocky islets? We would only be guessing, because we cannot see the underwater slope from where we sit on the cliff top, but the islands and the beds of seaweed extending to a considerable distance offshore suggest that the water is fairly shallow—that the slope of the sandy bottom becomes more gentle away from shore.

Thus with a little rumination we can ferret out the processes that are shaping the present coastline. One dimension we have omitted, the dimension of time. To reason like geologists, we must ask questions about the past and future. What will this pleasant scene look like ten years hence? A hundred years hence? Ten thousand years hence? Or go backward: was this same cliff here 100 years ago? 100 thousand years ago? 100 million years ago?

Fig. 1-4. Sea cliffs and marine terrace on the rugged coast of California. (David Muench)

In answering such questions we quickly pass beyond the point where common sense gives us much help. Ten years in the future, yes: we can say confidently that the cliff will still be here and that changes will be imperceptible—barring an exceptional storm that might cause some part of the cliff face to collapse. Our confidence depends simply on past experience with sea cliffs, or on pictures of cliffs from previous years, or on travelers' reports. We know that erosion of hard rock by waves is very slow, and that ten years is not long enough for change to be apparent. What about 100 years? Again we would not hesitate to say that the cliff would still exist, but we would grant that occasional rock falls might have changed many details. Ten thousand years? Here experience is no longer a sure guide. We recall, perhaps, that Greek and Roman writers 2000 years ago described cliffs around the Mediterranean (the rock of Gibraltar, for example, or the cliffs above Monaco) in language much the same as we would use today, and on this slender evidence we might guess that 10,000 years would not be long enough for wave action to destroy our cliff. It might be greatly altered, but about the extent of alteration we can only make wild guesses.

Can we do any better looking backward? We might expect to, because remnants of former aspects of a landscape can be preserved. Here on this bit of coast, for example, the reefs and islets along the shore represent former

Fig. 1-5. The city of Monaco, built on cliffs that have changed little since Roman times. (Courtesy of Princess Grace)

positions of the sea cliff. We have complete assurance that at some time in the past the cliff stood farther out to sea, at least as far out as the last of the reefs. But if we try to attach definite times to its positions, we are as helpless as with projections into the future. A century ago, on an average, the position of the cliff face was perhaps a few feet farther seaward. But was it in the position of the outermost reef 10,000 years ago? 100,000 years? a million years? From such questions the mind recoils, because ordinary experience gives us no basis for judgment. Yet these questions are of the essence of geology, and with sufficient study of landscapes we should be able to give them at least partial answers.

How does man change a landscape?

So far we have talked only about natural processes, uninfluenced by human activities. Ordinarily we think of natural forces as so powerful and on so huge a scale that man's best efforts can have important effects only locally and temporarily. Yet as machinery becomes more and more powerful and as mankind becomes more numerous, the influence of human occupancy grows increasingly noticeable. Along with the traditional questions of geology, another has become important in recent years: to what extent do man's activities, intentionally or not, play a role in the shaping of landscapes?

Looking at this query from our vantage point atop a sea cliff, we would first be struck by the human debris intermixed with rocks and driftwood along the beach—the cans and bottles and boxes discarded by picnickers and local residents. Certainly they make the landscape less pleasing, but do they have any geologic importance? It seems hardly likely. Such debris is slowly pounded to bits by the waves, and some of its fragments may ultimately be added to the accumulating sand of the beach and the offshore sediment. Except where the debris is dumped in enormous quantity, as from an urban area nearby, nature seems able to take care of the addition without modifying its accustomed pattern of behavior.

Perhaps at some point along the shore you note a pile of large angular blocks, bigger than the normal accumulation elsewhere and including blocks of concrete as well as natural stone. At the top of the cliff above, you see a house near the edge, and you guess that the owner has had several truckloads of riprap dumped on the shore to retard erosion of the cliff. This is deliberate human interference with a natural process: wave attack will indeed be slowed, and over a long period you can imagine the house and cliff being left as a small promontory while adjacent sections of the shore have been moved back by erosion. Yet this too would be ephemeral, unless the householder and his descendants renew the pile of riprap at frequent intervals. For any projection from shore is subject to concentrated wave attack, and eventually the riprap, like the natural rock, will be worn down to sand grains and clay.

You may notice a place where a gully cuts deeply into the cliff, a gully evidently fed by water draining cultivated fields behind the edge of the cliff. If the gully walls are steep and fresh you might guess that erosion has been accelerated by human occupancy—by intensive cultivation without precautions to prevent runoff from being concentrated at the head of the gully. Looking into the future you would expect the gully to continue to grow, and to supply mud and sand to the waves on the beach. Would the landward erosion of the cliff in general be affected by the local breach in its top?

Fig. 1-6. A sea cliff cut by a gully that drains cultivated fields. (Earl Brabb, U.S. Geological Survey)

Probably not greatly, because the soft material at the gully mouth would be protected from rapid erosion by the tendency of waves to concentrate their attack on the rocky projections at either side.

In an area of this sort, therefore, the effect of human interference is trivial. The geologic forces at work are so powerful that any changes man makes, deliberately or accidentally, are to a large extent obliterated in periods of time that are short by geologic reckoning. In places where natural processes are less aggressive—areas of broad plains or gently rolling hills, for example—the story may be very different. We will have occasion later to look at several kinds of environment where man and his machines have become important agents of geological change.

How does man use a landscape?

For his own survival man must adapt himself to his environment and find in it the materials necessary for living. As a general statement this is almost a truism, but the problems it entails can be sharply focused by looking at man's use of a small area. Suppose, for example, that the coastline we have been considering is a major part of a small principality (like the tiny European state of Monaco), whose inhabitants wish to be as self-sufficient as possible. What can they find in this sort of landscape to enable them to survive and enjoy a reasonable standard of living?

There is first and obviously the scenery. A rocky coastline is a fine place for recreation, and presumably the inhabitants would take full advantage of it. This will hardly help with survival, however, unless they can persuade foreign tourists to pay money for enjoying the recreational aspects.

Where could the inhabitants of our mythical princedom find food? The sea would provide a good deal, but to supplement a diet of fish they would need agricultural land. Here they might have difficulty, because inland from rocky coasts the land is often hilly, and flat areas with good soil may be scarce. Available land would have to be used intensively, soil would have to be carefully conserved, fertilizers would be needed in large quantity. Where would fertilizers come from? Some might be manufactured from air (nitrogen fertilizers), some could perhaps be obtained from seaweed or wood ashes (potash), but only by extraordinary luck would enough phosphate be found in the local rock. Presumably a large part of the fertilizer would have to come from the outside world.

Where would construction materials be found? Rock would be no problem, and timber might be ample on the hillsides. But if the inhabitants wanted to be part of the modern world they would need metals also, particularly iron. Where could these be obtained? The rocks would be searched for ores, but only rare good fortune would lead to discovery in so small an area. Even if ores of some one metal were found, the chance of finding supplies of all the metals needed in modern life is extremely small. Almost certainly imports would be the only possible source of most metals.

Even if the inhabitants solved their food problem and even if they chose to keep their requirements of metal to a minimum, they would face one major difficulty: how to acquire an adequate source of energy. They might look wistfully at the energy going to waste in the waves beating against the shore, but they would have no means of harnessing it. Windmills could be used for minor chores, but are hopelessly inadequate for modern households, transport, or industry. Coal and oil would become the objectives of intensive search, but the prospect of finding adequate amounts is just as small as it is for metallic ores. Again imports in quantity would be needed to maintain a reasonable living standard.

From the standpoint of man's use, the landscape we have described presents a bleak picture. So also would most landscapes on the earth. It is

Fig. 1-7. Florida as it looks from a satellite. The picture is taken looking south down the length of the peninsula from a point above Jacksonville. The distant horizon shows clearly the curvature of the earth. (NASA)

an inescapable fact of geology that most of the materials needed for modern living are not widely distributed, but are concentrated in only a few areas. The resources of almost any small area are limited. However enjoyable a landscape may be, life is generally possible only if an abundance of materials can be brought from elsewhere.

Planetary landscapes

We have sat long enough on our sea cliff. For a wider perspective, we now book passage on a satellite rocket circling the earth at an altitude of several hundred miles. The "landscape" now before us is a wide expanse of blue ocean, green and brown continents, and patches of white clouds, reaching on all sides to a circular horizon. From this distance we get no sense of difference of elevation, no feeling for mountains and valleys, plateaus and desert basins; all is reduced to a flat pattern of varied colors, highlands distinguishing themselves only by a difference in color from their surroundings (Plates 26 and 33).

We attempt to examine this planetary landscape with the same geologic eye we used for a short stretch of coastal scenery. Now we see no movement, no obvious sign of geologic processes at work. In successive circuits of the earth we may detect a gradual change in cloud patterns, but the continents and islands and oceans remain fixed. We note some oddities of distribution. As we circle the globe again and again we see that the continents are largely massed on one side of the planet, with the huge Pacific basin filling the other side almost wholly with water. All the continents together make up only about 30 percent of the total area, so that the scene below us is mostly water.

Perhaps we notice some regularity in the distribution of continental colors: green where a continent straddles the equator, generally broad patches of red and brown to the north and south, then more broad strips of green, and finally the white snow of Antarctica and the Arctic islands. We note further that clouds are more numerous in the belts where the continents are green. So we guess that the colors correspond with the earth's broad climatic zones: the rain forests near the equator, the strips of savanna and desert on either side, then the green temperate zones shading into ice-covered lands near the poles.

Then we begin to wonder about the shapes of continents: for the most part they seem irregular, but in places they look like pieces of a jigsaw puzzle that has been put together and then slightly disturbed. The Arabian peninsula, for example, looks as if it might be pulling away from Africa on one side and from Asia on the other. On a larger scale the similarity of coastlines on either side of the Atlantic Ocean suggests the possibility that the Americas were once closer to Africa and Europe. And the arc-shaped chains of islands along the west side of the Pacific, from the Aleutians south through Japan and the Philippines to Indonesia: have these been somehow pulled from the continent of Asia?

Now again we let our imagination roam, and start probing with questions. Why are there continents and ocean basins? Why are the continents largely restricted to one side of the planet? For such questions we have as yet no way even to approach an answer, as we could for the simpler questions about the origin of sea cliffs and beaches. Nothing in our experience gives any clue as to how continents might change, or how they might have formed if at one time in the earth's past they did not exist. The

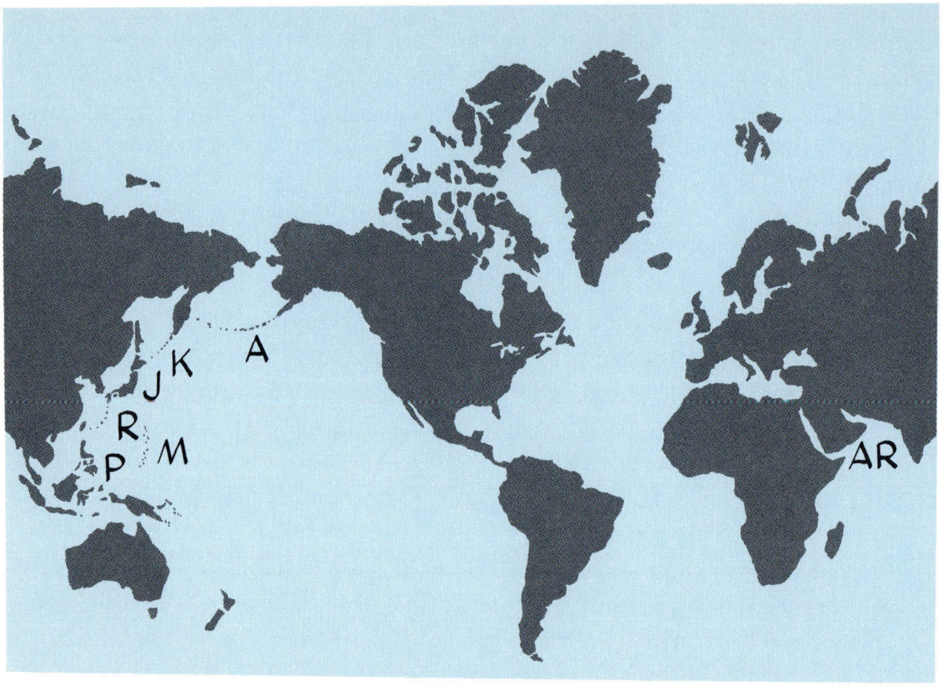

Fig. 1-8. Continental outlines suggest that even on a planetary scale changes may be in progress. Note how Arabia seems to fit into a corner of Africa, and how the eastern coast of the Americas would match roughly the western shore of Europe and Africa. (A-Aleutian Islands, K-Kuril Islands, J-Japan, R-Ryukyu Islands, P-Philippine Islands, M-Mariana Islands, AR-Arabia)

correspondence of opposed shorelines around Arabia and across the Atlantic gives a tenuous hint that continental masses might move about on the earth's surface, but the hint is exceedingly vague. The island arcs of the Pacific may once have been part of Asia, but they might have formed in a dozen other equally plausible ways. Are the questions then silly, or might they be answered with more information? We had best reserve judgment until we have looked at possible ways of obtaining additional data.

We could also try historical questions. If we had circled the earth in our satellite 100 years ago, would the continents have looked just as they do now? Or 10,000 years ago? Or a million years ago? We reflect that wave erosion is nibbling at continental margins, and that streams carry enormous quantities of soil and rock into the sea, so that given sufficient time we might expect to see at least slight changes in continental shapes. But the times necessary are overwhelming. If it takes, say, a thousand years for waves to make an appreciable change in a rocky coastline, surely it must take a thousand times as long for accumulated wave erosion to make a change in shape that would be detectable from a satellite. Or would it? Do such time figures have any meaning? At least we can say that changes in continental shapes would take far longer than perceptible changes in coastlines, provided the changes take place only by erosion and deposition of sediment. Could changes in continental shapes result from more rapid and violent activity? Perhaps it is best not even to guess until we have more facts at our disposal.

Then we look at questions of more human interest. Man has occupied this landscape below us for two or three million years, and in great number for a few hundred years; has his occupancy changed the appearance of the continents detectably? Not their shapes, certainly, but possibly their colors. We recall that large areas have been overgrazed and overcultivated, some of them for several thousand years, so that we might expect some of the brown patches of desert to be larger than they once were. The desert along the North African coast, for example, was once a fertile land that supplied abundant wheat to ancient Rome.

On the other hand, irrigation projects have brought water to parts of the desert, on a scale large enough so that some of the resulting green areas would be visible from a satellite. Perhaps the murkiness of the atmosphere over densely populated areas would also betray man's presence. But by and large the changes that man has produced are still minuscule on the scale of satellite observations.

Then we reflect on the suitability of this planetary landscape for supporting men in the style to which they have become accustomed. Places to grow food, availability of fertilizer, sufficiency of water, resources of metals and of energy—are these adequate for man's use? A possible reply is to say that the question is trivial: man has lived on this planet for at least two million years, he is completely adapted to his environment, obviously he must find in the environment all things needed for his survival. The situation is clearly different from that in the small coastal strip where we first asked this question, for now we have the whole planet at our disposal and scarce resources of all kinds can be distributed from the places where they are locally concentrated. Yet is the matter so simple? Humanity is increasing in numbers at a prodigious rate; the standard of living is rising, and more and more of the earth's resources are required to sustain it; even if we seek out all the points of concentration and perfect the distribution system, will there be enough food and metals and energy to go around?

This is a sobering thought for a satellite passenger. As you circle the earth time after time, reflect that the landscape below is all that mankind has to work with. Man has spread to the far corners of the landscape, and he is busily extracting metals and coal and oil wherever he can find them. When he has used up the local concentrations, where then does he go? In the coastal landscape we looked at first, resources were limited but necessities could be brought in from the outside. In the larger landscape resources are more plentiful, but once they are gone there is no outside place to turn for replenishment. This planetary landscape, too, is limited. It has become an important part of a geologist's function to seek out available resources and to give advice about their exploitation, so that planetary limits will not be approached so rapidly as to precipitate a crisis.

Beyond the planet

To appreciate just how finite the earth is, we change our point of view once again. From our satellite following an orbit around the planet we step now into a rocket heading out into space toward the nearest star. Through the back window of the rocket we watch the earth recede. Quickly we lose the sense of being poised above a flat, circular disk, and see our planet now as a sphere half illumined by the sun. For a time we can still make out the dim shapes of continents and the white patches of cloud, but as the sphere grows smaller all visible detail is lost. The sky now is completely black, studded with stars despite the brilliant sun shining in the distance. Our earth has become just a tiny bright spot against the blackness of space.

We notice another bright spot near the earth, considerably smaller, also apparently a sphere reflecting light from the sun. If we catch sight of it

Fig. 1-9. The moon as it would appear from a rocket. The left-hand side in this view is in full sunlight, but the right-hand edge is still in shadow. The large dark areas are *maria,* huge plains covered with lava. The lighter-colored areas are highland regions pockmarked by craters produced by meteor impacts. Note that the crater outlines are especially distinct near the right edge, where the rising sun casts long dark shadows. (Lick Observatory)

early enough we would recognize the familiar irregular surface features of the moon, but our rocket would carry us quickly beyond the point where surface details are visible. Thereafter we would see only two bright spots off in the distance; if we keep our attention on them, we note that the smaller one follows a path around the larger—circular or elliptical, depending on the angle from which our rocket has left the earth (Plate 25, Fig. 1-9).

But we have other things to watch besides the earth and moon. We assume that our rocket is equipped with a telescope, and with its aid we look at some other very bright objects in the sky around us. To our unaided eyes these are just bright stars, points of light like the fainter stars. But the telescope shows a sharp difference: most of the stars, despite the magnifying power of the instrument, remain only points of light, but these few bright ones show up as disks. From the shapes of the disks we can tell that these objects are spheres reflecting sunlight to us from their surfaces. They are evidently bodies much like the earth and moon, but more distant. As our rocket picks up speed on its journey, we can see that the earth and moon also will soon be mere points of light unless we use the telescope to make their surfaces visible.

The bright "stars" that turn out to be spheres when we examine them with a telescope are the planets. If we look at them closely we see faint points of light near some of them, points that follow paths around the planets, and we guess that these are smaller bodies associated with the planets just as the moon is associated with the earth.

With sufficient patience and mathematical skill we could work out from our planetary observations something about the sizes of these bodies, their motions, and their distances from us. But it is easier just to sit back and let the rocket carry us ever farther out into space, training the telescope occasionally on the earth and its sister planets. The light from the planets grows fainter, their distance apart seems to grow smaller, and even the sun itself now shrinks in size and brightness as our journey continues. At length we find we have outdistanced the planets: they are all behind us, all in the part of the sky near the sun; if we slow the rocket temporarily so that we can observe them over a period of time, we find that all the planets are in motion, following curved paths around the sun. We can no longer see them very clearly, but the amount of light they send us tells us something of their relative sizes. If we have kept track of how far we have traveled from the earth, the apparent distance of each planet from the sun can be translated into miles. So we see some order in the sun's planetary family: four small ones near the sun (the word "near" meaning within 150 million miles), then four big ones farther out and more widely spaced, and finally another tiny one some 4000 million miles from the sun.

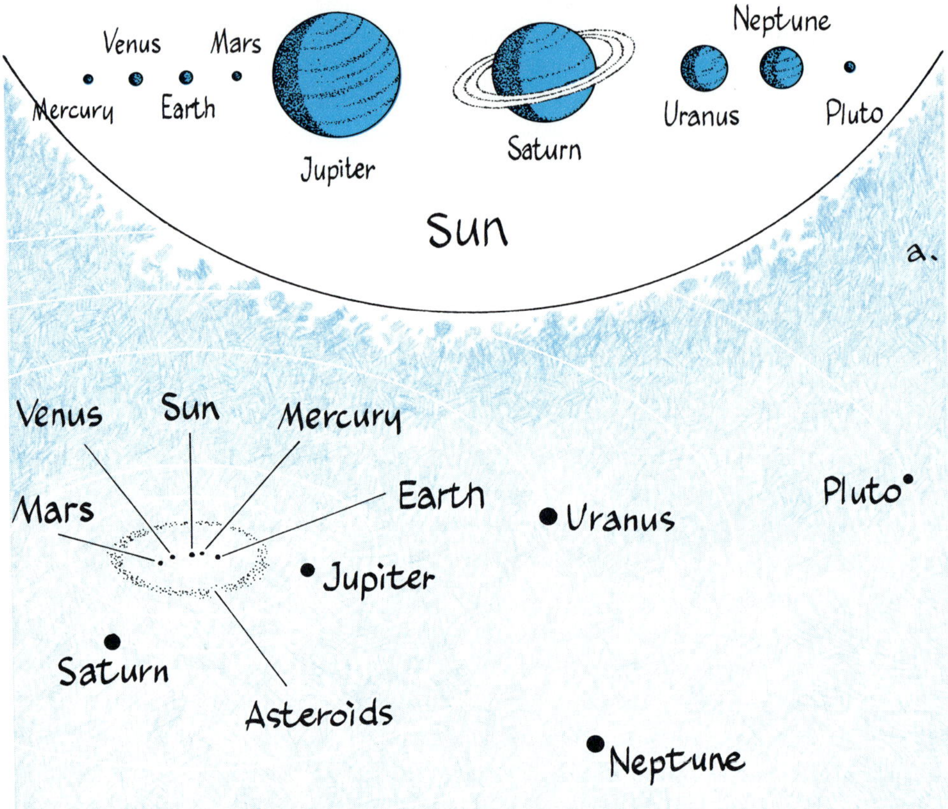

Fig. 1-10. The solar system, showing (upper) the relative sizes of planets and sun and (lower) the spacing of the planets in their orbits.

Looking from this distant point of view at the bodies circling the sun, we see little about the planet earth to make it distinctive. It is the third one out from the sun, and the largest of the four inner planets, but it is dwarfed by the giant planets beyond the orbit of Mars. Its moon is larger relative to the size of the planet than any other moon, but this hardly seems an important distinction. The earth is only one planet among many, below average in size and very tiny in comparison with the sun.

It is time that we pay attention to the navigation of our rocket, because if we really hope to get to the nearest star we will need all the speed we can muster. The outermost planet, Pluto, we have found to be 4 billion miles from the sun, and this seems a respectable distance. But we recall from our star map that our objective (a star called Alpha Centauri) is enormously farther, in fact more than 6,000 times as far. We can set down a number for the distance—25,000,000,000,000 miles, but the number means nothing to our minds. It becomes more vivid if we think of accelerating our rocket to the speed of light: even if we attain this maximum of all possible speeds for a material object, it will still take us four years to reach our destination. We can keep our telescope trained on the sun, which is fast shrinking to a mere point of light like the other stars, but soon even the telescope will lose track of the tinier points of light that represented the planets. We will be alone in space, separated by unimaginable distances from the stars on all sides, and we will have plenty of time to reflect on what we have seen, especially on the earth's position in all this vast emptiness.

In the course of reflecting we could try some of the questions we have asked about landscapes nearer home, but the questions no longer make much sense. Is this new landscape (or "starscape") changing? We have no way of knowing, no way even of approaching an answer. The only motion we have seen is the motion of moons around planets and of planets around the sun; but these motions seem just to repeat themselves, and as far as we can tell they may have gone on forever. Conceivably by studying the stars with more sophisticated instruments we might find hints of other motions, and perhaps even detect some non-repetitive changes that could give a clue to past and future. But such observations are far beyond the power of the naked eye and the telescope.

Has human activity had any effect on the sun's planetary system or on the larger universe of stars? The question is patently silly. Can humans make use of materials or energy from any of these objects, moons or planets or stars, beyond the earth's surface? To this question the answer is almost as clearly a flat "No"—except, of course, for the flow of energy from the sun in which the earth is continually bathed. With a vivid enough imagination, to be sure, we can dream of rockets traveling to the moon and bringing back gold or diamonds or uranium. But the expense of such an operation, with any presently imaginable source of fuel or energy for the rocket, is so outlandish that we can relegate such dreams to a distant science-fiction future.

As we sit twiddling our thumbs through the long and boring rocket journey among the stars, we will perhaps think fondly of the planet we have left, but we can hardly help thinking also of its extreme isolation. True, we can send men to the moon and television cameras to Mars, but any direct effect of these bodies on earthly life is hard to imagine. From the astronomer's viewpoint the earth remains a tiny planet, one among many, separated from its neighbors by tens of millions of miles and from all other objects in the universe by distances measured in thousands of billions of miles. On this isolated speck the whole drama of human history and geologic history has been played, and as far as we can see will continue to be played, with precious little interference from outside. The surface of this isolated speck will be our only home for the foreseeable future. From it must come all that we need for life, save only the light and heat supplied by the sun, and its limitations must set limits on both the quantity and quality of life that we can expect.

Back to the seashore

It is time we leave our rocket and come back to the rugged coast where we started. Once again we hear the waves hammering on the cliffs below, we watch the sand in its endless movement up and down the beach, and we see whitecaps on blue water extending before us out to the hazy horizon. Only a tiny patch on the surface of a tiny planet, from the perspective of our space wandering, but by human standards an impressive bit of scenery. To patches of the planet like this—patches that would include plains and mountains and deserts as well as seacoasts—our life is confined, and from such patches we must obtain all things necessary to our existence. Whether a bit of earthly scenery looks insignificant in comparison with starry distances, or all-important for the welfare of the human race, depends entirely on our point of view. And this point of view will change back and forth many times as we look more deeply into the meaning and history of such landscapes.

Here on the seashore, and also during our space journeys, we have asked many questions about processes taking place, about possible progressive changes, and about the history and probable future of the objects we were observing. For the seashore we could frame general answers from simple common sense, but the questions we asked on the rocket were largely unanswerable. On the basis of observations we could make, even observations aided by a telescope, we could get no further than idle speculation. More precise observations might have given us clues, especially if accompanied by measurements, but such activity would have required better instruments and more background knowledge. The necessary background can come only from study of parts of the universe that are close enough for detailed inspection—parts of the universe, for example, like our bit of coastline.

We have already seen how even elementary observations can lead to reasonable conclusions about changes that are going on and hence about probable past and future events. More sophisticated observations will certainly lead to deeper understanding, perhaps to conclusions that can be extended to other kinds of landscapes as well. So we may hope to build up a fund of information about the earth's surface as a whole and about the rocks under the surface, and with good luck we may even find some general rules that will help in understanding the moon and other planets also. This is always the general progression in science: to formulate questions, to make observations that apply to the questions, to seek general statements that explain the observations, and then to use the statements imaginatively to predict and explain observations elsewhere.

But it all begins with the part of the world nearby, in our case with the rocks and sand and sea water. Our questions will now have to be more penetrating. We have noted that details of the coastal scenery are determined by differences in hardness of the rocks undergoing erosion; but why should one rock be harder than another? What is it that determines how hard a given rock will be? Or we can go right back to fundamentals: what are rocks made of, and how are they put together? Again, we have observed that rock surfaces near the tops of the cliffs are colored in shades of brown, whereas surfaces lower down that are kept fresh by wave erosion are dark gray; does this mean that rock long exposed to air is somehow altered? What is there in a rock that would make its surface change from gray to brown? Questions like this could obviously be multiplied indefinitely.

Before starting to explore these questions, let us take one last general look at the shore below. We have perhaps not emphasized enough the multiplicity of living things clinging to the rocks, swimming or floating in the water, and attached to the bottom of the tidepools. If we climb down to the beach and focus attention on the living organisms, their number and variety are astonishing. Seemingly this should be a peculiarly difficult sort of environment for both plants and animals: continually pounded by waves, exposed to the air twice a day at low tide, subject to extremes of temperature and salinity as the sea sweeps in and out and fresh water is added from rain and rivulets. Yet the tidal zone teems with life. Species of most of the major groups of plants and animals can be found here, marvelously adapted to the vicissitudes of this constantly changing environment. No better proof can be found anywhere of the ability of living forms to adjust themselves to strange situations. If we are going to look into the history of the earth, this seashore scene should be a continual reminder that living things will play an important role in the history, since they are sure to flourish in great numbers on any part of the earth's surface where a bare minimum of necessary conditions is fulfilled (Plate 2).

In imagination—or by actual observation if we have diving equipment—we can picture the shallow water away from the beach as also a lively scene of organic activity. Remains of dead creatures cast up on the beach give us a sampling of the forms that live offshore. If we could extend our observations elsewhere, we would find in general that shallow marine waters, through which sunlight can penetrate to the bottom, are an environment in which organisms are particularly abundant. Then we think of fossils we have seen, either in museums or in rocks we have picked up in the field, and we recall how similar most of these remains of ancient plants and animals are to the creatures whose shells and bones we find on a modern beach. Evidently shallow parts of the sea have teemed with life in the past, just as they do today.

Again thinking of fossils, we recall that although many resembled closely the shells on a modern beach, others were distinctly different. Shallow parts of the sea have had myriads of organisms through much of the earth's history, but the character of the organisms has changed. Detailed study of fossils shows that assemblages of organisms can be arranged in sequence, those from older rocks showing greater differences from modern forms than those from younger rocks. This fact is the mainstay

Fig. 1-11. A record of an ancient shoreline—fossils in a rock that was deposited as sediment in shallow sea water more than 200 million years ago. If you look carefully at the shells you can distinguish several varieties, some of which are quite different from the shells you would find on modern beaches. (E.B. Hardin, U.S. Geological Survey)

Fig. 2-2. A handlens shows details that would be invisible to the naked eye. (E.F. Patterson, U.S. Geological Survey)

Fig. 2-3. Making a geologic map in the field. The table, which can be accurately leveled, is called a *plane-table,* and the instrument (for measuring distances and angles) is called an *alidade.* (L.C. Huff, U.S. Geological Survey)

But a modern geologist would tell you that this definition is too confining. He collects rock specimens with his hammer, he looks at them in the field with his lens—but then he brings the specimens back into his laboratory and subjects them to all manner of tests with a variety of sophisticated instruments. Even in the field he supplements hammer and lens with other gadgets: nearly always a compass, some maps and aerial photographs, sometimes instruments for measuring rock density, rock magnetism, or the composition of soil and water samples. Some of his field and laboratory instruments are built specifically for geological purposes, but many are devices that have wide use in other sciences. The increasing use of instruments from other sciences blurs the definition of geology. If a man spends most of his time in the laboratory analyzing rocks by using the techniques of mass spectrometry, atomic absorption, and neutron activation, is he practicing geology or chemistry or physics?

This depends on the kind of question that such an analyst is trying to answer. If his primary interest is in the process of analysis—in devising better ways to decompose rocks, to determine one element in the presence of others, to improve the speed or accuracy of his instruments—his activity is mostly that of a chemist or physicist. But if his major concern is the rocks themselves, if he uses his elaborate gadgetry to get data about the rocks that will help him to understand their relations in the field, then he is a geologist taking advantage of modern improvements on his handlens. The geologist's basic laboratory, we noted some pages back, is the outdoor world. He makes full use of indoor laboratories too, but always with the thought that he is learning something to help explain what he has seen in the field.

In the world of nature a geologist acts as a sleuth. He is forever trying to look behind the surface of things, to detect the processes that are working to change the appearance of rocks and landscapes. He wants to understand what is happening *now,* so that he can put together a reasonable story about what has happened *in the past,* and so that he can make predictions about what will happen *in the future.* He will use any means he can find to attain these ends, including his traditional hammer and handlens and then all the marvelous instrumentation that modern technology places at his disposal. But his focus remains fixed on what he has seen in the field. The questions he tries to answer are formulated in the field, and tentative conclusions from his laboratory work must be checked against field evidence.

Geology in action: a volcanic eruption

We have looked at the problems that arise in one kind of environment, the cliffs and beaches of a rugged seacoast. For a second example of the geologic approach to rocks and landscapes, we go now to another sort of environment where natural processes are obviously and spectacularly active, a volcano in eruption. This is one of the most awesome sights that nature provides. It has particular fascination for a geologist, because here he sees a landscape changing rapidly and rocks being formed before his eyes (Plates 4 and 5, Fig. 2-1).

We try to pick a "typical" eruption, but this is hard to do because volcanoes are notoriously individualistic. A good place to go would be Mexico or Central America, because here many of the active volcanoes exhibit a wide range of eruptive activity—explosions, quiet lava flows, and clouds that strew thick carpets of volcanic ash over the countryside. Parícutin and Colima in Mexico, Irazú and Santa María in Costa Rica are examples of such volcanoes that have erupted in recent decades. Before beginning our observations we shall have to wait for news reports of an eruption in progress, for major activity occurs only at intervals.

Assume that we have found a volcano in active eruption, and that we have stationed ourselves nearby to watch. We see a cone-shaped hill or mountain with a dense white cloud rising from its top and a stream of incandescent lava pouring out of a fissure in its side. As we approach we are deafened by the roar of escaping gas and frequent explosions as new puffs rise into the eruption cloud. For several kilometers around the base of the

Fig. 2-4. The Mexican volcano Parícutin. The cone-shaped hill is built of ash and larger fragments erupted from the crater at the top. The hill in the foreground is covered by ash about 3 meters deep, enough to kill the trees. The deep gullies carved in the hillside show how easily the ash is eroded during rains.

Fig. 2-5. Parícutin in another of its many moods. From a dark eruption cloud like this the fall of ash is very heavy.

mountain the ground is covered with a deposit of material that looks like fine gray ash, and near the volcano the ash is so thick that trees and bushes have died. When we have grown used to these strange surroundings, we start asking ourselves the usual geological questions. What is really happening, behind all the smoke and noise? What makes the volcano erupt? What changes is it causing in the landscape? What kinds of material is it producing? Have other eruptions occurred here in the past, and can more be expected in the future? Is the volcano destructive or dangerous to nearby settlements? What observations will be useful in framing answers to these questions?

This last query is a difficult one. Where do we start looking? A volcano is like a five-ring circus: there is always so much going on that it is hard to keep attention fixed on any one attraction. Perhaps a good place to start would be the most obvious site of activity, the eruption cloud that rises in a great billowy plume above the mountain. We note that the top of the mountain seems cut off, and that the cloud apparently originates in a series of explosions behind the summit ridge. It requires little imagination to see this ridge as the edge of a crater, and to think of the cloud as rising from an opening within the crater; if the explosions are not too violent, we may be able to verify this by climbing to the crater edge and peering within. In each explosion a quantity of gas seems to be projected up into the cloud, carrying with it gray dust and larger fragments, some of the fragments evidently red-hot lava. The fragments fall back into the crater or onto its rim, a few of them rolling down the slope of the mountain. The gray dust is carried a little higher, but it too in large part settles back onto the slopes. The upper part of the cloud, freed of the dust, is a brilliant white; we get some hint as to its composition from the rain showers that occasionally pelt the mountain. If we stay at the volcano for several days, we see the eruption cloud change its form: at times dense and dark, filled with dust which it spreads liberally over the mountain and its surroundings; at other times light and wispy, perhaps disappearing altogether for a few hours. And at night it is often a fine pyrotechnic display, the chunks of incandescent lava describing graceful curves against the dark lower part of the cloud as they are blasted out of the volcano's throat, and then leaving fiery trails as they roll down the steep slopes.

These elementary observations tell us immediately a good deal about the volcano. Molten lava must have risen into the throat, accompanied by enormous quantities of gas. The gas escapes continually, sometimes quietly but often in great explosions like bursting bubbles on an enormous scale. The explosions carry fragments torn from the dense, sticky lava, some of them large enough to remain incandescent as they travel through the air. Most of the lava is blown to bits by the explosions, producing tiny dust particles that cool rapidly and form the gray underpart of the cloud. The mountain itself is built from the falling dust and larger fragments, the angle of its slope representing the steepest incline to which rocks can cling. The crater is evidently kept open by the force of the explosions. The deposits of gray ash around the base of the volcano represent dust that has settled out of the eruption cloud at times when the activity was violent and the cloud was blown by the wind. The upper part of the cloud is largely water droplets, and we infer that water vapor makes up a large part of the gas that escapes from the lava.

Some of these conclusions we can verify with a handlens. The gray ash at our feet appears under the lens as an aggregate of tiny, sharp-pointed fragments of glass, the sort of thing we might expect from a sticky lava blown to bits by the explosive escape of gas. We call this material *ash,* or more precisely *volcanic ash,* but it is very different in composition from the ash produced by burning wood or paper. We climb the slopes of the mountain, and look carefully at a collection of the larger fragments. Some have the rounded irregular shapes and cracked surfaces that we would expect from chunks of still-fluid material that have traveled through the air and mashed themselves by impact with the ground; our lens will show us that they consist chiefly of dark glass. A few, evidently still more fluid

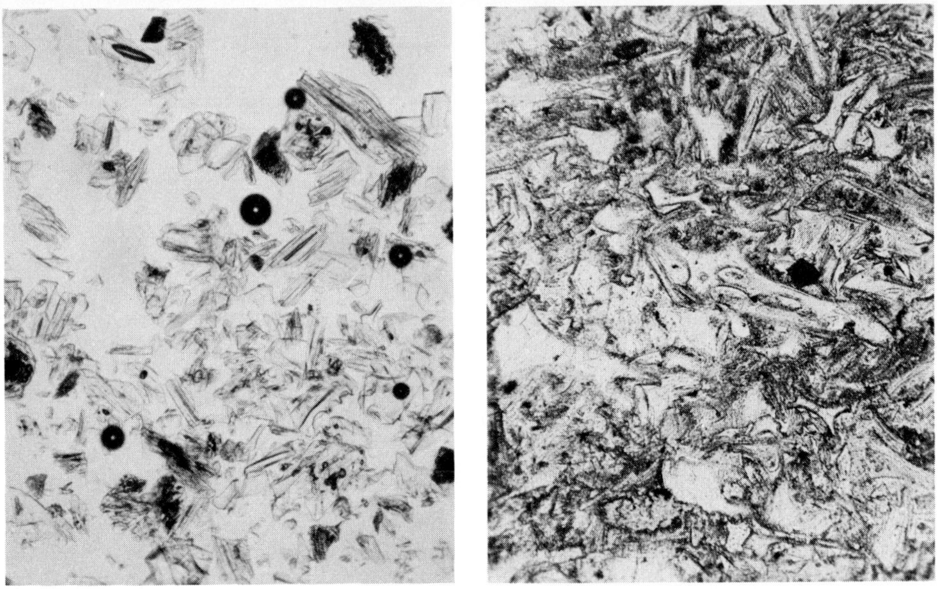

(a) (b)

Fig. 2-6. Volcanic ash as seen under a microscope. In (a) the fragments of glass are loose, and in (b) they are cemented together to form a rock (*tuff*). Both pictures show the sharply angular shapes, many with concave indentations, much like fragments produced when ordinary glass is shattered. Magnification about 34 times. (R.A. Sheppard (a) and M.N. Bramlette (b), U.S. Geological Survey)

during their flight, have the form of smooth spindles. Some of the fragments are conspicuously spongy, full of holes evidently formed by gas expanding as the fragments solidified. Other fragments, probably the majority, are simply angular pieces of gray lava. We call fragments in this collection *bombs, cinders,* or *blocks,* depending on their appearance; and because they make up so much of the mountain, we call the whole mountain edifice a *cinder cone.* So we see the volcano as a great pile of debris, much of it glassy, collected around an orifice where molten lava and gas are rising from somewhere beneath the earth's surface. The mountain is obviously being made larger by the present eruption, but it may owe its height to a long series of previous eruptions. To draw conclusions about its history we would need many more observations than the ones we have so far made.

We turn our attention now to the flowing lava. From a fissure far down the side of the cinder cone we see incandescent lava emerging from underground, accompanied by thin clouds of bluish smoke. As we approach the fissure we find the going difficult. We must scramble over piles of loose, sharp-edged lava blocks, the temperature grows more and more uncomfortable, and breathing becomes a problem in the suffocating acid gases. If we choose our route carefully and make sure the wind is behind us, we can

Fig. 2-7. Volcanic bomb, showing the spindle shape sometimes produced when a chunk of molten lava flies through the air. (A.M. Piper, U.S. Geological Survey)

work our way to a good vantage point. It is worth the effort, because the birth of a lava flow is a spectacular process to watch.

The glowing liquid, with about the consistency of toothpaste, oozes upward from the opening. On contact with the cold air its surface freezes to a black scum, which is broken up into angular blocks as the lava moves downhill away from the fissure. The lava contains dissolved gas, part of which escapes noisily at the point of emergence and part of which expands more slowly within the cooling blocks, blowing them full of holes. The gas evidently contains substances with powerful odors, and other substances that condense out as particles of the bluish clouds and as deposits of brown and white salts along the fissure walls. If we follow the flow away from the fissure, we find that its glowing surface is covered more and more thickly by dark, spongy blocks which jostle each other with sharp metallic sounds as they are carried downstream. Gas continues to escape from the lava in lessening amounts, in places reacting chemically with some of the blocks and turning them gray or red. In the lower part of the flow far from its source the incandescent lava is entirely concealed by piles of the varicolored blocks, and the only sign of motion is the occasional tumbling of one block over others as the pile is carried forward by the still-fluid lava beneath.

To learn more about what we are watching, we look at one of the cooled lava blocks with a handlens. We may see the flat surfaces of small crystals here and there, but most of the material is so fine-grained that we can make out no detail of its structure (Plate 23). Some of it has the sheen of glass, and we recall that the ash particles we have looked at before were largely glass. We note also that the lava as it emerges from underground has the same viscosity (or "stickiness") that we remember from ordinary glass heated to the softening point in the laboratory. If we have a device for measuring high temperatures we can take readings on the lava at different points along the flow: we find temperatures at the source of the order of 1000–1100° Centigrade, and farther down on the flow 800–900°. These are comparable to temperatures at which ordinary glass would become soft enough to flow readily, so our inference is strengthened that lava from a volcano has a strong resemblance to glass.

With these observations we can sit back and reflect a bit about the volcano's whole performance. There must be a column of liquid lava in the central throat of the cinder cone, and we can well imagine that this lava is connected underground to the body of lava that is the source of the flow. Evidently a volcano is a place where liquid rock comes from somewhere underground to the earth's surface; the liquid rock contains much dissolved gas, presumably dissolved under pressure somewhere below; the escaping gas blows some of the lava to bits and builds up a cinder cone with the debris, while another part of the lava flows out more quietly with only small amounts of residual gas. The different kinds of material found around a volcano—the ash, the vari-colored spongy cinders, the bombs, the blocks of solid lava, the lava flows—can all be explained as products of hot glass cooling in contact with air and expanded to varying degrees by the escape of dissolved gas.

Volcanoes and landscapes

Turning our attention from the materials of the volcano to its effect on the landscape, we note the characteristic conical mountain with a crater at its top, the long tongues of frozen or nearly frozen lava around its base, and the surrounding hills and valleys draped with deposits of ash. If the volcano is a large one, its present activity is only the latest in a long series of eruptions. The earlier events have left other lava flows, some of them now covered with vegetation, and perhaps pieces of earlier cinder cones which spoil the perfect symmetry of the mountain. Typically an erupting volcano is in an area where other volcanoes have been active in the past, so that the landscape roundabout will be dotted with other cinder cones and their accompanying lava flows.

Of this volcanic landscape we could ask questions similar to those we asked about the seacoast of the last chapter: what use can human beings make of the landscape, and what effect on the landscape would human activity have? Our first instinctive answer would be that such a landscape must be unfriendly to humans, that a volcano is necessarily dangerous, that it has the capacity of destroying works of man and even of causing great loss of life. But on second thought we recall statistics about the occupancy of areas where volcanoes are occasionally active, say Japan or Java or southern Italy: in general such areas are densely populated and are noted for their agricultural productivity. It is true enough that a volcano is always a source of possible danger, and that at rare intervals it can cause great damage. But deposits of volcanic ash after weathering make remarkably fertile soils, soils that are easily worked and retain moisture; and this characteristic of volcanic landscapes has made them throughout history irresistibly attractive, despite the danger ever lurking underground.

Clearly volcanic areas would be still more attractive if some means could be found for predicting when the next eruption will take place, and if some measure of control could be exercised when it does. The possibility of control is largely an engineering problem, and not much headway has been made toward solving it. The forces involved in a volcanic eruption are so enormous that all man's efforts seem puny in comparison. Occasionally a lava flow can be deflected away from a town or a road by a wall or an excavation, but the record of success is not impressive. The prediction of eruptions on the basis of increasing geologic knowledge seems a better possibility. Records of minor earthquakes, of changes in temperature and gas discharges from steam vents, and of minute changes in slope produced

Fig. 2-8. The famous Philippine volcano Mayon rises above a seaport town and cultivated fields. At intervals Mayon is violently active, but a large population continues to live near its base. (Perfecto Mary)

Fig. 2-9. A dome of viscous lava rising into the snow-filled crater of Great Sitkin volcano, Aleutian Islands, Alaska. (F.S. Simons, U.S. Geological Survey)

by the swelling of a volcanic cone as lava pushes up into it offer much promise that firm warnings of impending eruptions will eventually be possible.

The volcanic landscape we have been observing is a typical example, but it should be emphasized again that some volcanoes are very different in their behavior and in the kind of landscape forms they produce. The eruptions of Hawaiian volcanoes pour out much more fluid and more rapidly moving lava flows than the one we have looked at here, and the lava spreads over wide areas of gentle slopes. Hawaiian flows are often accompanied by great lava fountains, as gas escapes from the fissures through which the lava emerges (Plate 5). At the opposite extreme is the highly viscous lava typical of some volcanoes, lava so sticky that it cannot flow downhill but simply rises as a steep-walled dome in the crater or on the side of the volcanic structure. All volcanoes are places where molten glassy material approaches the earth's surface from somewhere underneath, but the way the glass behaves may vary greatly depending on its temperature, its chemical composition, and the amount of gas it contains.

A geologist looking at a volcanic landscape, then, asks the same kinds of questions he asks of a coastal landscape, and here also some fairly simple observations lead to an understanding of the major processes at work. Many questions remain unanswered, especially questions regarding the ultimate source of the lava, the reason for the high temperatures, and the factors causing an eruption to start and eventually to stop. Some progress can be made in answering these deeper questions, but to attack them requires more detailed observations made with more elaborate instruments than a hammer and handlens.

One final point about the volcanic landscape as compared with a seacoast. On the coast, from an historical perspective, the effect of geological processes is largely destructive, the wearing away of the land by waves along the shore and by streams from the nearby hills. Volcanoes, by contrast, build up the land. Lava, cinders, and ash are added to the surface, raising the general altitude. These are two clear examples of the oppositely directed geologic processes that have shaped the earth's surface, some of them tending to level the land and others to raise parts of the land above their surroundings. We will examine many of these processes, but few have such immediately obvious effects as coastal erosion and volcanic eruptions.

Chemistry and volcanoes

We have been looking at a volcanic eruption through the eyes of a geologist. Suppose now we try a different point of view. Imagine that we become deeply interested in the materials produced by the eruption, in the nature of the lava and gases and the changes they undergo as they cool off and interact with the atmosphere. We might well become so fascinated with this study that we would forget the volcano and focus attention on the nature of viscous liquids and acid gases in general. With this shift in viewpoint we change our approach to that of a chemist.

Our line of questioning might go like this: The behavior of lava is similar to that of glass; can this likeness be explained by a similarity in composition? Analysis would show that some elements—particularly silicon, oxygen, aluminum, sodium, calcium—are present in both glass and lava. Is this combination of elements necessary to give material with the properties of glass? Such a question suggests laboratory experiments with different combinations of these elements; how is the resulting glass affected by various proportions? Most lava specimens contain crystals as well as glass; what conditions of cooling or peculiarities of composition are necessary to produce crystals? Or to prevent crystals from forming? Is it possible to make glasses different from ordinary glass but with properties that would be useful for special purposes?

Or we might study the gases. What particular compounds give the gases their powerful odors and acid properties? To learn about the gases in detail, we would need to separate the compounds and study them individually. Are the different compounds affected by mixing with air? Do their compositions change as they cool off? Which ones dissolve in water to form acids? In general, what peculiarity of composition gives a gaseous compound acid properties? Some of the compounds are important industrial chemicals; would it be feasible to prepare them commercially from volcanic gases?

This sort of inquiry is clearly different from a geologist's approach. Note that the emphasis is on *the nature of materials as such,* not on the role they play in a natural process. Viscous liquids and gases are materials to be analyzed into their constituents, and their properties are to be explained as combinations of the properties of the elementary substances. These elementary substances are to be studied by trying new combinations in laboratory experiments, not only to better understand their make-up, but possibly to produce useful new materials not found in nature. A chemist has little interest in the origin or history of the lava and gases at a particular volcano, or in their possible effects on the landscape, or in their relation to the products of other volcanoes. He wants to bring these materials into the laboratory, completely apart from nature, and to find out as much as he can about them by breaking them down into the simple pure substances of which they are composed. The conclusions he reaches are timeless, applicable to lava and gases anywhere, at any time in the past or future.

Chemistry is a *laboratory* science, as opposed to a *natural* science like geology. The distinction is far from absolute. Some chemists have an interest in natural materials, and some geologists spend most of their working hours in laboratories. The difference is primarily one of emphasis: a chemist seeks understanding by analyzing things, by breaking down mixtures into simple substances, by studying the simple substances under artificial conditions where he can control such variables as temperature, pressure, and the kinds of materials added; a geologist looks for a different sort of understanding by accepting nature in all its complexity, by noting the influence of many variables but not trying to isolate or control them, and by sorting out the major influences on a natural environment from a host of minor ones.

A disquieting thought may have crossed your mind. Chemistry is a more fundamental science than geology, in that it delves into the nature of matter and its interactions; geology is concerned with matter in its various natural forms; so why should we not consider geology merely as a branch of chemistry? Or to put it more bluntly: we can never have a really thorough understanding of natural situations until we know all about the properties of matter, so why not just give up studying geology and concentrate on the more fundamental science? The answer is twofold. For one thing, natural materials and natural environments are so infinitely complex that chemical knowledge in the foreseeable future, no matter how it may increase, will be inadequate for handling many of the larger questions as to how our planet functions. The second answer is that geology differs basically from chemistry in its emphasis on the dimension of *time*: a chemist looks always for properties of materials that are independent of time, but a geologist is concerned with details of the development of the world through a long past and on into the future.

So geology is not a branch of chemistry, and presumably never will be, but the two sciences are, nevertheless, closely related. Very often the conclusions of a geologist depend on chemical data. A geologist would be foolish, for example, to try to understand a volcano without heeding what chemistry can tell him about the nature of the complex silicate liquid that constitutes a lava flow, or about the properties of the gases that escape from the liquid. The dependence of geology on a background of chemical knowledge has increased steadily over the past few decades. Parts of geology to which chemistry is especially applicable are often described as a separate subject, *geochemistry,* a "hybrid" or "interdisciplinary" science between the two traditional fields.

As an example of a typical geochemical problem, recall the gray and red and brown colors that many lava blocks assume on exposure to the air, and recall also the yellows and browns on the upper parts of the sea cliff of the last chapter. Clearly the colors represent some sort of alteration of the surface of the rocks, which in both cases are dark gray or black when fresh. But what makes the colors? This is a typical question for a chemist, and he would tell us after a bit of study in his laboratory that the colored material is iron oxide, the different shades representing different grain sizes and different amounts of combined water. We recall that we have seen such colors on other rocks, so we venture the geological conclusion that rocks exposed to the air often have a part of their contained iron oxidized to these brightly colored compounds. Thus a combination of geologic observation and chemical analysis leads to a useful geochemical generalization.

Fig. 2-10. A picture of Parícutin taken at night by leaving the camera open for several minutes, so that a number of explosions are recorded and the paths of incandescent chunks of lava appear as curves. Curves of this sort, produced by gravity acting on projectiles rising at an angle, are described mathematically as *parabolas.* (Carl Fries, Jr.)

Physics and volcanoes

There are still other ways of looking at a volcano. We might fix attention, for example, on the chunks of glowing lava blasted out of the crater by explosions in the volcano's throat. Each describes a graceful parabolic curve, some rising high in a narrow curve and others following broadly curved paths out to the side. How much force does it take to project these fragments high in the air? Why are their paths so perfectly parabolic? Does the nature of the curve depend on the size of the fragment? Such questions are related to the volcano, but they are really very general questions that could be asked about any chunks of matter thrown upward and allowed to follow a free path back to earth. We could best get answers by retreating to a laboratory where we could arrange to have chunks of different sizes and shapes projected upward at carefully measured angles by forces that we could control.

Or we might wonder about the rate of cooling of the lava in a flow. The molten rock starts to cool rapidly when it comes to the surface, and covers itself with cooled blocks which it carries downstream. The blocks partly insulate the flow from further cooling, and with this protection the lava can remain molten for distances of several kilometers from its source. How is the cooling rate related to the thickness of cover? Is there a limit on the length of the flow imposed by the rate of cooling? To what extent is the cooling accelerated by the escape of dissolved gas? These are difficult questions, but we could get at least partial answers by preparing artificial lava in a laboratory and measuring its rate of cooling under carefully controlled conditions. Again we start our questions at the volcano, but quickly find our attention shifting to the properties of hot viscous liquids in general.

Or perhaps as we make our way around the volcano we are caught in a sudden cloudburst coming from the eruption cloud high above, and we note that the rain is accompanied by frequent flashes of lightning. Why should electric discharges be produced in this kind of environment? Does the abundance of water vapor, or the rapid upward movement of gas, or the temperature variation in the eruption cloud, have something to do with it? Once more we are asking a general question, and once more we would need a laboratory study to answer it.

With questions of this sort we enter the domain of physics. Like a chemist, a physicist may start with questions prompted by a volcano, but he soon shifts his focus to the fundamental properties of the objects or the materials or the energy relations that he is observing. His answers, like the answers to chemical questions, come primarily from laboratory investigations where the variables can be rigidly controlled. The emphasis in physics, unlike that in chemistry, is on the properties and behavior of matter in general. A chemist is concerned with separating and distinguishing different kinds of matter, while a physicist is more interested in the

effects of force or motion or heat on matter of any kind. Obviously the distinction between physics and chemistry is far from sharp, and equally obviously the interests of a physicist and a geologist will have broad overlap.

One can argue cogently that physics is an even more fundamental science than chemistry, since the differences in properties with which a chemist deals must ultimately depend on the tiny particles which make up all matter. Thus a physicist can maintain with some reason that both chemistry and geology are really just specialized branches of his own science; and it follows logically that a would-be geologist should first master physics, then chemistry, and only with this background tackle the complex problems of the earth. The answer to this argument, like the answer to a similar argument from chemistry, is just that despite the logic our knowledge of physics is hopelessly inadequate to handle the problems of other sciences. It may even be downright misleading in geology, because the methods of a laboratory science are so different from the methods necessary to cope with the complex situations encountered in a natural science.

Applications of physics are vitally important for the solution of many geologic problems. *Geophysics,* the hybrid science that deals with these applications, is in fact one of the most active and rapidly growing of all the earth sciences. Conclusions from geophysics will have a prominent place in many of the following chapters.

Biology and geology

Biology is a science with two faces. In one of its aspects it is a natural science like geology, concerned with the forms and changes and development of the living part of the earth as geology is concerned with the non-living part. Its other manifestation is a branch of chemistry, a highly specialized and complex branch dealing with the structures of molecules that control life processes and particularly the mechanism of genetic inheritance. To the chemical side of biology, geology has only a distant relationship. But the more traditional, natural-history side of the subject is so closely allied to geology that a clean line between them cannot be drawn.

Biology and geology both take as their principal subject for study the solid materials of the planet. The inorganic part of the solid material—rocks and sediment and soil—changes only slowly, and is best attacked in the field with a hammer and a handlens. A biologist also may find a small lens useful in the field, but the objects of his interest are too delicate for a hammer, and they change so rapidly from day to day that the mechanisms of change become all-important for study. How does an organism grow and reproduce and die? What are the functions of its separate body parts? How

does it react to disturbance? What determines the kinds of organisms that will flourish in a given environment? These are questions that a biologist must keep uppermost in his mind, and they are obviously different from the questions that a geologist asks of rocks and landscapes. The differences in kinds of questions and in the tools used to seek answers account for the distinction between the two sciences.

Each science has much to learn from the other. The organisms that flourish in a given place depend on the food supply, and the food supply in turn is determined in large measure by the topography, soil, and underlying rock. Geologic events like a flood or the eruption of a volcano can wipe out whole populations of organisms in sudden catastrophes; yet a volcanic eruption or a flood can ultimately cause living things to flourish in greater numbers than before, as weathering of the volcanic ash or newly deposited silt enriches the soil. Plants contribute to the slow decay of rocks: recall the small plants growing in crevices on the sea cliff, and recall trees that you have seen whose roots have penetrated cracks in solid rock and are slowly widening them. Some rocks consist almost wholly of the remains of organisms: coal formed from plant material accumulated in ancient swamps is one example, and limestone made up of shell fragments is another. To understand the behavior of organisms a biologist must know something of geologic materials and processes, and a geologist is equally dependent on biology for details of both the wearing away and the formation of rocks.

The two sciences are even more intimately intertwined in an historical context. Biology and geology, unlike the laboratory sciences of chemistry and physics, are deeply concerned with development through time—with the evolution of species of organisms on the one hand, and the evolution of rocks and landscapes on the other. A biologist studying past evolution is completely dependent for his material on remains (fossils) found in rocks, and the remains mean little unless he knows something about how the rocks were formed. A geologist trying to reconstruct past environments of the earth's surface finds his most valuable clues in the fossils entombed in ancient rocks, but the fossils betray their secrets only if he examines them in part with the eyes of a biologist.

This is far from the whole story. Back toward the beginning of the last century the discovery was made that rock layers can be recognized and can be traced by the kinds of fossils they contain, and furthermore that fossil assemblages change regularly from each layer to the overlying layer in a sequence. Here was the key that enabled geologists ultimately to arrange rocks over the entire earth in order from the long distant past to the present—and also the key that made it possible for biologists to follow the evolutionary sequence from primitive organisms to the complex forms of the present day. Small wonder that the borderland between biology and geology is a flourishing science in its own right, a specialty that goes by the name of *paleontology* (which in Greek means "study of ancient life").

Other sciences

Anyone who asks questions about the solid earth finds himself exploring not only the borderlands of biology, chemistry, and physics, but large parts of other sciences as well.

Soil science, for example, in a logical classification should be part of geology, since it concerns a solid, inorganic portion of the earth. The origin of soil, the recognition of ancient soils, the incorporation of soil material into rock—these are indeed geologic problems. But the major emphasis in soil science lies in different directions: what makes a soil suitable for agriculture, how is a soil changed by organic activity, how does a soil respond to application of fertilizer, what determines the availability of nutrient elements. Such questions lie outside the usual range of geologic thinking. A hammer and a handlens would help little in attacking them, and a geologist's usual questions about natural directions of change and historical development would be only marginally useful. So the study of soil is properly a separate science, but one with close ties to geology.

The science of water, *hydrology,* is likewise a very close relative. The movement of water in streams and glaciers, the mysterious ways of water underground, the composition of water in lakes and swamps and rivers—these are matters of obvious concern to a geologist, for they affect and are affected by the solid materials over which and through which the water moves. Yet much of an hydrologist's interest is in details of fluid movement, a subject requiring mathematical tools that are familiar to an engineer or physicist but not ordinarily to a geologist. So a student of hydrology must be half geologist, half engineer; and whether his subject is a branch of geology or a separate science is purely a matter of definition.

Study of the oceans is also a study of water, but problems of the oceans are so distinct that they are grouped into another science, *oceanography*. Oceanography requires exploration by ships, hence is distinct from most geological work in its basic techniques. Much of oceanography, too, is concerned with movements and mixing of water masses, problems in which a geologist has only marginal interest. Yet in other senses the oceans are necessarily part and parcel of geology: in them most sedimentary rocks have formed and most of the organisms preserved as fossils have lived, and the waves and currents along ocean coasts are major agents in the shaping of landscapes. Even more important is study of the solid materials of the ocean floor, a study which in recent years has revolutionized geologic thinking about the earth's behavior and history. Is oceanography a branch of geology? This is yet another question to which a yes-or-no answer would be pointless.

Somewhat more distinct is study of the earth's other major fluid, the atmosphere. Two terms are used to describe this study: *meteorology,* if emphasis is on the day-to-day changes that make up weather, and *climatology,* if emphasis is on the long average trends that distinguish different climates. Both depend on analysis of large-scale movements of gases on a rotating and differentially heated globe—a highly mathematical undertaking that is hardly relevant to most geologic inquiries. Geologists are interested in air insofar as it attacks the surfaces of rocks, or moves particles of soil and sediment, or is changed in composition by addition of gases from volcanoes; and one of the goals of geologists in reconstructing earth history is to visualize the changing climates of the past. But the major current problems of meteorologists and climatologists are outside the mainstream of geologic thinking.

Astronomy, although by definition having its major concern outside the earth, is still a subject to which geologists must pay heed. As the third planet of the solar system, the earth must be somehow related to other bodies in this family of planets and satellites, and working out details of the relationship can shed light on the earth's structure and history. Recent close-range exploration of the moon and Mars has made this study more pertinent than ever, for already we have learned much from these nearest neighbors that helps in understanding the behavior of our own planet. Then, too, the ultimate origin of the earth must be essentially an astronomical question. As astronomers change their ideas about the most

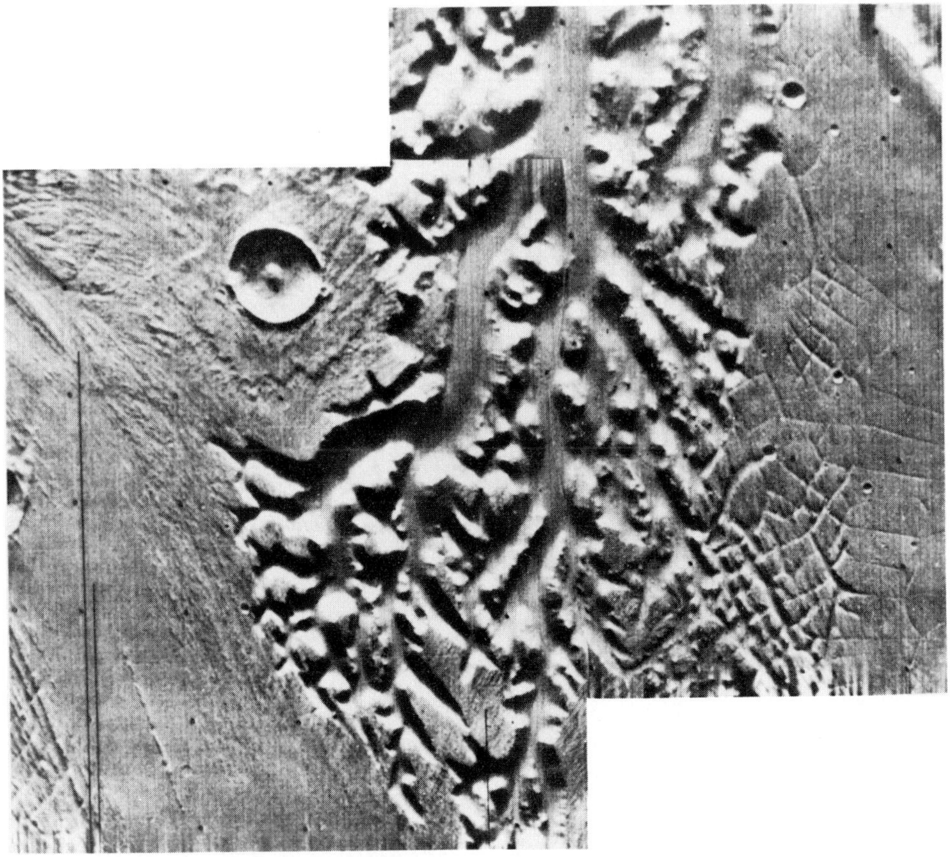

Fig. 2-11. Part of the surface of Mars, representing an area about 50 km wide. The landscapes of Mars have many features that can be recognized, including craters formed by volcanic activity and others formed by meteor impact like those on the moon. Other features remain mysterious. The branching valleys in this picture, for example, resemble stream-carved valleys on the earth, yet Mars has far too little water to maintain flowing streams. Does this mean that the planet had much more water at some time in its past history? (NASA; courtesy of D.J. Milton, *Journal of Geophysical Research*, vol. 78, p.197, 1973, copyright by American Geophysical Union)

probable way for a planetary system to form, so must geologic thinking shift with respect to events in the earth's earliest days.

What is a science?

We call geology a science, then, because it is a body of logically related information about our natural environment, information obtained by application of a reasoning process called the scientific method. Use of the scientific method in geology differs somewhat from its use in a laboratory science, because geologists must apply it to complex and large-scale natural situations rather than the controlled experiments of a laboratory, and because geology is concerned with historical questions as well as questions about contemporary processes. In one sense the laboratory sciences may claim to be more fundamental, since they use the scientific method to study the basic properties of matter and energy. But in another sense geology is equally fundamental, for its concern is with the raw materials from which the refined substances of the other sciences must come, and with the problems that these materials pose in all their pristine complexity both in the present and through their long past history. The scientific method has proved as successful in attacking one of these kinds of problems as the other.

Generalizations like these are useful, but they leave one question unanswered. How do we restrict the subject matter of a science, so as to delimit it from others? In this chapter we have rapidly surveyed a group of sciences, making the customary distinctions among them and examining particularly the points of overlap with geology. The distinctions and the points of overlap have changed during the last century or so, and doubtless will continue to change in the future. Is the definition of a science, then, only temporary, changing as the years go by? At any point in history, what sets off a particular kind of inquiry so that for convenience we give it a special name?

Trying to draw strictly logical boundaries is useless, because the different sciences overlap so extensively. But when we look at any one science, we find a community of interest in particular kinds of questions, in generally accepted methods of attacking the questions, and in a roughly delimited kind of subject matter. Science is a community enterprise, impossible unless its practitioners stimulate one another. In each field scientists must be able to talk with one another, to understand each other's questions, and to criticize each other's methods. It is often said half-facetiously that chemistry is what chemists do, geology is what geologists do, and so on. Such definitions have more than a germ of truth, for the essential characteristic of a science is enough community of interest to make understanding and arguments possible.

What about the borderline fields, subjects like geophysics and geochemistry? Are they sciences themselves, or branches of the major sciences? Here our easy definition gets fuzzy. Parts of geophysics and geochemistry are close enough to geology so that questions and methods are mutually understandable. Other parts have become so specialized, so different from the ways of thinking in ordinary geology, that geochemists and geophysicists can scarcely communicate with their geological colleagues. The separation of the border fields becomes more pronounced as time goes on—in terms of our definition, they become more and more like sciences in their own right.

In a sense this tendency for science to split apart into specialties is only an extension of an historical process that has been going on a long time. All science a few centuries ago was encompassed in the term "natural philosophy," and divisions among the traditional sciences were set up only when knowledge grew to the point where a single individual could no longer hope to master the entire field. With continued expansion of knowledge it is only natural that more and more divisions should seem convenient. Human capacities are limited; a scientist finds that he can accomplish more if he concentrates attention on a single narrow field. The growth of many separate sciences is a natural development, and it has produced a phenomenal increase in scientific knowledge.

Yet one may reasonably question this tendency to extreme specialization. Is it wholly desirable that scientists should separate themselves into little groups that have trouble communicating with each other, and even more so with people in general? This is a good question to keep in mind as we look at recent developments in the science of geology.

Fig. 3-1. The Grand Canyon of the Yellowstone River. A spectacular V-shaped gorge cut by the swift-moving stream. (U.S. Department of the Interior, National Park Service)

3

Reading Landscapes

Take any landscape you are familiar with, and reflect a moment on what it has meant to you at different times of your life. As a very young child you were hardly conscious of the landscape: it was just out there, a fixed part of your childish universe, a distant backdrop for your more immediate concerns with nearer objects that you could touch and taste. Later, when you could move about the landscape without strict parental supervision, it became something to be explored, a world where new objects and new experiences awaited you at every turn. Some of the new experiences were pleasant, some unpleasant, so that parts of the landscape took on emotional coloring. Most of us think of cool woods and stream banks as pleasant parts of a landscape, flat expanses of dust or mud as neutral or mildly unpleasant, areas covered by rubbish as distinctly unpleasant—largely because of experiences remembered from childhood.

Then perhaps you tried to change parts of the landscape for your own purposes—to use its materials for building castles or digging caves, to clear an area for a playing field, to dam a stream for a swimming hole—and the landscape took on a more practical meaning. Later you sought to reproduce parts of the landscape with water colors, charcoal, or oil paints, and your esthetic appreciation deepened. You watched a violent summer storm sweep across the landscape, or you tried to force your way into the teeth of a January blizzard—and the landscape became part of your feeling of awe at the power of natural forces. From memories of this sort come our ordinary reactions to the scenery around us.

A geologist starts with such memories, but he adds to them a special way of looking at landscapes. His way of looking goes back to a particular set of questions: what natural changes are taking place in the landscape? How has the landscape developed in the past? What will it be like in the future? How is man's activity affecting the landscape? What parts are especially suitable for man's use and what parts offer hazards to human occupancy? The questions are seldom stated explicitly in this form. They may not be part of a geologist's conscious thinking, but they describe his approach to a landscape. Answering the questions requires discipline, long practice, development of an habitual way of seeing landscapes—and the answers have deepened our knowledge about processes that shape the earth's surface and about ways in which landscapes can be used most profitably and safely by mankind. The answering of questions by disciplined thinking is what we mean, of course, when we say that the study of landscapes is a science.

Already we have taken a preliminary look at two landscapes from a geologic point of view, a stretch of rugged seacoast and a volcano in eruption. We will now extend this reasoning to other samples of the earth's surface, seeking more general conclusions about how landscapes are formed and how they change. This is a good way to start the study of geology because no sophisticated techniques are required. The questions are simple and straightforward, and many can be answered with little more than use of common sense. The subject of landscapes could be made highly technical, but some satisfying general answers are obtainable with reasoning of a fairly simple sort.

Fig. 3-2. One of the steep Appalachian ridges rising above the gently rolling Piedmont plateau. Delaware Water Gap, Pennsylvania-New Jersey. (John Shelton)

Fig. 3-3. The Front Range of the Rocky Mountains in central Colorado, rising out of the Great Plains on the left. The city of Boulder is in the middle distance, Pike's Peak on the far skyline. (John Shelton)

Why are there hills?

In an area where hills rise abruptly out of a plain, a common query by laymen is, "Why are there hills here?" Along the eastern seaboard, in Virginia and the Carolinas, for example, as one travels west the flat coastal plain gives way to the low rolling hills of the Piedmont, and then quite suddenly one is in the midst of the steep Appalachian ridges. Why do these ridges exist? Even more spectacular are the high Rockies of central Colorado, looming above the flat plain that stretches eastward from Denver. Somehow the hills and mountain peaks demand an explanation. One never hears a similar question about plains—plains are natural, taken for granted. But hills, especially if they are high and steep, are sufficiently striking to arouse curiosity.

What, really, does the question mean? There are several possibilities: Why is there a difference in elevation? Why is the slope so steep? Why does the hill have this particular shape rather than another? Why is the hill here rather than somewhere else? Usually the person asking the question has not tried to make it precise. He has noticed an abrupt change in topography, he has vague memories of reading about sudden dislocations of the ground during earthquakes, and he wonders whether the hill might have been pushed up in some cataclysm during an earlier and more violent stage of the earth's history.

To a questioner with this kind of imagination, a geologist's approach seems disappointing. The geologist would start by looking methodically at the processes going on now which might be changing the hill's shape. The most obvious of these processes is erosion by streams; either in the Appalachians or in the Rockies, or on the sides of hills almost anywhere, the geologist's eye would note the tell-tale effects of stream erosion. In particular he would see the steep-sided V-shaped valleys, ranging from small gullies to valleys a mile or more across, formed by the down-cutting of streams at their bottoms and the slumping and slope-wash on their sides. If he had any doubt that such valleys are the work of streams, a few observations on floors of the larger ones would show the sand and gravel bars and the rounded boulders that are typically moved and deposited by streams in time of flood. So the geologist could assert triumphantly, "The shape of the hill is a consequence of erosion of its sides by streams."

But this answer would hardly satisfy the questioner. "Yes," he would say, "but that's a mere detail. I want to know how the hill itself got here."

So the geologist tries another tack. Stream erosion is clearly not actually producing the hill at the present time; if anything it is wearing the hill down, because sand and gravel coming out of the V-shaped valleys represent material of the hill that is being carried out and deposited on the adjacent plain. Evidently a more detailed examination is needed. The geologist picks up his hammer and handlens and climbs the hill, looking at rocks wherever they crop out on its sides. Study of the rocks leads him to a tentative conclusion. The rocks of the hill and the rocks under the adjacent plain are of many different kinds, but on an average the ones making up the hill are harder than those he finds on the plain. If streams are eroding both, the origin of the hill is clear: it stands above the plain because its rocks are less easily eroded; over a long time the material above the plain

Fig. 3-4. A hillside gully, steeply V-shaped because of rapid cutting by a stream at its bottom and slumping on its sides. (John Shelton)

has been worn away and the hill has been left standing. Thus the hill exists because of *differential erosion.* It stands here because when erosion began the rocks at this particular spot were harder than those of adjacent areas.

By this time the questioner will probably throw up his hands in despair. "But you haven't explained the hill at all! The hill stands *high,* and all you can talk about is erosion, which makes things *low.* How did the rock of the hill get to its present elevation? Surely it must have been thrown up in some ancient upheaval?" The questioner, in spite of himself, is refining his earlier query. What he really wants to know is why the top of this hill is so far above sea level. The elevation is impressive because of its contrast with the plain, but the essential point is that at some time in the past a part of the earth's surface must have been raised far above the level of the sea.

Now the geologist sees the difficulty. "So *that's* what you want to know. Of course, you are right that at some time the whole land surface here must have been elevated so that erosion could take place. It was probably not a sudden upheaval, but this is a part of geology that is still very theoretical. We know that parts of the earth's crust have been raised and other parts lowered at various times in the past, but we are uncertain as to the causes. Now if we could measure the gravity anomaly here . . ."—and the geologist would be off on a technical lecture in which he would lose his listener after the first sentence or two.

Conversations like this are all too common when geologists try to interpret landscapes for their laymen acquaintances. The layman is idly curious about the origin of unusual features in the scenery around him, and he thinks in terms of sudden, catastrophic movements of the earth's surface at some hazy time perhaps a few thousand years ago. The geologist, trained to think of very slow changes taking place over periods of time measured in millions and even billions of years, sees erosion as the principal shaper of most landscape features. For erosion to have been active, some parts of the earth's surface must of course have been higher than others, but from a geologist's point of view differences in elevation are nothing to be marveled at. He knows that the earth's surface is in a constant state of very slow motion, some parts moving upward and some downward at any given time, so that erosion is being accelerated in some areas and slowed in others. Very locally a steep mountain front may be formed directly by earth movement, by a succession of small earthquakes over a very long period of time—and in such places the layman's picture of an "upheaval" is accurate, provided that the upheaval is regarded as a slow process rather than catastrophic. But by far the great majority of hills and mountains, from a geologist's standpoint, are more accurately described as originating by differential erosion.

And erosion, if not otherwise qualified, means stream erosion. Along coastlines waves are a powerful erosional force, in high mountains glaciers play a role in shaping peaks and valleys, where wind is exceptionally strong it may polish and groove rock surfaces, but the great carvers of landscapes are streams. This is abundantly evident from the shapes of valleys—steeply V-shaped where streams are cutting rapidly, more gently V-shaped where gradients are low and the cutting is slower. Even the dry mountains of desert areas, where rain falls but seldom, are carved with V-shaped valleys formed by infrequent but often torrential downpours.

What happens on plains?

A river that comes down from hill country onto a plain changes its character. Think of the Platte River in Nebraska, for example. From its headwaters in the Rocky Mountains of Colorado and Wyoming, the Platte flows out across the high plains, and by the time it reaches Nebraska it has become a broad stream moving sluggishly through a belt of flat land with low hills on either side. Its valley no longer has a V shape, and the material in its bed is now mud and sand rather than gravel and boulders. To all

Fig. 3-5. A gently sloping, stream-carved valley in the Great Smoky Mountains, Tennessee-North Carolina. (W.B. Hamilton, U.S. Geological Survey)

Fig. 3-6. A stream-carved valley with nearly vertical sides—the gorge of the Uncompahgre River near Ouray, Colorado. Why should the slopes in this picture be so much steeper than those in Fig. 3-5? (L.C. Huff, U.S. Geological Survey)

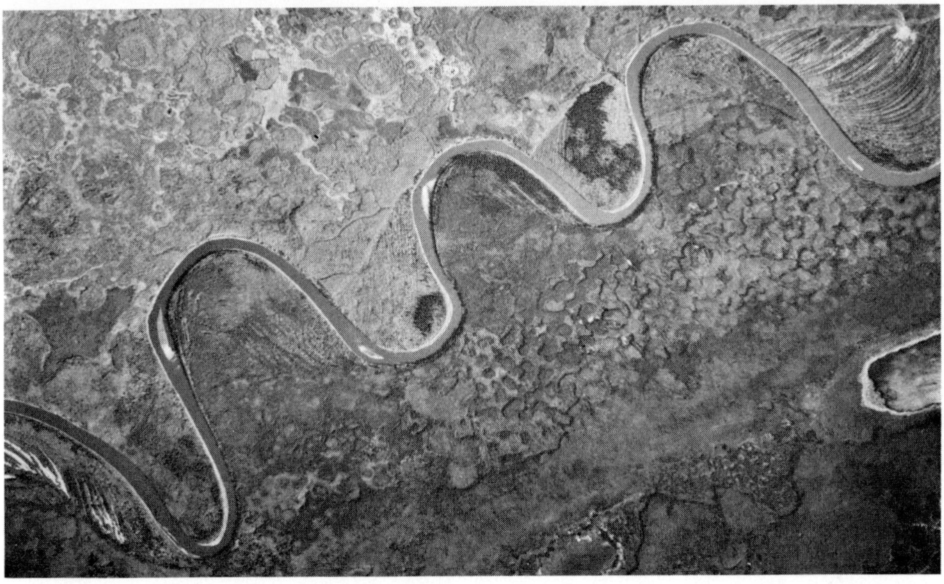

Fig. 3-12. Meanders of the Hay River in a very flat area in northern Alberta, Canada. What details of the picture indicate progressive changes in the course of the river? Can you tell from these details in which direction the river is flowing? (The flow is from left to right.) (Department of Energy, Mines and Resources, Canada)

stream courses, and the scars of recently abandoned stream channels are good signs that deposition is dominant over erosion. In areas of intermediate topography, nearly flat but with perceptible differences in elevation, the overall effect of stream activity may be uncertain. The existence of floodplains along the main streams suggests that deposition may occur at least locally and temporarily, but stream gradients may be high enough to permit slow erosion some of the time. The balance between deposition and erosion is a complex matter, depending on such things as the amount of water, the particle size and amount of sediment, the gradient or slope of the stream course, and the shape of the channel. Small wonder that a single stream may change its character from season to season or even from day to day in stormy weather, sometimes actively eroding and sometimes piling up debris in its channel or over its surroundings.

From a very long perspective—if we think now in millions of years rather than a few hundreds or thousands—the changes that streams produce in landscapes must tend toward leveling of the earth's surface. Active erosion in hilly areas is all in the direction of reducing high elevations, and deposition on floodplains tends to build up low areas. Ultimately, if streams had their way, the continents should be reduced to nearly flat plains across which rivers could wander without accomplishing much of either erosion or deposition.

Volcanoes again

The work of streams is so pervasive that a question about the origin of landscape features can usually be answered by the phrase "stream erosion and deposition"—whether you have seen the landscape or not. At least ninety percent of the time you will be right. But obviously the leveling activity of streams must somehow be opposed, for otherwise no scenery would be left on the earth's surface. Running water has been cutting valleys and flooding plains for untold millions of years, and despite the slowness of these processes they should long since have leveled the highest of the earth's mountains.

One of the opposing forces we have already looked at—the activity of volcanoes in pouring out lava flows and piling up mounds of cinders and ash. In some places this kind of de-leveling activity is the dominant agent in shaping a landscape. Go to the high plateau of northern Arizona near Flagstaff, for example, or central Oregon along the east front of the Cascades, or better yet the central part of Mexico west of Mexico City. In

Fig. 3-13. A volcanic landscape north of Flagstaff, Arizona. Cinder cones of various ages dot the landscape, the most recent one accompanied by a fresh lava flow that emerged from a fissure and moved about 11 km. Much of the area is underlain by older flows, but they are obscured by weathering and growth of vegetation. (John Shelton)

these areas the most conspicuous landscape features are isolated cone-shaped hills, ranging from small symmetrical hills with craters at their tops to high mountains of more irregular shape. Closer observation will show the shapes of old lava flows—tongue-like masses a few tens of feet thick, with steep sides and rough, hummocky surfaces, some so fresh that the jagged lava blocks are still visible and others covered with soil and vegetation. Stream erosion has left its mark on both hills and lava flows: steep gullies carved in the sides of cinder cones, larger valleys cut into the mountains, occasional steep-walled gorges through the older lava flows. But streams are clearly not responsible for the principal landscape shapes: streams do not ordinarily produce isolated hills and mountains or flat tongue-shaped areas raised above the general level. The scenery is chiefly the result of volcanic activity, only slightly modified by the work of streams.

Farther east in Oregon, or across the Columbia River in Washington, you would see a different kind of volcanic scenery. The landscape is volcanic in the sense that all the rocks you see are of volcanic origin—horizontal layers of the dark gray lava called *basalt,* piled one on top of another and easily visible where the Columbia and its tributaries have cut down

Fig. 3-14. Columbia River basalt, near Palouse Falls, Washington. This part of Washington is underlain entirely by basalt in huge, nearly horizontal flows, one on top of another. (F.O. Jones, U.S. Geological Survey)

Fig. 3-15. Two ancient basalt flows exposed in a quarry near Orange, New Jersey. Both show conspicuous columnar jointing, but the joints are more widely spaced in the lower flow. These flows are roughly 200 million years old, in contrast to a scant 20 million for the flows in Fig. 3-14. (J.P. Iddings, U.S. Geological Survey)

through them. The upland surface between the rivers is a monotonous grass-covered plateau; there are no isolated hills and no recognizable tongues of lava. The climate is semi-arid so that permanent streams are scarce, but the networks of branching valleys in parts of the plateau show that running water is active during storms. How do we interpret a landscape of this sort?

The absence of cinder cones and fresh lava, the depth of soil on the plateau, and the well developed system of stream valleys all indicate that volcanic activity here is older than in the areas we looked at before. Furthermore, the activity was of a different sort—not much explosive activity, but huge lava flows that spread in flat sheets for tens and even hundreds of miles over the countryside. The last of the flows was formed long ago, and since then the area has been subject to stream erosion. To interpret this landscape, then, we would have to describe it as partly the work of volcanic action, partly the work of running water. All the details of present valley patterns and valley shapes betray the activity of streams, but the huge pile of material into which the streams are nibbling is the product of volcanic eruptions in a long-distant past.

On the Columbia River plateau the volcanic activity, although ancient by human standards, occurred recently enough so that separate lava flows are still easily distinguishable on the sides of stream-cut gorges, and the inner structures of the flows are beautifully preserved. In other areas where lava is found, the volcanic activity that produced it was still more ancient, so ancient that individual flows are hard for even a trained geologist to trace and most details of structure are obliterated. Parts of the Connecticut River Valley, for example, are underlain by lava, and also much of the region around Lake Superior; yet the landscapes in these areas are typical products of stream erosion, with no detail in the topography that would suggest the former existence of volcanoes. The lava in hillside outcrops is proof positive that eruptions once occurred here, but they were so far back in earth history that all topographic evidence has been obliterated by streams.

Thus volcanic activity opposes the leveling work of running water by piling up material on some parts of the earth's surface. If the activity is continuing today, or if it ceased in the recent geologic past, the isolated hills and lava flows betray the volcanic origin of the landscape. Streams begin their relentless work of erosion as soon as the hills and flows are formed, and if volcanic activity ceases the landscape is progressively altered until eventually all sign of its volcanic origin is lost. How long does the transformation take? By ordinary observation there is no way to assign a time in years, except to say that the times are long in comparison with recorded human history. More sophisticated methods suggest that the volcanic features of northern Arizona were formed during the past few hundred thousand years, the Columbia River plateau roughly ten million years ago, the lavas in the Connecticut River valley about 200 million years ago, and those near Lake Superior at least 1000 million years ago. We will see later how such estimates are made.

Movement of solid rock

Impressive as volcanoes may be, they are clearly not the only agents that work against the leveling tendencies of streams. Volcanic rocks make up only a small proportion of the materials exposed in most of the earth's high mountains. Somehow there must be other forces capable of moving rocks up and down, hence of maintaining highland areas for streams to erode. Because much of the rock in mountains shows no sign that it was ever very hot, these forces must operate on solid rock rather than the liquid rock we see in volcanoes. It violates our instinctive sense of the stability and immovability of ordinary rock to think of it being pushed around, crumpled, broken, and elevated, but the simple geologic evidence is unarguable.

What evidence do we find that solid rock is moving, or can move, in the present-day world? The most spectacular evidence is the movement sometimes observed in earthquakes. Some earthquakes result in visible breaking and displacement of the earth's surface, one side of a crack moving up or sideways with respect to the other side; such movement of the surface must mean that the solid rock beneath has also broken and shifted. But earthquakes are rare, and displacements are seldom more than a few feet. What connection can they have with the formation of mountains?

For answer we again look at landscapes, this time landscapes of a quite different sort. We can make the necessary observations in any one of a number of places in the huge area of the American West called the Great Basin—the arid region between the Wasatch Mountains on the east and the Sierra Nevada on the west, a region from which streams never flow to the sea but lose themselves in dry basins between steep mountain ridges. We choose the front of one of these ridges, preferably one that is conspicuously straight and steep, and sit down in the sagebrush of the adjacent flat valley to think about it (Fig. 3-17).

Is this the kind of landscape that would be produced by stream erosion? We see steep V-shaped valleys carved into the mountain face,

Fig. 3-16. A fresh scarp produced by the Hebgen Lake earthquake in southwestern Montana in 1959. During the earthquake the land on the right-hand side moved down with respect to the left-hand side, leaving exposed the scratched and polished rock surface. (I.J. Witkind, U.S. Geological Survey)

Fig. 3-17. Part of the steep west side of the Panamint Range, southeastern California. Note the straightness of the mountain front and the abrupt change in slope at its base. (B.M. Page)

many with fan-shaped deposits of sand and gravel (*alluvial fans*) piled up at the points where they emerge onto the flat floor of the basin. We can visualize torrents of water pouring down these valleys during occasional desert cloudbursts, eroding their channels and then depositing the debris when their speed suddenly slackens at the mountain front. In a sense, then, this is indeed a stream-produced landscape.

But look at the larger picture: is it usual for streams to form such a straight mountain front, rising so abruptly out of a flat basin? Note that all the mountain valleys are short, none of them coming from far back in the range; note too that ridges between the streams are cut off at the main valley, none of them extending out farther than others. Think of other mountains you have seen: foothills in front of the main range, some ridges projecting far out into the adjacent valley, some streams heading well back in the mountains. An irregular mountain front is what you would expect from long-continued stream erosion on rocks of different hardnesses. But this desert range we have been looking at—how do we account for its straightness, its steepness, its lack of projecting spurs and long stream valleys?

One possibility is to suppose that the major valley where we are sitting—the flat, sagebrush-covered basin between this range and the next one—is itself the work of stream erosion, erosion by a mighty river which once flowed between the ranges but has long since disappeared. The straight, steep mountain front would then be the side of a stream-carved valley. How reasonable is this hypothesis? We can dismiss it quickly by noting some characteristics of the basin: it lacks the V shape of a stream-made valley, it has no large tributary valleys such as we might expect if the main valley had been once filled by a river, and its central part lacks the huge boulders we would expect to find in the bed of a large river flowing through mountainous terrain.

To explore all possibilities we would also wonder if the basin might have been scooped out by a glacier, or perhaps by violent winds at a time when the atmosphere was more turbulent than at present. Yet the slightest acquaintance with glacier-carved scenery or the work of winds at the present time would show that the characteristics of this landscape are very different from the features produced by these agents.

So we think back to what we know about earthquakes, and ask ourselves if this steep mountain wall in front of us might be the result of a huge fracture in solid rock during a particularly violent earthquake. The rock of the mountains might have moved up suddenly several thousand feet with respect to the basin, and ever since then ephemeral streams would have been eating into the uplifted block and filling the basin with their debris. This hypothesis would account nicely for the straightness and steepness of the range and for the short stream valleys carved into its front. The difficulty with the hypothesis is that no earthquake in recorded history has done anything like this. We need here an uplift of thousands of feet, but displacements in the greatest recorded earthquakes are measured in tens of feet.

There is a possible way out if we imagine that the rise of the mountain block resulted not from a single cataclysmic earthquake but from a succession of small ones. Suppose that once in a century, or once in several centuries, an earthquake occurred that increased the height of the mountains by a few feet. Then in a few hundred thousand years the observed difference in elevation would be attained. Erosion, of course, would begin as soon as the uplift started, and would be continuous down to the present, but it is so slow that a few hundred thousand years would be only enough to produce the short, steep valleys we observe. As so often happens in geology, by expanding our time scale we find that processes similar to those going on at present will reasonably account for apparently enormous changes that happened in the past.

If the mountain front has been caused by a long succession of small earthquakes, might we expect that the process is still going on—that tomorrow, or ten years from now, or a century from now, an earthquake will occur that will produce a fissure at the base of the range along which the mountains will be raised another ten feet or so? The Great Basin is one of the parts of North America where earthquakes are most frequent, so that renewed elevation of at least some of the desert ranges seems likely. With this thought in mind we might inspect the base of our particular range more carefully. If we are lucky we will find evidence that earthquakes have indeed occurred here in the not very distant past. The most common kind of evidence is a break in slope of the alluvial fans near the mountain front: an alluvial fan as it is deposited has an evenly sloping surface, but sometimes at the base of desert ranges one finds a shoulder, a sudden steepening, in all the fans for a long distance along the range front. Such a change in slope means that a crack has opened in the fans and the side toward the mountains has moved up, so recently that erosion has not yet had time to obliterate the small cliff (or *scarp*) even in the loose stream-deposited sediment of the alluvial fans (Fig. 10-12). Where this kind of evidence exists, the presence of a still-active break or *fault* in solid rock (beneath the fan) along the mountain front is demonstrated, and future earthquakes can be confidently expected.

In this look at a Great Basin landscape we have assumed that our choice of mountain range was peculiarly fortunate—a range where the geologic relations are ideally clear and simple. Such ranges exist, but they are the exception rather than the rule. Most of the desert ranges show complications: some have an intricate pattern of faults rather than a single break along the mountain front, and some have irregular or deeply indented fronts because fault movement has ceased long ago and erosion by streams is far advanced. Yet even in the more complicated ranges, the working out of geologic details often shows clear evidence that movement along faults in solid rock over times of the order of hundreds of thousands or a few millions of years is responsible for much of the present topography.

Back to the sea coast: terraces and estuaries

Evidence for up-and-down movements of solid rock is also provided by a further look at the sort of coastal scenery we studied in the first chapter.

Waves at work on a steep sea coast, we have already noted, characteristically produce a cliff-and-terrace profile: they cut actively at the base of the cliff, the slope of the cliff is maintained by undercut rocks falling from above, and currents generated by the wave motion spread the debris of erosion over a smooth underwater surface sloping very gently seaward. As wave attack continues, the cliff is pushed farther into the land and the underwater terrace is broadened. Waves erode the rock selectively, wearing away soft parts faster than hard; remnants of particularly hard material may be left as islets (called *sea stacks*) rising out of the terrace offshore.

Now take a longer perspective, and consider how this kind of erosion and deposition would affect the form of a coastline over distances of many miles. Wave attack is selective, but once an area of soft material is eroded into a cove or bay the force of the waves is blunted because they must travel over a longer stretch of shallow water to reach the point of attack. Areas of harder rock, on the other hand, are left as bold cliffs with deep water just offshore, and here the waves can concentrate their attack without hindrance. The tendency of a shoreline to become indented, then, is counterbalanced by the more violent attack of waves on headlands and the diminished erosion in bays. Deep indentations are prevented also by the ability of currents (set up by prevailing wave motion at an angle to the shore) to carry sand and gravel into bays, building up wide beaches or offshore bars. Thus wave erosion on a rocky coast tends to produce a shoreline that is irregular in detail—with bays and promontories of the

Fig. 3-18. Sea stacks on the coast of Oregon. These remnants of harder rock are left projecting above the flat coastal terrace which the waves are eroding. (David Muench)

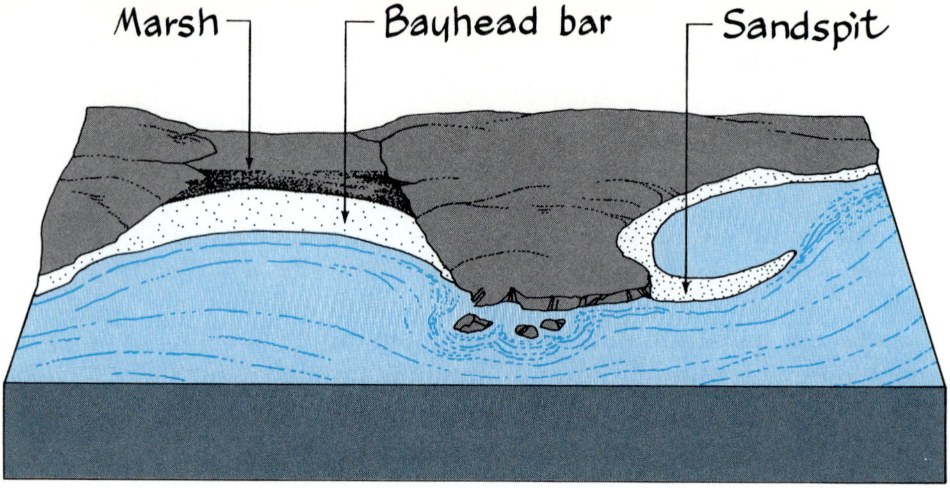

Fig. 3-19. Wave attack is concentrated on headlands, and sediment accumulates in the quieter water of bays.

order of tens or hundreds of feet long—but on a larger scale is straight or nearly so. Deep bays and long promontories cannot be the direct result of wave erosion, but betoken a more complex geologic history.

Long stretches of the Pacific coast of the U.S. show the characteristics of wave-eroded shorelines, but in places the geologic development has been more complicated. One kind of complication is the common occurrence above the present sea cliff of a conspicuous flat or gently sloping area, up to several hundred yards wide and from a few tens to several hundred feet above sea level. This flat, or *terrace,* ends abruptly against the steep slope of the adjacent hills, and its surface is often dotted with rocky hillocks or pinnacles. Where gullies cut through the terrace one may be able to find exposures of sand and gravel beds a few feet or tens of feet thick resting on a smooth surface of bedrock. How are these details of the landscape to be interpreted? (Fig. 1-4)

The terrace clearly has no connection with the wave erosion going on at the present time. The waves nibbling at the base of the cliff, in fact, are actively destroying the terrace. Nor can the terrace be a product of present-day stream erosion; streams have no way of producing continuous flats in hilly terrain except by deposition, and streams today are obviously dissecting the terrace rather than building it up. Again we must consider an hypothesis of large-scale movement of solid rock: we wonder if the terrace might have resulted from uplift of the land relative to sea level. Suppose, for example, that the steep slope where the terrace ends against the hills is an old sea cliff, with its shape modified by erosion; suppose that the terrace surface represents the one-time gentle offshore slope, first eroded by waves and then covered with a veneer of sand and gravel; and suppose further that the isolated rocky hillocks are the remnants of old reefs and sea stacks that existed just offshore when waves were pounding at the base of the hills. The hypothesis fits the observations beautifully. It could be made even more convincing by a successful search for fossils of marine organisms in the sand and gravel layers just under the terrace surface.

One serious question about the hypothesis remains. The terrace is nicely explained by a difference in sea level between the present and some time in the past, but does the difference mean an uplift of the land or a withdrawal of the sea? In other words, has the land in this area actually moved upward, or has the sea level dropped? This would be a hard question to answer in a single locality, but if we travel along the Pacific coast we find that the terrace changes its elevation from place to place, and in many places is absent altogether. In some areas a trained geologic eye can pick out more than one terrace, and the various levels are far from constant. Even if a drop in sea level can account for some of the observations, the changing levels can hardly be explained without movement of the land.

Like the straight, steep fronts of mountain ranges in the Great Basin, marine terraces along the Pacific shore speak eloquently of large-scale

Fig. 3-20. An upraised terrace on the California coast. The flat terrace was eroded when sea level was higher and waves were pounding at the base of the steep slopes. (John Shelton)

Fig. 3-21. Estuaries of the Delaware and Susquehanna rivers. The shapes of these long bays suggest that they are "drowned" river valleys—valleys eroded by the major rivers and their tributaries at a time when sea level was lower and now invaded by salt water as a result of sinking of the land.

movement of solid rock. The association of the coastal movements with earthquakes is less clear than in the Great Basin, for some segments of the Pacific coast have only slight present-day earthquake activity. It may well be that the rock movements here recorded take place slowly and continuously over long periods of time, without the sudden fracturing and release of accumulated energy that characterize earthquakes.

Next we cross the continent for a brief look at a very different kind of coastline, the Atlantic shore from Virginia north to Nova Scotia. On a large scale this coast is exceedingly irregular, each major river ending in a huge estuary that reaches far into the land. How do we read such a landscape? Rivers are incapable of eroding estuaries, for their currents are stopped when they enter the sea. Ocean waves are equally powerless; as we have seen before, waves cannot cut deeply into the land because their energy is dissipated by moving long distances in shallow water and because entrances to bays are commonly blocked by spits and bars. The Atlantic estuaries are being slowly destroyed, not formed, as streams fill them with sediment at their upper ends and as waves and currents pile up sand at their mouths. To account for their origin we appeal once more to movement of solid rock. The pattern of tributaries leading into the estuaries suggests that they were once stream-eroded valleys, cut at a time when sea level was lower than it is today and now drowned by the advancing sea. Again we may wonder whether the change in sea level means an actual lowering of the land or an increase in the amount of water in the sea, and again we suspect that both have played a role. At least some movement of the land is suggested here, as on the Pacific side of the continent, by the fact that the amount of change in sea level differs from place to place. This movement is probably a very slow sinking without accompanying earthquakes, for earthquake activity on the Atlantic coast is very slight.

As we try to read one landscape after another, the conclusions we can reach become more general. The chief agent in shaping details of the land surface is running water, which operates by cutting valleys of characteristic shape in highland areas and depositing the debris in lowland areas. Along coastlines, and to a lesser extent on lake shores, the work of streams is supplemented by waves and wave-generated currents. The leveling activity of streams and waves is counteracted by the escape of liquid rock to the earth's surface during volcanic eruptions and by large-scale movement in solid rock, sometimes accompanied by earthquakes and sometimes not. In statements of this kind we can summarize broadly the shaping of the earth's surface features, but we obviously leave many details and many questions untouched.

Exceptions

Some landscapes don't fit the generalizations we have made. By way of concluding this discussion, we look at a few examples.

Take first the Grand Canyon of the Colorado River in northern Arizona. You approach the canyon by driving for miles over a nearly flat plateau. You come to the rim, and it seems as if the earth has broken open in front of you. You stand for a minute, trying to take it all in: the immense distances, the brightly colored layers of rock, the isolated crags and pinnacles that look like ruins of ancient castles perched on the top of steep hills (Plate 9). Soon someone will tell you that this is all the work of a river, and he will point to the apparently tiny stream of which you can see a short segment far below. But you recall other stream-eroded landscapes, and you object that this one doesn't entirely fit your expectations. A stream-carved valley is supposed to be V-shaped in cross-section; this one has sides of very irregular shape, vertical cliffs alternating with broad platforms of gentle slope. Perhaps overall, on a very large scale, you can discern the shape of a

Fig. 3-22. The Grand Canyon of the Colorado River in northern Arizona. For other views of the great canyon see Plate 9 and Fig. 5-25. (U.S. Geological Survey, oblique air photo No. GS-OAM 1-70)

broad V, but the sides of the V are so intricately stepped with the cliff-and-terrace pattern that the general shape is obscured.

By now your first amazement has worn off, and you begin to piece things together. You note that the rocks in the canyon form thick, nearly horizontal layers. Any given layer is recognizable on both sides of the canyon and on the walls of the pinnacles between. The outcrops of some layers stand in vertical cliffs wherever you see them; especially conspicuous are a light-colored cliff-forming layer (the "Kaibab Limestone") just under the canyon rim, and a bright red layer (the "Redwall Limestone") farther down. Small outcrops on some of the gentle slopes suggest that the rock here is softer and more easily eroded. So the idea takes shape that the cliff-and-terrace profile results from stream erosion in horizontal beds of differing hardness, the resistant beds being left as cliffs and the softer beds being worn down to gentle slopes. To verify this guess you would want to follow a trail down into the canyon so that you could examine the characteristics of the different layers more closely.

The Grand Canyon is the most spectacular example, but the same cliff-and-terrace profile can be noted in many other valleys. Rates of erosion are very sensitive to slight differences in rock hardness, and wherever rock layers are roughly horizontal the eroded edges of the more resistant beds will form steeper slopes than the softer ones. This effect of differential erosion is not limited to the work of streams. In Chapter 1 we noted the ability of waves to carve more rapidly into the softer parts of a sea cliff, leaving the harder rocks projecting; where a sea cliff is made up of horizontal layers, the etching out of the softer beds is especially conspicuous.

For a second example of a stream-eroded landscape with forms different from the usual ones, suppose you are in an airplane flying east to

Fig. 3-23. Ridges of the Appalachian Mountains near Harrisburg, Pennsylvania, as seen from an airplane. Note the general straightness of the ridges and the sharp bend in one of them. Fig. 3-2 is another air view of some of the Appalachian ridges. (John Shelton)

west across Pennsylvania, at an elevation low enough so that the Appalachian ridges are conspicuous features of the landscape. Perhaps at first they look like any other mountain ridges, but if you keep watching you will be struck by their extraordinary straightness and continuity. In a few places a large river cuts through a ridge, but the gaps are narrow and the ridge continues on both sides of a gap. Some of the ridges are symmetrical, some conspicuously steeper on one side than the other. Over long distances the ridges are remarkably parallel; if the ridges of a parallel set are asymmetrical, the steeper slopes of all of them face the same direction. Once in a while you will note a sharp turn in a ridge, the ridge doubling back into a direction almost parallel with itself, and adjacent ridges on either side following the same sharp turn. Shapes of valleys indicate that the landscape has been carved by streams, but the pattern of ridges seems odd. In most stream-eroded areas the principal ridges have a branching, irregular pattern, and their heights are conspicuously variable; here we see even-topped, parallel, continuous ridges that extend for tens and hundreds of miles.

If now you abandon the airplane for a car, and stop to look at roadcuts as you drive through Pennsylvania, you will find that here also, as in the Grand Canyon, the rocks are formed in layers. The difference is that in northern Arizona the layers were horizontal, whereas now in almost every outcrop you will see layers steeply tilted. Each ridge is underlain by a layer of hard rock, its steep side being the eroded edge of the layer and its gentle side the tilted surface of the layer. Erosion of a thick series of tilted beds of alternately hard and soft rock gives the parallel ridges, with valleys eroded into the soft layers between. The beds are not only tilted, but folded on a large scale; and the places where the ridges turn abruptly are the eroded edges of folds.

Fig. 3-24. Tilted beds (limestone and shale) along a highway through one of the Appalachian ridges. The tilt of beds underlying these ridges is also well shown in Fig. 3-2. (W.B. Hamilton, U.S. Geological Survey)

The pattern of parallel ridges and valleys so beautifully shown in the Appalachians is duplicated on a smaller scale in many other places. It is a perfectly general rule that streams carve an area of tilted and folded beds by concentrating their activity in the softer layers, leaving the hard rocks as ridges whose pattern reflects the pattern in which the rock layers were deformed.

From these two examples we can generalize further that any deviation from the kind of landscape normally produced by streams may find a simple explanation in the structure and differences in hardness of the rocks beneath. The two examples we have looked at are ideally simple cases: thick sedimentary beds lying horizontally in the one case, deformed by tilting and folding in the other. More complicated situations can readily be imagined, but the simple rule holds almost universally that the harder rocks of an area betray themselves by standing above the landscape as hills and ridges, or as cliffs on valley sides, while the softer rocks underlie the valleys and gentle slopes. Differences in rock hardness are always the first guess in trying to account for seeming peculiarities of a stream-eroded landscape.

But this kind of explanation will not suffice everywhere. As a third example, which we will not develop here but will save until later, suppose that you undertake a canoe trip in the lake country of northern Minnesota. The work of streams is everywhere evident in the present landscape, but what about the lakes? Can they be produced by normal stream erosion? Streams cut down high places and fill low places; hence streams should destroy lakes rather than form them. That lakes have indeed been destroyed in Minnesota is evident in the very numerous swamps, which represent old lake basins that have been nearly filled by stream sediment. Look also at the pattern of the streams: they wander aimlessly over the landscape from lake to lake, now sluggish and now tumbling over the rapids that make your canoe trip exciting. Look at the ridges: for the most part low, isolated, rounded, with no discernible pattern. This is hardly a normal stream-carved landscape, and differences in rock hardness are only locally helpful in explaining its peculiarities. What sort of geologic history can we read from the lakes and swamps and odd drainage pattern? We will leave this question until a later chapter.

Fig. 4-1. Sandstone. This specimen is more decorative than most sandstones, because it is irregularly splotched with red iron-oxide cement. (Joachim Hampel)

Fig. 4-2. Basalt. The right-hand specimen shows holes left by expanding gas (*vesicles*) and tiny white crystals of feldspar. The left-hand specimen is so full of vesicles that it looks like a sponge. (Joachim Hampel)

Fig. 4-3. Granite. The light-colored minerals are quartz and feldspar, and the dark-colored grains are black mica (biotite). (Joachim Hampel)

4

Rocks, Minerals, Atoms, and Ions

Reading landscapes with little background beyond common sense is a pleasant diversion. It can give us useful geologic insights, but the depth of insight is limited. Details of many landscapes depend on characteristics of the underlying rocks, and we need some understanding of the rocks before the details will make sense. Then, too, landscapes at best are temporary features. From a perspective of millions of years, the positions of hills and valleys, plains and broad seas have changed so radically that we can see little of the earlier history in present-day landscapes. Records of the early history are preserved not in landscapes but in rocks, so again a deeper insight requires that we take a closer look at what rocks are made of and how they are formed.

The word *rock* is so familiar that it hardly needs defining. One of the hard solid inorganic materials that make up substantial parts of the earth—something like this would express the generally accepted meaning. A nitpicker can find fuzzy spots in the definition, as he can in most geologic definitions. Coal, for example, is generally considered a rock, even though its substance is organic rather than inorganic; a pile of decaying vegetation is certainly not a rock; yet we find all gradations between decaying vegetation and coal. Again, sand and mud freshly deposited by a stream would not usually be called rock, but there is no sharp line dividing such material from the consolidated sediments that go by the familiar rock names of sandstone and shale. Despite such borderline uncertainties, the usual understanding of "rock" is clear enough for all ordinary geologic purposes.

Fig. 4-4. A sandstone outcrop. Beds of different thickness are etched out by weathering of softer material between. The beds have a conspicuous tilt; could they have been deposited originally at this angle? (W.R. Dickinson)

Most of the rocks you pick up are heterogeneous, meaning that they consist of two or more different kinds of material visible with a handlens if not to the naked eye. The separate constituents are called *minerals.* Most minerals are definite or fairly definite chemical compounds, which means that their compositions can be expressed by formulas showing the relative numbers of different kinds of *atoms* present. The atoms in most minerals are not neutral but electrically charged, some positive and some negative, the charges always balancing so that minerals themselves are electrically neutral. The charged atoms are called *ions.* It is our purpose in this chapter to look first at rocks, then at minerals, then at the atoms and ions which are the building-blocks of all the earth's materials.

Sandstone

The rock whose origin is most obvious to casual inspection is *sandstone.* It consists of sand grains cemented together—a fact that was just as clear in ancient times as it is in our modern science-saturated world. Sandstone is one of the commonest of rocks, forming conspicuous outcrops in many parts of the country and often used as a building stone. It is so very commonplace and ordinary-looking that when you pick up a piece you are tempted to toss it aside with the thought, "Oh, that's only sandstone."

But let's hold on to the piece for a minute and look at it more closely. A sandstone actually is a mute record of a good deal of geologic history. The sand grains must have formed at some time in the dim past by the weathering and erosion of earlier rocks. Then they must have been carried from their place of origin and heaped together—perhaps by a stream, perhaps by waves and currents along a beach, perhaps by wind on a desert plain. Next the deposit of sand must have been covered by other sediments and slowly cemented into a coherent rock by material deposited from percolating water (water moving slowly through open spaces between the sand grains). And finally the rock layer must have been uncovered by erosion and partly disintegrated to produce the piece that you now hold in your hand. At least some of this history should be decipherable by close inspection of the grains and the cementing material, supplemented if possible by study of the outcrop from which the piece has broken. A sandstone is always an invitation to a bit of geologic sleuthing.

We look first at the outcrop. Typically the sandstone surface will be stained in many colors, mostly dirty yellows and browns. Patches of moss and lichen may give a greater variety of color. These colors are distractions and tell us nothing useful about the sandstone; they are surface features resulting from the action of air and rainwater and the growth of organisms. To get a fresh sandstone surface we must break the rock—and this, of course, explains why a hammer is such an indispensable tool to a geologist. Freshly broken surfaces often have colors quite different from the weathered surface and are generally more uniform from place to place on the outcrop.

Before examining a fresh surface in detail we look at the outcrop for indications of bedding. Since the sand of the rock was originally deposited as a sediment, the layering of the sediment might well be preserved in the rock. If layering is indeed present it will show up as parallel bands of sand with different colors or different grain sizes or different degrees of hardness, and the differences will be visible also in the interior of the rock as more and more fresh surfaces are exposed by hammering. Bedding can easily be confused with some of the surface staining, but the streaks of stain generally do not penetrate more than a few millimeters into the exposed surface. Bedding may also be hard to distinguish from cracks (*joints*) in the rock, but the cracks in general do not separate sand layers of different character. The bedding may be horizontal, or tilted, or vertical; if it is not horizontal or nearly so, the mere fact of tilt means that the solid rock has been deformed or moved after its consolidation. Some sandstone outcrops show no sure sign of bedding even after careful inspection, apparently because the sand grains accumulated in deposits of nearly uniform material in which the original layers are indistinguishable from one another.

Details of the layering—thickness of individual beds, variation in grain size within each bed, continuity of beds—can give important information about the agent that deposited the sand, for the bedding produced by streams and waves and wind has different characteristics. The method of deposition of the sand is also sometimes betrayed by the kinds of fossils found in the rock, although good preservation of fossils is not common in sandstones. We defer the interpretation of bedding and fossils to a later

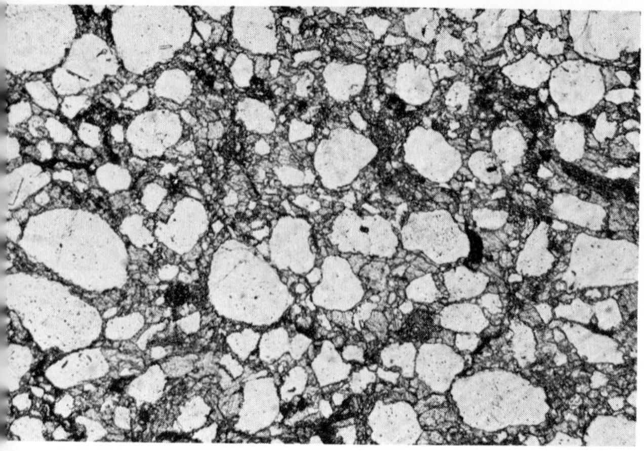

Fig. 4-5. A photomicrograph of a thin slice ("thin section") of sandstone. The magnification is a little greater than you can get with a handlens, and the grains stand out more clearly because light is transmitted through them. In this specimen the grains are largely quartz and the cementing material is calcite. The grains would be described as slightly rounded and poorly sorted. Is there a difference in degree of rounding between the large and small grains? (C.D. Rinehart, U.S. Geological Survey)

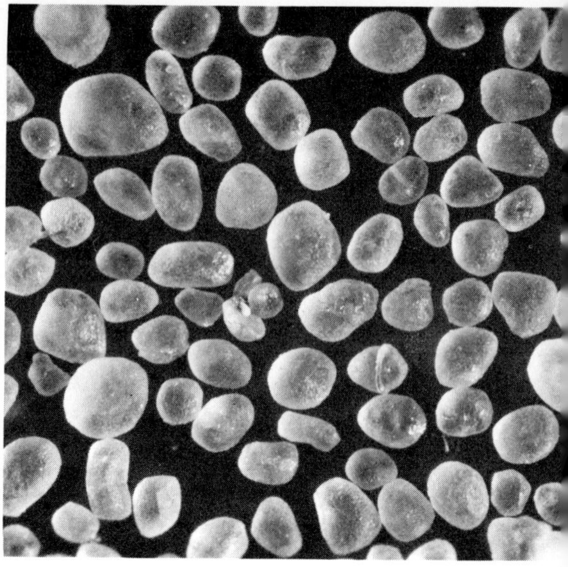

Fig. 4-6. Grains separated from a pure quartz sandstone. In contrast to Fig. 4-5, these grains are very well rounded and well sorted. Keeping in mind that quartz is the hardest of common minerals, what conclusion might you draw from these characteristics about the history of the sand? (K.B. Ketner, U.S. Geological Survey)

chapter, and concentrate here on what we can learn from study of fresh surfaces of the rock itself.

With the aid of a handlens we inspect a freshly broken surface. If the sandstone is typical, we see sharply defined grains set in a matrix (or background) of fine-grained material. The grains can be described in various ways. If the average diameter is greater than about 0.5 millimeter (mm) we describe the sand as coarse, and if less than 0.2 mm we describe it as fine. The grains may range in size from coarse to very fine, in which case we call them poorly sorted; or the size may be nearly uniform, and we say the grains are well sorted. In shape the grains may be sharply angular or smoothly rounded or anything between. The grains may consist mostly of a single mineral or they may obviously represent many different minerals. The most abundant grains of most sandstones are a light-colored, glassy, hard mineral called *quartz*; other light-colored grains that are cloudy and have flat surfaces are *feldspar;* and most sandstones have a scattering of dark grains that represent a variety of other minerals. Recognition of the minerals in sandstone with a handlens is not always easy, but it is generally a safe guess that the light-colored grains are either entirely quartz or a mixture of quartz and feldspar (Plates 16, 18, 21).

Handlens observations of this sort lead to several elementary interpretations. Coarse grains obviously needed stronger currents for transportation than fine grains. Poorly sorted grains were not subjected to the winnowing action of currents for as long a time as well sorted grains. Rounded grains have been abraded for longer times than angular grains, hence presumably were carried farther from their source. Quartz and feldspar are the dominant minerals of sandstone grains because these are the hardest and among the most abundant of minerals in ordinary rocks. A sandstone that contains other minerals in large amount, particularly soft minerals, must represent a sand deposit formed rapidly in an unusual environment. Such qualitative statements are easy to make, but it is far less easy to translate them into quantitative estimates of just how far a given assortment of sand grains was transported or how long a time was involved.

Now we turn from the grains to the cementing material that holds them together. Ordinarily not much detail in the cement is visible under a handlens; its study requires chemical tests or observation with a high-powered microscope. If the cementing material is very hard, so that it cannot be scratched with a knife-blade, it is probably *quartz.* If it is fairly hard but scratchable, and if it fizzes when a drop of acid is placed on the surface, it is *calcite.* If it is red-brown or yellow-brown it is a form of *iron oxide.* And if it is soft and light-colored it is probably *clay.* These are the common cements, but they by no means exhaust the possibilities.

The nature of the cement in large part determines the character of a sandstone. The color of the rock is usually the color of its cement: thus a red or red-brown sandstone is most commonly an aggregate of colorless quartz or quartz-plus-feldspar grains cemented by abundant iron oxide. The hardness of sandstone has an extreme range, depending largely on the amount and nature of the cement: a sandstone consisting of quartz grains cemented by quartz is one of the hardest rocks known, while a sandstone with grains held together only by a little clay may fall apart when rubbed between your fingers. The hardness of sandstone, let it be emphasized, has little to do with the hardness of its individual grains. The grains of almost all sandstones are dominantly hard minerals, but the hardness of the rock depends on how well the grains are cemented together. The permeability of a sandstone to fluids is determined largely by the amount of cementing material between its grains, a fact of central importance in studies of sandstone as a reservoir material for petroleum and groundwater. Permeability, like hardness, has an extreme range, so that some sandstones contain much oil or water and permit them to flow freely, whereas well cemented sandstones may be substantial barriers to all fluid flow.

Sandstone is representative of a large group called *sedimentary rocks,* for the obvious reason that they consist of material originally deposited in the

Fig. 4-7. Beds of conglomerate interbedded with sandstone. From the nature of the bedding, could you make a guess about how the original gravel was deposited? (W.G. Pierce, U.S. Geological Survey)

Fig. 4-8. Steeply dipping beds of shale in a roadcut.

Fig. 4-9. Crystals of kaolinite, the most common of the clay minerals, enlarged 15,400 times with an electron microscope. The crystals are tiny hexagonal plates. (K.M. Towe, U.S. National Museum)

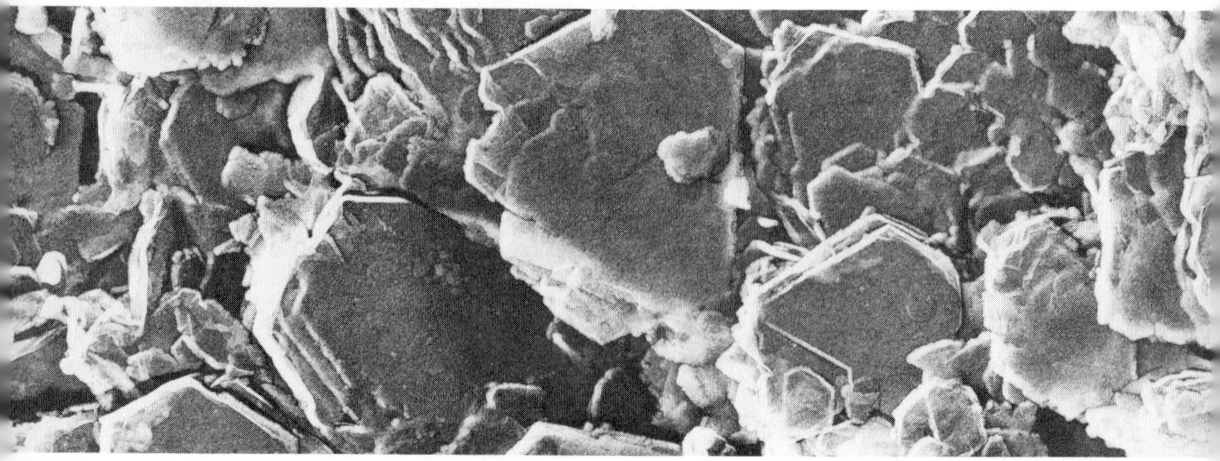

form of sediment. We look now at a few other important members of this group.

Other sedimentary rocks

Cemented sand grains form sandstone, and cemented gravel forms a rock called *conglomerate.* There is no sharp distinction between the two. They have the same kinds of cementing material, and their properties depend in large measure on the nature and amount of cement. The source of original sediment is easier to work out for a conglomerate than for a sandstone, because pebbles show more clearly than sand grains the kind of rock from which they came.

Sedimentary particles smaller than sand grains make up the material called silt or mud, and the general name for rocks formed when such material hardens is *shale.* The tiny particles of shale may consist in part of the same minerals found as grains in sandstone—chiefly quartz and feldspar—but a considerable fraction is the material known as *clay.* Clay is a general name for a group of minerals formed during rock weathering, with the typical properties of softness, very fine grain-size, and plasticity when wet. The individual grains of clay are far too small to be seen with a handlens, but under a powerful microscope they appear as flakes, usually with irregular shapes but sometimes crudely hexagonal. When pressed together the clay flakes stick to one another, so that a mud compressed by the weight of overlying sediments becomes a coherent rock just by the ability of clay particles to cling together. Cementation with quartz, calcite, or iron oxides may aid the process, but many shales have little cementing material beyond the clay minerals themselves. Because clay particles tend to settle on their flat surfaces during sedimentation, the grains of shale are roughly oriented parallel to the bedding; the rock breaks more easily parallel to the orientation than across it, so that typically fragments of shale are thin, flat slabs. This characteristic distinguishes most shales from other rocks: if an outcrop shows thin beds and a tendency to break into flat fragments along the bedding planes, it is very probably a variety of shale. Of all sedimentary rocks shale is by far the most abundant, but it is less conspicuous than others in outcrop because it is generally soft, easily eroded, and subject to sliding on steep slopes.

Another important sedimentary rock consisting of fragments cemented together is consolidated volcanic ash, or *tuff.* This is simply the fragmental material (called *pyroclastic* material) blown out during a volcanic eruption, and deposited in layers after traveling through the air. Most tuff consists chiefly of tiny, sharp-edged particles of glass (Fig. 2-6); if it is deposited in

Fig. 4-10. Limestone containing abundant fossil fragments of marine organisms. Shells of clams (pelecypods) and snails (gastropods) can be recognized. (Joachim Hampel)

thin layers, as tuffs often are, it may resemble shale in appearance, but can often be distinguished by the sharp "feel" of the powder when a little is rubbed between the fingers. Tuffs commonly contain fresh, angular fragments of lava and pumice, which aid in their recognition. All gradations exist between unconsolidated ash and tuff that has been hardened by any of the common cementing agents. The pyroclastic material of some tuffs was so hot at the time it was deposited that the glass fragments have partly fused together, giving a hard variety of tuff called *welded tuff.*

A very different kind of sedimentary rock is the one called *limestone*—which we have met before in two of the common cliff-forming beds of the Grand Canyon (page 81), and which is a common, useful, and often conspicuous rock in other parts of the country. Limestone is clearly sedimentary because it shows bedding, in many places contains fossils, and is interstratified with other sedimentary rocks. But in its structure limestone contrasts sharply with the other rocks we have considered. A handlens does not help much in seeing the structure, because most limestones are very fine-grained, but under stronger magnification the rock appears to be made up of grains with irregular shape and complexly intergrown with one another. These are not detrital grains, as sand grains are detrital; in other words, they are not fragments of other rocks cemented together, but grains that have grown in place and interfered with each other's growth. The kinds of fossils that are fairly common in limestone show that the original sediment must have accumulated under water, so we infer that the material of the grains must have formed as a precipitate from something dissolved in the water—or at least that after deposition their shapes were altered by something deposited from water. Thus we describe limestone as an example of a rock formed from a *chemical* sediment, in contrast to the *detrital* sediments which are cemented together to form shale, sandstone, and conglomerate.

What is the material that is chemically precipitated to form limestone? We examine the rock for a clue, and note first that it is fairly soft—easily scratched with a knife or hammer. When grains are visible with a handlens, as the larger ones often are, they have flat surfaces showing many parallel cracks. A drop of acid on the rock surface causes lively effervescence. These are characteristics we have noted (p. 89) for the mineral *calcite,* one of the common cementing materials of sandstones and conglomerates. When pure, the mineral is white or colorless. Some limestones are almost pure calcite, but most have enough impurities to give the rock color, commonly light gray but locally almost any light color and sometimes even dark gray to black. Thus we conclude that the original sediment that hardened into rock must have been largely calcite precipitated from water.

Because the formation of limestone requires precipitation from water, calcite must be somewhat soluble. The solubility is not evident in simple experiments with calcite: if you put chunks of limestone in water and stir, as you would do to demonstrate the solubility of sugar, the chunks remain obstinately undissolved. If you take the trouble, however, to make delicate chemical tests on the water after stirring, you will find that a little calcite has indeed dissolved, much more than would be dissolved in similar experiments with quartz or feldspar. Also, if you examine a limestone

Fig. 4-11. Beds of limestone exposed in an old quarry in Missouri. (S.N. Davis)

Fig. 4-12. Photomicrograph of a thin section of limestone. The rock consists of crystals complexly intergrown. (M.G. Dings, U.S. Geological Survey)

Fig. 4-13. A weathered limestone outcrop. The steeply dipping beds are etched out, and the surfaces are grooved and roughened, by slow dissolving of the slightly soluble rock in rain water.

Fig. 4-14. Limy mud exposed at low tide along the Persian Gulf. The numerous holes are made by boring shrimps. The scale is given by the rod, on which each black bar is 10 cm long. (G.M. Friedman)

Fig. 4-15. Beds of chert separated by thin interlayers of shale. The irregular pinching and swelling of beds is common in chert outcrops. (E.H. Bailey, U.S. Geological Survey)

outcrop you will find evidence of slight solubility. In contrast to most rocks, an exposed limestone surface is conspicuously rough, commonly grooved and furrowed, with cracks deeply indented—evidently the work of rain-water slowly dissolving the surface material. By ordinary standards calcite does not appear very soluble, but over the long time periods considered in geology its slight solubility is sufficient to account both for the formation of limestone and then later for its slow dissolving when exposed to rain and running water.

If limestone results from precipitation of calcite, where can we look for such a process taking place in the modern world? The coast of Florida is one such area; even better is the shallow sea water of the Bahama banks. Here fine-grained particles of calcite (and a mineral of similar composition called *aragonite*) are accumulating in enormous quantity. Much of the material—what proportion is not certainly known—consists of shells and fragments of organisms, for many marine creatures use calcite and aragonite as material to build their shells. The particles accumulate as loose sediment, but as they are buried by continual deposition they are slowly converted into solid rock. The consolidation is more complicated than the cementation of sandstone, because it involves the dissolving of some grains and the growth of others, eventually leading to the pattern of intergrowing grains we find in older limestones—a process that goes by the general name of *recrystallization.*

Thus limestones—or at least many of them—can be accounted for by the formation of limy mud, in part by direct precipitation and in part by the activity of organisms, followed by slow recrystallization after burial. Such processes must have been commonplace in the geologic past, for limestone is a very abundant rock. From a human standpoint this is a fortunate circumstance, because limestone is the most commercially valuable of all ordinary rocks—for building-stone, road surfacing, manufacture of cement and other building materials, smelting of ores, and a host of other industrial processes.

Several other sedimentary rocks form by consolidation of chemical precipitates, but for the moment we will note only one—the rock geologists call *chert.* In popular language chert is known by many familiar names signifying particular colors or patterns, for example flint, agate, jasper, carnelian. In any of its varieties chert is extremely hard, tough, fine-grained, breaking like glass on rounded concave surfaces that leave very sharp edges. These are the properties that account for its wide use in primitive societies to make arrowheads and cutting tools. Because of its hardness and toughness chert survives long battering during erosion, and it

Common Sedimentary Rocks

Rocks formed from detrital sediments (fragments of other rocks):

Rock	*Name of loose sediment*	*Size of principal fragments*
Conglomerate	Gravel	Diameter more than 2 mm (0.1 inch)
Sandstone	Sand	Diameter 0.1–2 mm (0.005–0.1 inch)
Shale	Clay or mud	Diameter less than 0.1 mm; larger grains barely visible with a handlens
Tuff	Volcanic ash	Mostly fine-grained, may include large fragments

Rocks formed from chemical precipitates:

Limestone: consolidated limy mud; chief mineral is calcite, chemical composition is calcium carbonate

Chert: chief mineral is quartz, in very fine grains; chemical composition is silicon dioxide (silica)

Other sedimentary rocks, somewhat less abundant:

Siltstone is a detrital rock with fragment sizes between those of sandstone and shale. **Dolomite** is a rock much like limestone, but formed of the chemically precipitated mineral dolomite (calcium magnesium carbonate) rather than calcite. **Rock salt** is formed as a chemical precipitate of salt (sodium chloride or halite) when a concentrated brine evaporates. **Coal** is formed by the partial decay and consolidation of plant fragments.

is accordingly the commonest rock variety among the pebbles of gravel and conglomerate—although in general, as a constituent of earth materials, chert is not abundant. If you pick up from a gravel deposit pebbles that have interesting color patterns, and that break to give smooth surfaces with a somewhat "greasy" appearance, you have probably selected pieces of chert. Your handlens will tell you little about the rock, but greater magnification would show that it consists of very tiny intergrowing particles of quartz. The existence of chert as a sedimentary rock must mean that quartz, like calcite, is slightly soluble in water; but its solubility is very much less, and the formation of chert must take correspondingly longer times or must require more unusual conditions.

Common Igneous Rocks

Usual color:	*Light*	*Medium*	*Dark*
Minerals:	**Quartz*, feldspar,** dark minerals	**Feldspar,** dark minerals	**Dark minerals** feldspar
Coarse-grained (intrusive or plutonic)	**granite**	**diorite**	**gabbro**
Medium-grained			**diabase**
Fine-grained (extrusive or volcanic)	**rhyolite**	**andesite**	**basalt**

Other igneous rocks: **Granodiorite** and **dacite** resemble granite and rhyolite, respectively, and cannot easily be distinguished without a microscope; they have plagioclase in excess of orthoclase. **Peridotite** is a coarse-grained rock consisting entirely of dark minerals, with no quartz or feldspar. **Obsidian** and **pumice** are glassy varieties of rhyolite or dacite.

*NOTE: Boldface in the heading indicates the most abundant minerals. The feldspar of granite and rhyolite is chiefly orthoclase (K-feldspar) with minor plagioclase (Na-Ca-feldspar). The feldspar of other rocks in the table is chiefly plagioclase.

Igneous rocks

When we watched earlier the eruption of a volcano, we witnessed an entirely different method of rock formation—the cooling of molten material to form an *igneous* rock, in contrast to the *sedimentary* rocks we have been discussing here. The kind of volcano we were watching, characterized by the building of cinder cones and free-flowing lava, most commonly produces a variety of igneous rock called *basalt.* This is a hard, fine-grained, gray rock, generally containing a few crystals large enough to be visible with a handlens, often with rounded holes representing bubbles of gas trapped as the hot lava froze. By far the commonest of volcanic rocks, it is the rock which in common speech goes by the name of "lava." Other volcanoes produce different kinds of lava in great variety. We mention here only two fairly common ones, *andesite* and *rhyolite,* and describe them very roughly by noting that typically—but far from universally—rhyolite is a white or light-colored rock and andesite has shades intermediate between those of rhyolite and basalt.

Fig. 4-16. Rhyolite. Note the small crystals (quartz, feldspar, black mica) set in a light-colored, very fine-grained groundmass. (Joachim Hampel)

Lavas are sufficiently hard and resistant to survive erosion for long times, long even on the expanded time-scale of a geologist. Lavas, in fact, are found among the earth's oldest known rocks, and their widespread occurrence with rocks of different ages is a sure indication that volcanoes have been active at many times and places during the earth's history. Commonly lava occurs in flows buried beneath sedimentary layers. Often from a distance an old basalt flow in an outcrop may look like a sedimentary bed, just another layer in a sequence consisting otherwise of sandstone, shale, and limestone. Close examination is generally sufficient to show the igneous origin of the basalt—the characteristic small, straight-edged crystals, the bubble holes, the reddened spongy rock at the top of the flow, perhaps the baking of a thin layer of sedimentary rock beneath the flow. But sure identification is not always easy, as is evident from the fact that the early history of geology was made colorful by a long controversy over the question of whether buried basalt is an igneous or a sedimentary rock.

Under a microscope basalt proves to be made up of several kinds of minerals, the grains intergrowing in much the same way as do calcite grains in limestone. In contrast to limestone, some of the grains in basalt have straight edges, the edges being parallel or at sharp angles to one another (Plate 23). The intergrowing grains seem reasonably explained as a consequence of rapid growth in a fluid medium, the grains touching and interfering with each other's growth as they get larger. Thus a pattern of grains with irregular, intergrowing edges may form either by precipitation out of solution in water or by freezing of a melt. This pattern is hammer blow opens a cavity in a rock outcrop. What connection is there between these geometric crystal forms and the rounded or shapeless mineral grains in rocks?

The geometric form of a crystal, if you think about it a minute, must be a consequence of some inner structure of the crystalline substance. This is strikingly shown by the growth of a crystal in the laboratory: suspend a tiny grain of alum in a saturated alum solution, and let the solution evaporate slowly; over a period of days the alum crystal grows to a large size, maintaining the same shape during its growth; evidently something in the inner structure of the solid determines how new material is added from the solution to preserve the flat crystal faces and the sharp angles between them. The alum crystal maintains its shape because nothing in the liquid interferes with its growth. If the evaporation takes place rapidly, however, so that many small crystals start to grow in all parts of the solution, no one crystal can grow large without interference from its neighbors and the result is a mass of irregularly shaped and partly intergrowing grains. But each grain in the mass has the characteristic internal structure of alum, and we speak of alum as a crystalline solid even though separate crystal forms may not be visible.

So, in nature, perfect crystal forms develop only when crystals grow freely into cavities or cracks or soft material, and perfect crystals are rare because such situations are uncommon. The same minerals, however, when they occur as constituents of ordinary rocks, have the same internal structure that exists in the well-developed crystals, and in this sense we speak of them as crystalline solids. The internal structure may be revealed in a number of ways, particularly by the effect of mineral grains on polarized light and on X rays. For detailed study of the minerals in rocks the use of these two forms of radiation has proved indispensable, but their effects are observable only with special equipment.

Fig. 4-20. Calcite has good cleavage in three directions at oblique angles (about 75°).

The inner structure of many minerals betrays itself in another property that can be observed more directly. This is the property called *cleavage,* the ability of a mineral to break or cleave in certain preferred directions. It is well illustrated by a large crystal of rock salt: if the crystal is struck with a hammer it breaks into many fragments, most of the fragments being cubes or at least having flat sides that meet at right angles. The tendency to break into right-angled fragments is evident in the sets of parallel cracks that traverse any large chunk of rock salt. Formally, we describe this characteristic of salt by saying that the substance has *good* cleavage in *three* directions at *right angles* to each other. A crystal of calcite breaks similarly under a hammer blow, but its fragments are rhombic, with oblique angles between the flat sides; hence calcite has good cleavage in three directions at angles of about 75° with each other (Plate 20). Mica has a single direction of excellent cleavage, which accounts for its ability to break into thin sheets. Quartz, by contrast, has no direction of good cleavage, so it breaks to give rough surfaces like the surfaces of broken glass.

Cleavage is a very useful property for recognizing minerals, but it is unhappily the property that means most trouble for a beginner. In different minerals cleavage may be good, poor, or completely absent; it may be in one direction, two directions, three directions, or (rarely) several directions; the angles between cleavage directions may be anywhere between 45 and 90°. When a piece of rock is broken, each mineral grain (if it has cleavage) tends to break along a cleavage direction. The flat or stepped surfaces so often visible on mineral grains are cleavage surfaces. A mineral that has more than one cleavage direction will show its other cleavage or cleavages by parallel cracks on the flat surfaces. With a little experience these various aspects of cleavage can be easily recognized.

One frequent source of confusion is the distinction between cleavage surfaces and crystal faces. Remember that cleavage can be found (or can be developed by hammering) in any part of a crystalline solid, whereas crystal faces appear only on the outside of a crystal that has grown freely. Nearly all the flat surfaces that you see in broken rocks are cleavage surfaces rather than crystal faces.

To use cleavage confidently in identifying minerals requires considerable practice, but other mineral properties are less troublesome. Color is an obvious characteristic of some minerals—the black of biotite (black mica), the brassy yellow of pyrite ("fool's gold"), the red of hematite (an iron oxide), the green or blue of many copper minerals. The shades of light-colored minerals may sometimes be confusing, because minerals that are white or colorless when pure may contain small amounts of other materials that color them strongly; for example, quartz is commonly glassy or white, but some varieties are purple (amethyst), pink (rose quartz), or nearly black (smoky quartz). The hardness of a mineral is easily tested by trying to scratch it with a knife, or by scratching the knife-blade with the mineral. Some minerals are notably heavy (formally, have a high *density* or a high *specific gravity*); thus garnet is conspicuously heavier than most ordinary rock minerals, and gypsum is conspicuously lighter. These properties—cleavage, color, hardness, density—are the commonest mineral characteristics that a geologist depends on for identification in the field.

Mineral displays in large museums include hundreds of different varieties, but fortunately the minerals that are abundant in most ordinary rocks are limited to a dozen or so. We will cut down this number even further by lumping some varieties into mineral groups and by limiting the present discussion to minerals in the most common sedimentary and igneous rocks.

Quartz (Plates 11, 16). The hardest of ordinary minerals: scratches steel or glass easily. Colorless, white, any light shade, rarely dark; in rocks, most commonly colorless and glassy. No cleavage; broken surfaces rough, like broken glass. In igneous rocks occurs generally as grains of irregular shape intergrown with others. Crystals lining cavities are characteristically six-sided prisms surmounted by six-sided pyramids. In the form of chert, quartz occurs as microcrystalline masses of almost any color, breaking to give smooth curved surfaces with a waxy appearance. Chemical composition: silicon dioxide, SiO_2.

Feldspar (Plate 18). Nearly as hard as quartz. Colorless or any light color, most commonly white or pink; in contrast to quartz, generally milky rather than clear. Good cleavage in two directions, at right angles or very nearly so. Feldspar grains in igneous rocks generally have a roughly rectangular shape with conspicuously flat cleavage surfaces, and traces of the other cleavage are generally visible as cracks parallel to the long dimension of the rectangular faces. Feldspar grains in sandstones are often recognizable by their flat cleavage surfaces. Chemical composition: silicate of aluminum, calcium, sodium, and potassium. (Example of a feldspar formula: $KAlSi_3O_8$.) Feldspar is technically the name for a group of closely related minerals; the two major varieties are *plagioclase,* an aluminosilicate of sodium and calcium, and *orthoclase,* an aluminosilicate of potassium. The distinction is not easy without a microscope.

Dark-colored minerals (ferromagnesian minerals). The black or dark greenish grains of igneous rocks include a number of minerals with widely differing properties. Distinctions among them are not important here, although we will single out a few for attention in later chapters. One easily recognizable dark mineral is *biotite* or black mica, conspicuous by its softness, excellent cleavage in one direction, and shiny surfaces. Other abundant dark minerals are harder; some have conspicuous cleavage in two directions, so that flat cleavage surfaces (in contrast to biotite) have parallel cracks. Chemical composition: silicates of magnesium and iron; in particular ones, also calcium, aluminum, potassium. (Examples: olivine, $(Mg,Fe)_2SiO_4$; biotite, $K(Mg,Fe)_3AlSi_3O_{10}(OH)_2$; pyroxene, $Ca(Mg,Fe)Si_2O_6$.) Dark colors are characteristic of minerals containing magnesium and iron, hence the name "ferromagnesian."

Calcite (Plate 20). Soft enough to scratch easily with a knife, but too hard to scratch with a fingernail. Colorless, white, any light shade, rarely dark; in the form of limestone, most commonly light gray. Excellent cleavage in three directions at oblique angles; rhombic fragments of broken calcite and the rhombic pattern of cracks on calcite cleavages are diagnostic. Effervesces in a drop of acid. Crystals of calcite on the walls of cracks and cavities have a variety of shapes, some rhombohedral and some roughly hexagonal, superficially resembling quartz. Chemical composition: calcium carbonate, $CaCO_3$.

Clay minerals (Fig. 4-9). Soft enough to scratch with the fingernail, light in weight. White or any light color, rarely dark. Always in aggregates of microscopic or sub-microscopic grains; large crystals and cleavage surfaces never observed. The clay minerals are a large group with a variety of properties; some, for example, show much more plasticity than others when moistened. Chemical composition: various; the commonest clays are silicates of aluminum and hydrogen, many containing magnesium also. As an example, the simplest clay is *kaolinite,* $H_4Al_2Si_2O_9$.

Atoms and ions

Minerals are chemical compounds, some with very definite compositions (quartz, calcite) and others with compositions variable over a considerable range. The compositions of the simpler minerals can be shown conveniently by chemical formulas, as illustrated in the last section. What do these formulas mean?

Take calcite for an example. A chemist would name the compound calcium carbonate, and write its formula $CaCO_3$. Either the name or the formula would tell him that the compound is composed of three elements: calcium, carbon, and oxygen. The last two are not explicit in the name, but the word *carbonate* always refers to the combination carbon-plus-oxygen united with a metal (in this case calcium). Each of the elements consists of

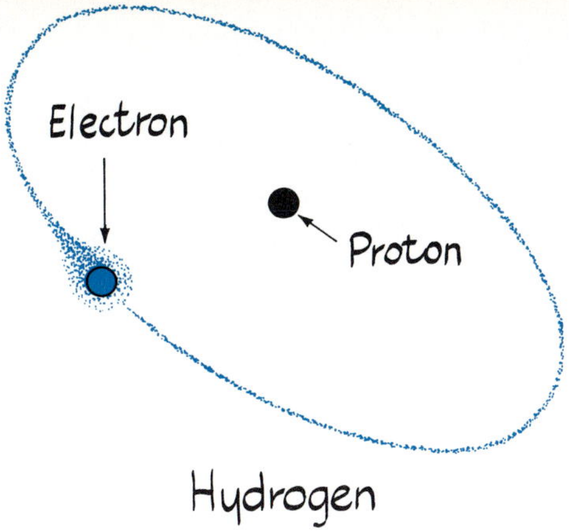

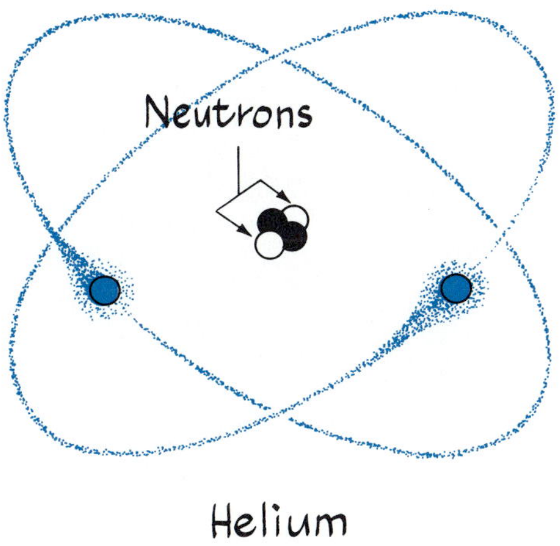

Fig. 4-21. Symbolic drawings of hydrogen and helium atoms. The H atom consists of a single proton and an orbiting electron; the He atom has a nucleus with two protons and two neutrons, and two orbiting electrons. Atoms of other elements have more complex nuclei and additional electrons.

atoms, the atoms of any one being all alike but differing from those of the other two. The chemist uses symbols for the elements, e.g., Ca, C, and O. Each symbol has two meanings: (1) it is an abbreviation for the name of the element, and (2) it represents a single atom of the element. With a combination of the symbols the chemist can show the proportions in which the atoms combine; thus he writes the formula $CaCO_3$ to indicate that in calcite every atom of calcium is accompanied by an atom of carbon and 3 atoms of oxygen. (Note that the subscript "3" refers only to the symbol just before it, O.) This kind of shorthand is a convenient way of condensing a good deal of chemical information about minerals.

How does the chemist know that the formula of calcite is $CaCO_3$ rather than, say, $CaCO_4$ or Ca_2CO_3? This is a matter of analysis, of breaking down the compound in the laboratory and measuring the relative weights of the three elements, and then comparing these weights with the relative weights of the atoms. It is a seemingly complex procedure, but one with ample justification that has been standard practice in chemical laboratories for more than a century and a half.

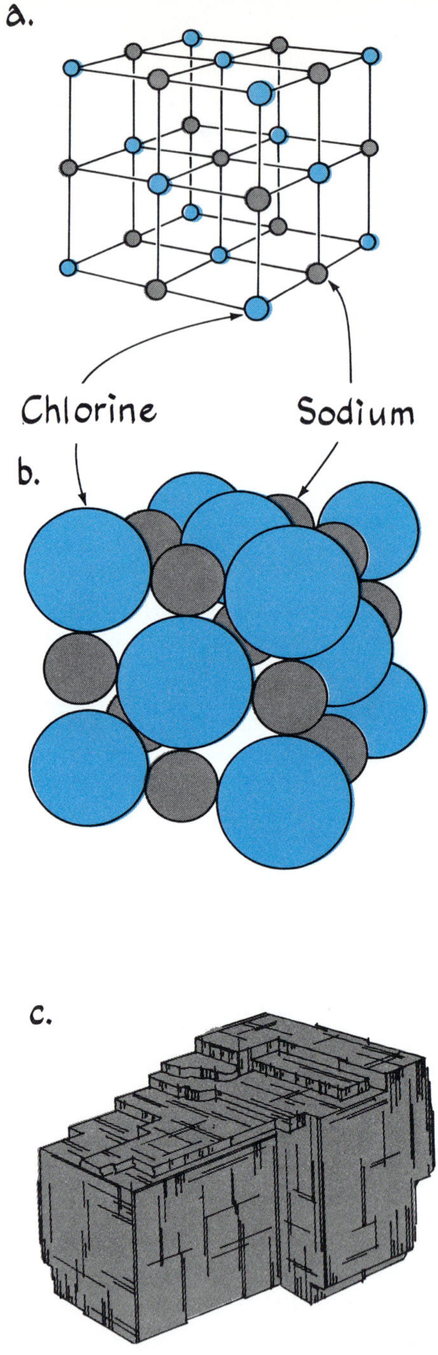

Fig. 4-22. A crystal of rock salt (halite) consists of sodium ions (Na^+) and chloride ions (Cl^-) in a rectangular array of alternating Na^+ and Cl^-. The upper drawing shows the arrangement in space; the middle drawing shows the ions drawn to scale and touching each other, as they would in a crystal; and the lower drawing shows how the right-angled cleavage of the mineral is related to the crystal structure.

Minerals are often grouped according to their chemical compositions. Thus calcite is representative of a group of minerals whose chemical formulas all include the combination CO_3, and which are therefore called the *carbonate minerals.* Quartz, with the formula SiO_2, may be classed with *oxide minerals,* those which consist of oxygen united with a single other element; the two common oxides of iron, hematite (Fe_2O_3) and magnetite (Fe_3O_4), would be other members of this group. Another important class consists of elements that occur in nature all by themselves (as "native" elements), for example gold (Au), diamond (C), sulfur (S). The most numerous mineral group of all is the *silicates,* represented in the above descriptions by feldspar, ferromagnesian minerals, and clay minerals. The word "silicate" means a compound containing silicon, oxygen, and one or more metals. The formulas of most silicate minerals are forbiddingly complex; as fairly simple examples, the orthoclase variety of feldspar would be written $KAlSi_3O_8$, a simple member of the ferromagnesian group (pyroxene) as $Ca(Mg,Fe)Si_2O_6$, and the simplest of the clay minerals as $H_4Al_2Si_2O_9$.

Thus most minerals are chemical compounds, meaning that they consist of elements whose atoms are combined in specific numerical ratios. Since atoms are the building blocks of elements, and since minerals have internal structures as shown by their crystal form, their cleavage, and their effect on X rays and polarized light, it is a reasonable inference that the internal structure consists of a definite geometric arrangement of the atoms.

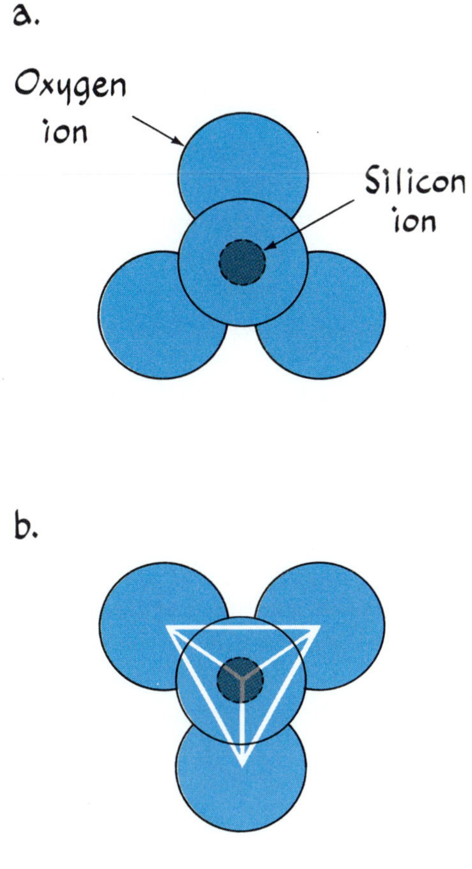

Fig. 4-23. The silicon-oxygen tetrahedron, consisting of a Si^{4+} ion surrounded by four O^{2-} ions.

One can even go further with this sort of speculation. If atoms are pictured as small hard spheres of different sizes, perhaps the structure is determined by the way these different-sized spheres can arrange themselves to fill space.

It turns out that this speculation is pretty close to reality, but it needs one major modification. Atoms are not hard spheres but aggregates of electric charges, the negative charges (*electrons*) of an atom moving in orbits around the tiny concentration of positive charge at its center (*nucleus*). When atoms combine to form chemical compounds, the outermost electrons in their structures are disturbed. One form of combination involves the loss of an outer electron (or more than one) by one kind of atom and the gain of an electron by another; in different language, chemical combination is effected by the transfer of an electron from one atom to another. The first atom, then, is left with an excess of positive charge, and the second atom has acquired an excess negative charge. These atoms with electric charges are called *ions;* the kind of chemical combination involved is described as the formation of an *ionic bond,* and the compound formed is an *ionic compound.* The reactions involved in forming most minerals are of this type, and most minerals are therefore ionic compounds.

Thus for most minerals the units of structure are not atoms but ions. The number of charges on the positively charged ions must balance those on the negatively charged ions, because minerals themselves are electrically neutral. It is the ions that can be pictured as small hard spheres (although the "hardness" is just a crude way of saying that other particles cannot easily penetrate the powerful electric fields in an ion's electronic structure), and it is the sizes of the ions that determine how a crystalline solid is put together. This is an over-simplification, because some ions can distort others out of their spherical shape (in other words, the bonds are not entirely ionic), but to a surprising degree the properties of crystalline solids can be predicted simply from knowledge about the kinds of ions present and their relative sizes.

The science of crystal structure ("crystallography") is a fascinating subject, but one that would take us far afield from our central purpose. We mention just one detail that is important in understanding the chemistry and mineralogy of the silicates, the most abundant of all mineral compounds. In any silicate the silicon ions (Si^{4+}, an Si atom that has lost 4 electrons) and oxygen ions (O^{2-}) are arranged in a particular pattern, four oxygen ions forming the corners of a tetrahedron around a central silicon ion. This combination is very stable, and may itself act as an ion (SiO_4^{4-}) in mineral structures. The SiO_4 tetrahedra can be linked together in various ways (Fig. 4-24), and the different possible linkages account for the

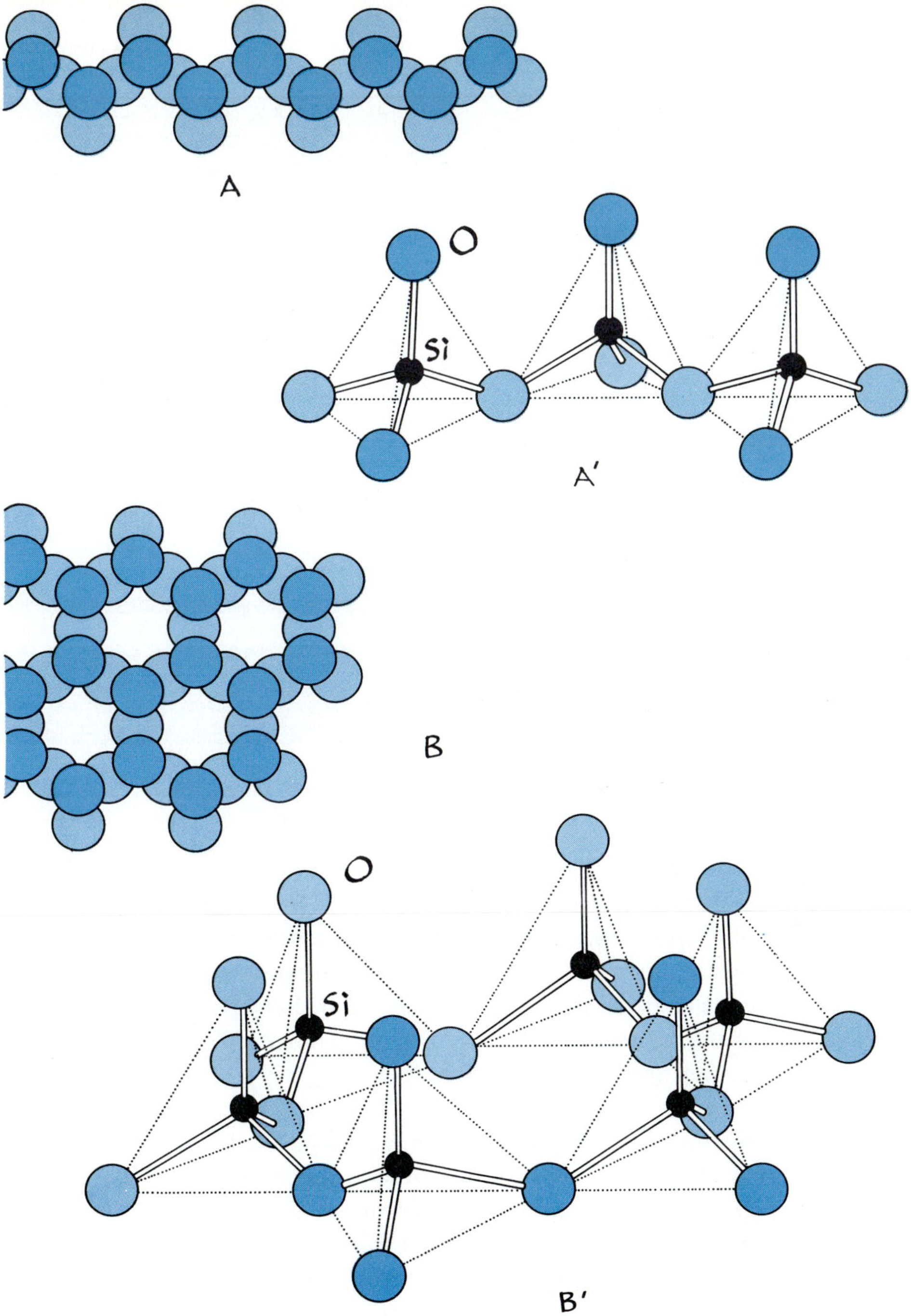

Fig. 4-24. Examples of the linkages of SiO_4^{4-} tetrahedra in silicate mineral structures. The linear arrangement shown in A and A′ is found in the ferromagnesian mineral called pyroxene, and the arrangement in sheets with a hexagonal pattern (B and B′) is characteristic of the micas and clay minerals.

This "landscape" is a slab of sandstone with colored bands produced by deposition of small amounts of iron oxide. (Edwin B. Eckel)

enormous diversity of silicate structures found in minerals. For some minerals the correspondence between inner structure and outer form is fairly clear; a particularly good example is mica, whose structure consists of two-dimensional sheets of linked tetrahedra parallel to the conspicuous cleavage. For many minerals, however, the outer form shows little obvious relation to the inner arrangement of ions.

By way of summary: In this chapter we have started with the solid materials that make up the crust of our planet and have torn them apart, first into the elementary substances that compose them and then into the particles responsible for their inner structures. We could have used the opposite sequence, building complex structures from simple units. Thus we might have begun by noting that all the earth's materials consist ultimately of electrically charged particles—the positive nuclei of atoms and the negative electrons around them. Nuclei and electrons are aggregated into atoms and ions, and these larger particles join together in infinite variety to form chemical compounds. Naturally occurring compounds, made up chiefly of ions, constitute minerals, the characteristics of different minerals depending in large part on the geometric arrangements of their ions. And finally, aggregates of minerals are the constituents of ordinary rocks.

Not only rocks on the earth, but samples brought back from the moon and rocks that have fallen from space as meteorites, consist for the most part of familiar minerals. Scanty information obtained by instruments about the surfaces of other planets indicates that on these bodies also many familiar substances are present. The same pattern that we have worked out on earth, from rocks to minerals and then to atoms and ions and electrons, is very probably the pattern on which most of the solid material of the universe is constructed.

Fig. 5-1. Dinosaur tracks in sandstone, near Rocky Hill, Connecticut. (Peabody Museum of Natural History, Yale University)

5

The Record in the Sediments

In the Connecticut River Valley a prominent rock is red sandstone. The sandstone beds are flat or gently sloping, and in places erosion has exposed large areas of the bedding surfaces. The exhumed surface of a bed is the record of an instant in geologic time, the brief period just after this sediment had been deposited and before other sediment had been laid on top of it. Most of the sandstone surfaces in the Connecticut River Valley show little detail; they are simply flat expanses of well cemented sandstone, typically ending in a small cliff on one side and overlain by soil or other sandstone beds on the other. A few of them, however, are startlingly different, for they are criss-crossed by the tracks of animals embedded in the solid rock. Most of the footprints are three-toed, ranging from tiny prints a centimeter or so long to huge ones with lengths of a few tens of centimeters. They are the kind of tracks we would identify as bird tracks on a modern mudflat—but what enormous birds some of these creatures must have been! Actually they were not birds at all. From similar tracks preserved elsewhere, in places where bones also are found, we know that these are prints left by giant reptiles of species that have long been extinct. Thus a sandstone surface gives a vivid picture of what happened on a single day in the earth's long-distant past, a day when lizards of many kinds and sizes walked across a moist surface of freshly deposited sand and left tracks to be covered by the next layer of sediment.

The record in the sediments is particularly striking when they contain fossils, as they do near the Connecticut River, but a great deal can be learned from the nature of the sedimentary rocks themselves. It is the inorganic rock structures that will occupy us in this chapter, and we will come back to the fossils later.

Where do sediments come from?

You look at a sedimentary rock, and ask what sort of history it records. By this question you mean that you are going to try to reconstruct the conditions that existed when the original sediment was deposited. To approach an answer you will need to know something about the kinds of sediment that are being deposited in various environments at the present time. This involves an unconscious assumption—you guess that sediments of the past were laid down under conditions like those that are producing similar sediments today.

Where does the material of sediments come from? Clearly, from the weathering and erosion of pre-existing rock. That rock slowly weathers, or decays, is evident from the softened, pitted, discolored surface of almost any rock that has been long exposed to wind and rain. Not only does the rock surface disintegrate, but weathering penetrates deeply along cracks and loosens blocks of partly weathered material. The products of weathering—angular blocks, grains of resistant minerals, and new-formed clay and iron oxide—may fall from steep rock surfaces to accumulate as piles of *talus,* or, on flatter surfaces, remain in place to form the raw material of soil.

The debris of weathering is easy prey to erosion, especially to erosion by running water. Watch, for example, the work of a stream in time of flood: the moving water picks up loose material from its bottom and its

Fig. 5-2. Talus accumulating at the base of steep slopes in the Sierra Nevada, California. (C.W. Chesterman, California Division of Mines and Geology)

Fig. 5-3. The wall of a gravel pit, showing the typical rough sorting and uneven layering of stream-deposited sediment. (John Shelton)

banks, material that includes soil and weathered rock fragments and the more rounded grains and pebbles of earlier stream deposits. If bedrock is exposed the water may add to its load of debris by actively cutting into the rock, using pebbles and sand grains to wear down the rock surface. The grains and pebbles have their sharp edges and corners smoothed off in the process. Because the water is turbulent the fragments become roughly sorted according to size, the coarser pebbles being moved only by the swiftest flow and dropped wherever the speed slackens, while the fine-grained material is winnowed from them and deposited in quieter water. As the flood subsides and the stream returns to its usual tranquil state, the material it was carrying is dropped to form sand bars, gravel bars, and mudbanks. If you imagine these deposits to be buried, hardened, and converted into sedimentary rock, the rock should betray its origin by its rough layering, crude sorting of coarse and fine material, and partial rounding of the pebbles and coarser sand grains.

Such structures in a rock may suggest that it originated as sediment laid down by a turbulent stream, but somewhat similar structures can be produced also by waves and currents on a beach. Here again the moving water handles most easily material already loosened by weathering, but is capable of wearing away the solid rock of a sea cliff by driving sand and pebbles against it; again the turbulence sorts debris roughly into coarser fragments that accumulate where movement is most violent and finer particles that are carried out into quieter water offshore; and again the long continued motion smooths and rounds the edges of sand grains and pebbles. Evidently the distinction between stream sediments and shallow-water marine sediments is not always easy, and must be based on something more subtle than the mere existence of sorted and rounded grains.

Fig. 5-4. A sandstorm on the plain west of the Salton Sea in southeastern California. The wind is blowing toward the camera, lifting dust high in the air and sweeping sand along the ground. Note the crescent-shaped dunes at the left. (John Shelton)

Still another agent that can produce rounding and sorting is the wind. Winds are seldom strong enough to move particles larger than sand grains, but they can pile up sand into ridges (*dunes*) by rolling it along the ground and sweep the fine particles of clay high into the air. Once again we are dealing with currents in a turbulent fluid, the shifting strands of current having different speeds and hence different carrying capacities; the stronger currents carry the larger particles, rounding and polishing them by driving them together or against rock surfaces.

One can imagine other ways for sediments to form that do not involve turbulent currents. Glaciers, for example, push ahead of themselves mounds of debris (*moraines*), scraped by the ice from its bed and dumped without sorting and with little rounding of fragments. Some sedimentary deposits require no active agent of transportation and deposition at all, but only the force of gravity: talus piles at the base of steep slopes are the most common example, but the debris of landslides and mudflows has a similar origin. Some of the sediment at the bottom of the deep oceans and large lakes consists of dust particles brought long distances by wind and then deposited by settling slowly through the water; some volcanic ash is a similar sediment that has settled quietly out of the air. To complete the list we would have to add the chemically or biochemically precipitated sediments that ultimately form limestone, chert, and salt deposits, and other sediments that consist largely of remains of organisms.

Such a survey of sediments forming in the world around us gives an idea of the kinds of structures that can be looked for in trying to read the history of older rocks. The survey also suggests how difficult the interpretation may be. A major stumbling block is the fact that similar sediments may form in very different environments: a bed of sandstone, for example, may represent an ancient deposit formed in a stream bed, in the ocean, or in desert sand dunes. To distinguish between such sediments requires a good deal of ingenuity, and is not always possible even for experts. But the information to be gained is important enough to repay a closer look at the kinds of sediment and sedimentary rocks formed in particular environments.

Lake sediments.

The sediments that accumulate in lakes are good ones to start with because of their relative simplicity.

Compared with the ocean, lakes are bodies of quiet water. Currents are generally feeble, limited to transporting only the finest mud. The kind of mud deposited may differ from one season to another, particularly if organisms grow luxuriantly for only part of the year. The mud deposited during a major storm may be slightly coarser than normal. Such differences produce recognizable beds, typically very thin beds of remarkably uniform thickness. If seasonal differences are pronounced, the beds may show regular alternations of color. In permanent lakes of the present time these

Fig. 5-5. Material pushed into a ridge by an ancient glacier in Wisconsin. The fragments range in size from fine clay to large boulders, heaped together with no sign of sorting or layering. A low ridge of this sort is called a *moraine,* and the material is called *till.* (W.C. Alden, U.S. Geological Survey)

Fig. 5-6. The thin, very uniform beds of clay suggest deposition in quiet water, probably an ancient lake. (F.O. Jones, U.S. Geological Survey)

characteristics of bedding can be seen only in cores brought up from the bottom, but they are often conspicuous in shallow cuts through sediments left by ephemeral lakes (*playa lakes*) in the semi-arid region of Nevada and Utah.

The typical thin, uniform bedding of lake deposits is formed only on the level bottom away from shore. Closer to shore, where the activity of waves and wave-generated currents is stronger, sand may accumulate in considerable quantity. The shores of large lakes may be broad sandy beaches, much like beaches along the seashore; the famous beaches around the south end of Lake Michigan, for example, have sand in such quantity that wind heaps it into dunes that cover broad areas. The mouths of streams that enter lakes are also places where coarser sediment can be deposited, for any sand and pebbles that a stream carries must be dropped when the moving water is suddenly slowed as it enters the lake. Thus lake sediments may show considerable variety: thin parallel beds of fine-grained material are characteristic, simply because currents in lakes are generally too weak to transport larger fragments, but such beds grade into coarser debris locally near the shore.

Where would one look for lake beds preserved from the past? Particularly clear examples can be seen in the plains and foothills around the Great Salt Lake in northern Utah. The present salt lake is the last remnant of a larger lake which once covered much of the northwestern part of the state, and slowly evaporated as the climate grew drier. Old shore lines are visible as horizontal terraces on the mountainsides near Salt Lake City, each terrace representing a time when evaporation slowed and the lake level remained constant for an extended period, long enough for waves to cut a cliff in the mountain slopes and for currents to deposit the derived sand and gravel in an offshore terrace. The mud deposited in deeper water is now visible in places where gullies and roadcuts expose material just under the sagebrush-covered surface of the lower slopes and flat areas around the present lake; the uniformly fine grain size and the thin beds are good indications that this material is sediment deposited in quiet water. These beds date from only a few hundred thousand years ago, and for the most part they remain nearly horizontal, just as they were deposited.

Then go east in Utah, through the hilly country south of the Uinta Mountains. Occasionally along the highway, both in the valleys and in the higher hills, you will see spectacular roadcuts through many tens of feet of shale in the thin, uniform layers that are characteristic of lake deposition. This material is harder than the lake beds of the Salt Lake valley, and the beds in some roadcuts are tilted at moderate angles. If you are lucky enough to find fossils, and if you give them to a paleontologist to identify, he will tell you that they are remains of fresh-water organisms—which confirms your guess that the beds were deposited in a lake—and also that

Fig. 5-7. The conspicuous flat terraces along the base of a ridge south of Salt Lake City represent ancient shorelines of Great Salt Lake, at a time when the lake was much higher than at present. The terrace materials are sand and gravel deposited by waves and currents in the lake. (John Shelton)

Fig. 5-8. Lake beds of the Green River formation, near Castleton, Utah. The rock consists of alternating shale and very fine-grained sandstone. Note the extraordinarily uniform thickness of the layers, here tilted at a slight angle.

they are species that flourished 60 or 70 million years ago. You are looking, then, at the record of a much older lake, old enough so that its beds have been tilted, folded, and broken by movements of the earth's crust.

If your interest is aroused by the old lake, and if you try to determine its original size by following its beds from place to place, you will find yourself embarked on a very ambitious project. For this was a huge body of water, extending over much of what is now eastern Utah, southwestern Wyoming, and northwestern Colorado. In Wyoming some of the beds are impregnated with salts, indicating that the lake at times during its long history partly evaporated. In places the salts are abundant enough for commercial exploitation. Potentially more valuable commercially are some of the lake beds in Colorado, which contain large quantities of organic matter from which the constituents of gasoline and lubricating oils can be distilled; these are the famous *oil shales,* a potential source of petroleum fuels that will become increasingly important as petroleum itself grows scarcer. Geologists call these old lake beds collectively the "Green River formation," the name coming from a town in southwestern Wyoming where the beds are well exposed.

Still in pursuit of lake beds, you might next turn attention to some of the older rocks of eastern Utah and western Colorado, rocks that lie beneath the Green River shales. You will find a succession of shale and sandstone beds, many of them with bright red colors indicating the presence of iron oxide as an important cementing material; these are extensions of the beds that form the cliffs and gorges and arches of the beautiful canyon country of southern Utah and northern Arizona. Among these beds is one group of red shales and fine-grained sandstones that can be followed over an area of many hundreds of square miles, and that shows by its very uniform bedding and its fossils that it represents sediments deposited in a still older lake. Among the fossils the most spectacular are the bones of dinosaurs, giant lizards that lived along the shore or wallowed in the mud of shallow areas or swam in the deeper water of the lake. Geologists call these beds the "Morrison formation," and assign them an age of the order of 150 million years.

Utah and its neighboring states have unusual quantities of sediments laid down in ancient lakes, but similar deposits are not uncommon elsewhere. The outstanding characteristics of all such sedimentary rocks are the fine grain size and thin, uniform beds. These features are not sure proof of deposition in a lake, because similar sediments can be laid down in quiet water in the sea, but the evidence becomes certain if fossil remains of fresh-water organisms can also be found.

Stream sediments

Streams are the great carriers of sediment, the movers of debris from highlands to lakes and oceans. As sediment is carried downstream, some of it may be dropped as sandbars and piles of gravel and expanses of mud. Such deposits are ephemeral, generally picked up and moved along at the next time of high water. Yet in some places, particularly on floodplains, stream-deposited material may accumulate and remain long enough to be buried by other sediment. If such material becomes compacted and cemented into solid rock, what characteristics would enable you to recognize it?

Think first of the debris deposited just below a hillside gully, or dropped by a rain-swollen rivulet that courses down a fresh roadcut or the side of a pile of dirt. Currents in the moving water are rapid and shifting; particles are picked up in one place, deposited in another, then picked up once again. Sand is the dominant material in one miniature bar, mud in another. As sand sweeps over a tiny bar, it is deposited on the lee side on a curved and sloping surface. If the current shifts and a new bar is deposited on the remnants of the older one, the slope of the surface of deposition may be different. If a sudden shower increases the flow, a channel may be cut through earlier deposits. If the flow diminishes, mud may be deposited on top of earlier-deposited sand. In contrast to the quiet environment of a lake bottom, all is activity and rapid change.

Now, in imagination, magnify this tiny turbulent scene to the scale of a large river emerging from mountains onto a plain, and a time of years or centuries rather than minutes. Debris of all grain sizes, from pebbles and boulders to fine mud, will be dropped in roughly sorted piles. Where main channels have been cut through earlier deposits, the sides of the channels will show cross-sections of old sandbars and gravel bars one on top of another; each bar will be roughly *stratified,* the beds showing the slight curvature of the bar's one-time surface, and being cut off at an angle by the curving beds of the next bar. This kind of stratification goes by the name of *cross-bedding,* and is characteristic of situations where grains accumulate in piles as they are swept along the surface of deposition. Depending on the size of the stream and the distance the debris has traveled, the sand grains and larger fragments may have any shape from angular to well rounded.

This imaginary reconstruction suggests the chief characteristics of rapidly deposited stream sediments: rough sorting of coarse and fine fragments, beds differing widely in thickness and individual beds thinning and thickening erratically, much cross-bedding, generally only partial rounding of larger fragments, local cutting off of beds by lens-shaped

Fig. 5-9. Cross-bedding in stream deposits. The thin beds at an angle to the main bedding planes are parts of old sand bars, formed by a stream with shifting currents on an ancient floodplain. Supai Sandstone, Grand Canyon. (E.D. McKee, U.S. Geological Survey)

Fig. 5-10. Cross-bedding in marine sandstone. Structures very similar to those in Fig. 5-9 are produced by the accumulation of sand in shifting sand bars in shallow sea water. Tapeats Sandstone, Grand Canyon. (E.D. McKee, U.S. Geological Survey)

Fig. 5-11. Cross-bedding in wind-deposited sandstone. The wedges of thin beds at various angles are cross-sections of ancient sand dunes, piled one on top of another as the wind changed its direction. Note that cross-bedding in itself does not permit a distinction between wind-deposited, stream-deposited, and wave-deposited sand, but cross-bedding on a very large scale, as in this picture, is a good indication of the work of wind. Navajo Sandstone, southern Utah. (W.R. Hansen, U.S. Geological Survey)

deposits of gravel and sand representing old stream channels. Sedimentary features of this sort are commonly seen in the alluvial fans formed where streams descend from steep mountains into desert basins in the western states. They are not common in older rocks because such accumulations of the past have generally been destroyed by continued erosion of the rugged country in which they formed.

Where streams flow more sluggishly across plains, and deposit material more widely over the plains in time of flood, the sediment has more chance of survival. The manner of deposition and the kinds of sediment differ only in degree from those of alluvial fans. Coarse material is absent, because the streams flow too slowly to move it. In and near the main channel of a stream, sandbars and mudbanks may be conspicuous. When a stream overtops its channel during floods, the coarser-grained part of its load (sand) is dropped near the channel, on the low levee that generally borders it, and the slow-moving water that advances over the plain deposits only a layer of fine mud. The sorting of coarse from fine is much better than in an

Fig. 5-12. Stream sediments deposited on an ancient floodplain, showing a conspicuous channel scoured by the stream and then filled with coarse sand. These deposits date from the Eocene Epoch, about 60 million years ago. (Warren Hamilton, U.S. Geological Survey)

alluvial fan, but still imperfect; the bedding is more even, especially in the fine-grained material away from the channel, but cross-bedding on a small scale is prominent in the sand of the levees and the main channel. Thus in older deposits of this kind, now buried and solidified into rock, the tell-tale characteristics that would betray floodplain origin are the thin beds of fine-grained material grading into cross-bedded sandstone and in places cut by shallow channels filled with sand. Sedimentary beds of this kind are common among the rocks that underlie the coastal plain of the southeastern states and the states bordering the Gulf of Mexico.

Perhaps a difficulty has occurred to you. A sedimentary layer that represents an old lake deposit, we have concluded, should consist largely of fine-grained material with thin, uniform beds, but the beds may grade into sand where the ancient shoreline was located. Now we describe floodplain deposits in much the same language: fine-grained, evenly bedded material grading into the sand of old levees and stream channels. How can we be sure of telling them apart? The difficulty is compounded when we reflect that floodplains are often the site of large shallow lakes, dammed when a shift in the stream channel creates a low barrier; hence floodplain deposits may be expected to include lake deposits locally. Making the distinction is not always easy, especially in places where outcrops are few, and can lead

Fig. 5-13. The delta of the Mississippi River.

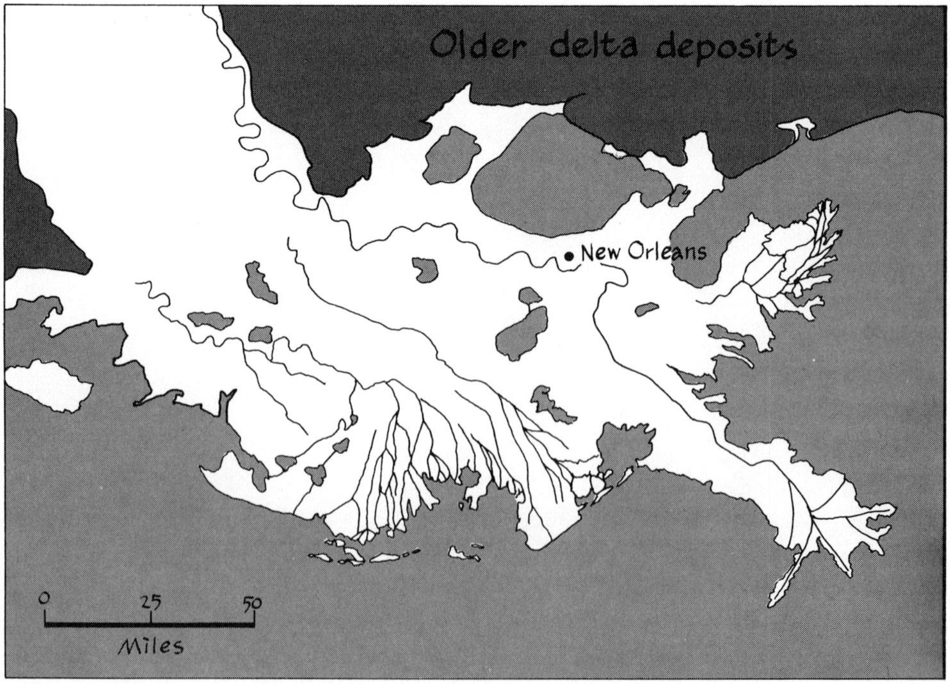

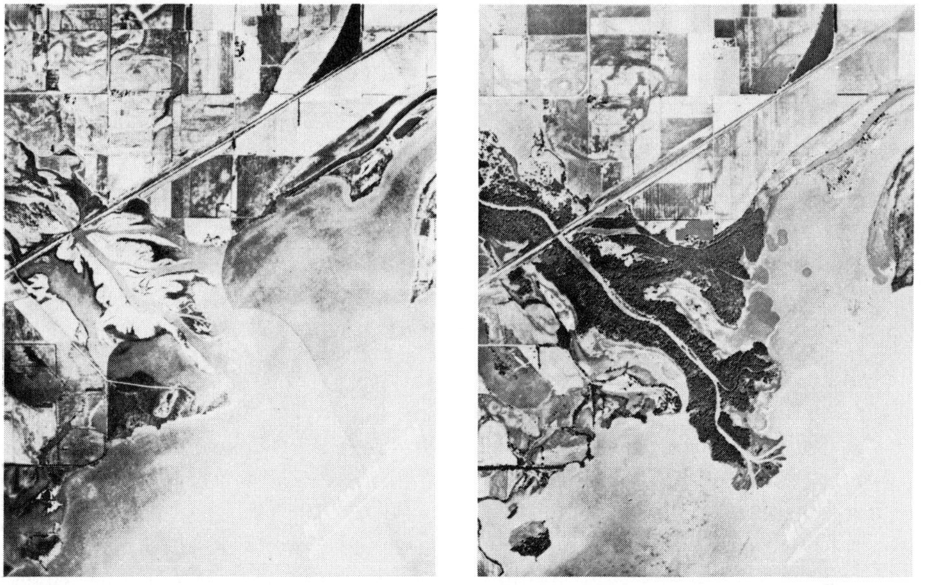

Fig. 5-14. The growth of a small delta. The two pictures are vertical air photos taken 26 years apart, showing a delta being built by a small tributary bringing sediment into a quiet part of the Mississippi River in Iowa. (H.P. Guy, U.S. Geological Survey)

to much geologic arguing. Typical lake beds are more uniform over long distances, grading into sand in only a few places, and lacking old stream channels; floodplain deposits generally have less uniform material, are much more commonly interbedded with and gradational into sand, more commonly show cross-bedding, and are interrupted by stream channels. But differences of this sort are fairly subtle and not everywhere readily distinguishable.

Floodplain deposits may be gradational into lake deposits, and also into marine deposits. Think, for example, of the broad floodplain of the lower Mississippi Valley: at its mouth the river is building out the shoreline by depositing material brought down from its huge drainage basin, and the deposited material is in part being worked over and redeposited by the waves and currents of the Gulf of Mexico. Technically we say that the Mississippi is building a *delta* into the Gulf—a delta being a deposit formed where a stream moves into a body of water and drops its load because its speed is suddenly reduced to zero. Material of the delta, then, should show some characteristics of a stream deposit and some of a marine deposit. The intermingling of stream activity and the activity of waves and currents is shown not only on the present Mississippi delta, but also by structures observed in drill cores obtained from sedimentary rocks underlying the floodplain far upstream from the present Gulf. Evidently the great river has been pushing its delta out into the Gulf for a very long time.

(Text for Chapter 5 continues on page 140.)

Legends for Plates 1 through 26

Plate 1. Edge of a continent. A beach and sea cliff at a time of low tide and calm water. During storms, waves pound at the base of the cliffs, and this constant undermining maintains the nearly vertical slopes. The waves sweep back over the shallow, gently sloping terrace, which is here partly exposed seaward of the beach. Cliff-and-terrace topography like this is a sure sign of wave activity on a steep coast. (David Muench)

Plate 2. Life in a tide pool. The camera is pointed straight down at a small rock basin, where plants and animals grow in profusion despite the rugged environment. Twice each day the basin is exposed to air at low tide, and then at high tide is covered and pounded by waves. The green flower-like creatures are animals called sea anemones, the bright orange is a sponge, and the brown and magenta colors are algae. (California Academy of Sciences Picture Collection)

Plate 3. Hot spring and fumarole. Near Lassen Peak in northern California volcanic rocks at depth are still hot enough to heat and partly vaporize groundwater that comes in contact with them. The warm water rises to the surface in the blue pool at the lower right, and gases escape in the fumarole above. Acid constituents of the gases produce the white and brownish alteration of the volcanic rock nearby. (John Shelton)

Plate 4. The volcano Helgafell. This volcano, near the outskirts of a prosperous town on a small island off the south coast of Iceland, came to life suddenly in 1972 after long dormancy. The town was temporarily abandoned when it was inundated by ash and threatened by moving lava. (G. Hannesson)

Plate 5. Fire fountain and lava flow. The very fluid basalt lava in Hawaiian eruptions often emerges from fissures as fountains of incandescent fragments and as thin flows that spread rapidly down even the gentlest slopes. (National Park Service)

Plate 6. Valley glaciers. Two glaciers in adjacent valleys in the Canadian Rockies come together in the middle of the picture, with a dark train of rock debris to mark their junction. More rock debris is piled into a small moraine at the end of the glacier. The rounded valley bottom and steep walls in the foreground are clear evidence that glacial activity was more extensive in the past. (John Shelton)

(Legends for these plates continue on page 137.)

Plate 1. Edge of a continent.

Plate 2. Life in a tide pool.

Plate 3. Hot spring and fumarole.

Plate 4. The volcano Helgafell.

Plate 5. Fire fountain and lava flow.

Plate 6. Valley glaciers.

Plate 7. Limestone cavern.

Plate 8. Bryce Canyon.

Plate 9. The Grand Canyon of the Colorado River.

Plate 10. Sandstone arches.

Plate 11. Geode with quartz and agate.

Plate 12. Petrified log.

Plate 13. Oak to opal.

Plate 16. Vein quartz with gold.

Plate 17. Pyrite crystals.

Plate 18. Feldspar.

Plate 21. Sandstone under a microscope.

Plate 22. Granite under a microscope.

Plate 14. Galena crystals.

Plate 15. Gold from the Mother Lode.

Plate 19. Fluorite.

Plate 20. Calcite cleavage.

Plate 23. Basalt under a microscope.

Plate 24. Schist under a microscope.

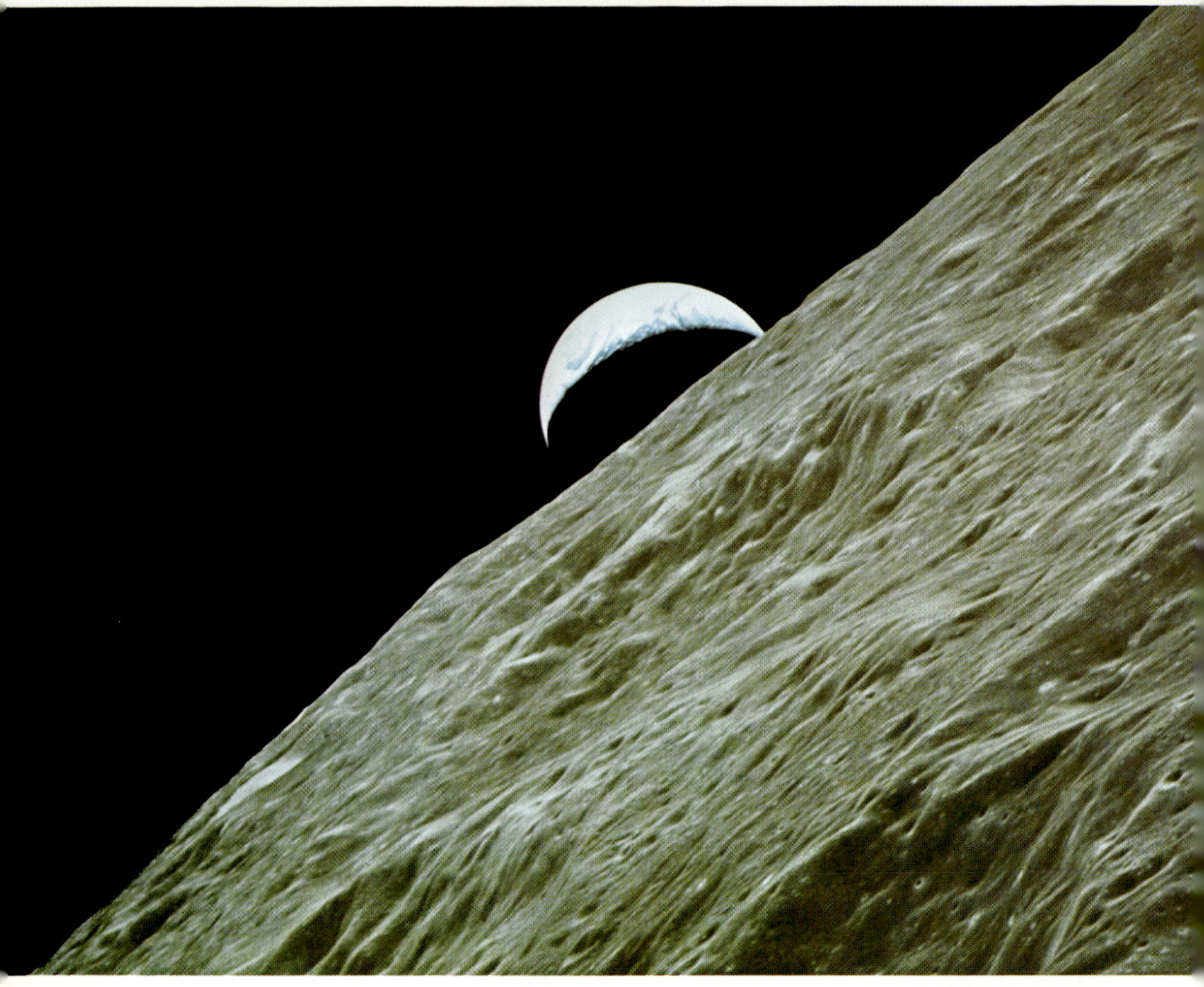

Plate 25. Earthrise seen from the moon.

Plate 26. The earth from space.

Plate 7. Limestone cavern. The huge cave at Carlsbad, New Mexico, has many rooms like this, carved by groundwater out of a thick limestone bed and then decorated by the slow deposition of calcium carbonate from water that drips from the ceiling. The curtains, pedestals (stalagmites), and rock icicles (stalactites) all result simply from this very slow accumulation of material that was once dissolved in drip water. The colors come from traces of ferric oxide deposited with the calcium carbonate. (National Park Service)

Plate 8. Bryce Canyon. Erosion of horizontal layers of soft shale in the semiarid climate of southern Utah has left this brightly colored panorama of spires and pinnacles. (Utah Department of Development Services)

Plate 9. The Grand Canyon of the Colorado River. A breath-taking panorama of light, color, and geologic history. The record of a billion years is displayed here, from the ancient dark schists of the Inner Gorge in the foreground through a succession of horizontal beds to the uppermost Paleozoic on the rim in the far distance. The lowermost of the prominent cliffs is the Tapeats Sandstone (Cambrian), the bright red one in the middle distance is the Redwall Limestone (Devonian), and the white one near the top is the Coconino Sandstone (Permian). (David Muench)

Plate 10. Sandstone arches. Erosion by water and wind has left more resistant parts of thick sandstone beds in the form of huge arches, brightly colored because of ferric oxide in the cementing material of the rock. Arches National Park, southern Utah. (Utah Department of Development Services)

Plate 11. A geode partly filled with quartz and agate. A geode is a cavity lined with material deposited from solution. The material often consists of well formed crystals, but in this example it is chiefly the microcrystalline variety of quartz called agate. (E.B. Eckel, Geological Society of America)

Plate 12. Petrified log. The end of the log has been cut and highly polished to bring out the colors of the opal (noncrystalline silica) and chalcedony (microcrystalline quartz) which replaced the original wood when the log was buried. The minerals themselves would be white or colorless, but they contain enough impurities (chiefly ferric oxide) to give a variety of colors. (Petrified Forest Museum Association)

Plate 13. Oak to opal. In contrast to the log in Plate 12, the petrified wood in this picture shows replacement of the finest details of the old organic structures. The replacement is a very slow process, accomplished by groundwater which removes constituents of the wood and simultaneously deposits silica from solution. Although usually the wood structure is in large part destroyed, under the right conditions the replacement can faithfully duplicate details. (National Museum of Natural History, Smithsonian Institution)

Plate 14. Crystals of galena. A mineral that is easy to recognize because of its bright silver color, its cubic crystals, and its excellent cleavage in three directions at right angles. It is the chief ore mineral of lead. (Joachim Hampel)

Plate 15. Gold from the Mother Lode. Different varieties of gold from mines in the foothills of the Sierra Nevada, California. The large specimen is an unusual occurrence, a "sandwich" of gold between layers of quartz. (Bank of America)

Plate 16. Vein quartz with gold. White quartz like this is a material with which gold is often found associated—but *very* seldom in the amount shown here. (Joachim Hampel)

Plate 17. Crystals of pyrite. This is the commonest of the sulfide minerals, FeS_2. The brassy color makes it easy to mistake pyrite for gold, hence the common name "fool's gold." The two minerals can be distinguished on the basis of the slight difference in color and the much greater hardness of pyrite. (Joachim Hampel)

Plate 18. Feldspar. The two lower specimens are cleavage fragments of large crystals, showing the characteristic two directions of cleavage approximately at right angles. The small upper specimen is a crystal, showing several crystal faces in addition to two that are parallel to the cleavages. (Perfecto Mary)

Plate 19. Fluorite. Calcium fluoride, CaF_2. A mineral that often appears in varicolored, well formed crystals which make handsome museum specimens, and seemingly should be prized for jewelry—except that it is much too soft and too common. (Joachim Hampel)

Plate 20. Calcite. Three cleavage fragments and an aggregate of small crystals (left side). The cleavage fragments show by their flat surfaces and abundant parallel cracks their characteristic three directions of cleavage at angles of about 75°. The crystals have faces in many other directions, but if broken would form fragments of the same shape as the larger specimens. (Perfecto Mary)

Plate 21. Sandstone under a microscope. This is a very thin slice (about 0.03 mm) or *thin section* of sandstone enlarged about 10 times. The light-colored grains are quartz and feldspar, the brownish green ones ferromagnesian minerals and rock fragments that have been somewhat weathered. The grains would be described as fairly well sorted (most of them roughly the same size) and subangular (predominant straight edges and sharp corners, but some grains rounded). (Joachim Hampel)

Plate 22. Granite under a microscope. The colors in this thin section are not natural, but are produced by interference when the section is put between two polarizing prisms in order to make grain boundaries more distinct. The white, black, and blue-gray colors are mostly quartz and feldspar; the greens, browns, and brighter blues are ferromagnesian minerals. Note that the grains are intergrown, not separate as in a sandstone. (Joachim Hampel)

Plate 23. Basalt under a microscope. As in Plate 22 this thin section has been put between polarizing prisms, so that the colors are not realistic. The larger straight-edged crystals, in colors from white to light gray-blues, are plagioclase; the pinks, browns, and bright blues are ferromagnesian minerals (chiefly pyroxene). This specimen is a *porphyritic* lava, meaning that it has crystals of two conspicuously different sizes: the large plagioclase crystals (*phenocrysts*) set in a fine-grained *groundmass* consisting of a mixture of plagioclase and pyroxene. (Joachim Hampel)

Plate 24. Schist under a microscope. Again as in Plate 22, the colors are due to interference. The rock has been cut perpendicular to the foliation, so that the mica flakes are seen edgewise as colored streaks. The small intergrowing crystals with colors from white to blue-gray to black are mostly quartz. Note the conspicuous alignment of the micas, which gives the rock its foliation. (Joachim Hampel)

Plate 25. Earthrise on the moon. The illuminated crescent of the earth, bright because the sun is shining on its cloud cover, is on the horizon of the moon's bleak gray landscape. In one sense the title "Earthrise" is not accurate. The earth cannot "rise" for an observer on the moon's surface, because the moon keeps the same face always turned toward the earth. The earth would appear to rise only for an observer in a rocket moving around the moon, like the observer who took this photograph. (NASA)

Plate 26. The earth from space. The picture is taken from a point over the equator in the eastern Pacific. The outline of South America is clear on the right-hand side, and toward the top North America is half hidden by clouds. The brown desert areas of Mexico and western South America contrast with the dirty green of equatorial Brazil and the dimly visible eastern part of the United States. (NASA)

Marine sediments

The sediments of floodplains, of lakes, of alluvial fans and deltas are all temporary, in the sense that they represent material which is dropped locally by rivers on their way to the sea and which presumably will sometime be picked up and moved farther as the rivers continue to erode the land. Most such sediments never become rock, because they are eroded before the slow processes of compaction and cementation can harden them. In places they do become buried and lithified, forming rock layers that may be preserved for millions of years before erosion cuts into them. The lake beds of the Green River formation and the red sandstones of the Connecticut River Valley are examples. But among sedimentary rocks in general such layers are exceptional. The great bulk of the material preserved in sedimentary layers is debris that has been carried to the sea and deposited in salt water. This would certainly be expected, because the ultimate goal of all erosional processes is to transport material from the continents into the ocean.

What, then, are the characteristics of these most abundant of sediments and sedimentary rocks? If we think back for a moment to our

Fig. 5-15. Long-shore drift of sediment, produced when waves driven by prevailing winds strike the shore at an angle.

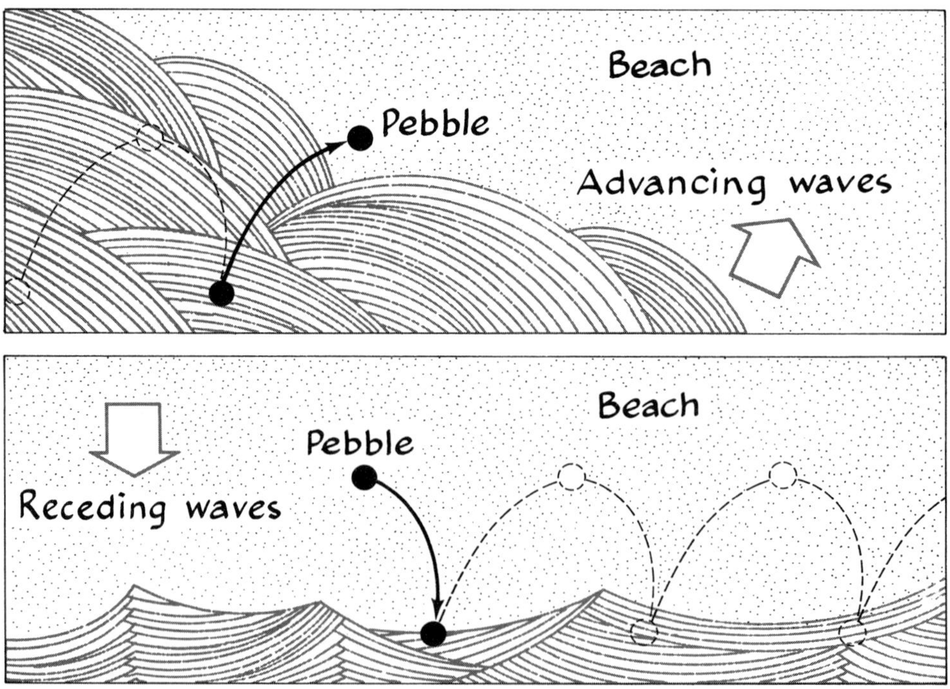

Fig. 5-16. Muddy water brought by a river into the ocean drifts to the right (south) with the long-shore current. Mouth of the Russian River, central California. (John Shelton)

ruminations on a sea cliff in Chapter 1, we can make some plausible guesses. Sediment is obviously accumulating offshore from the cliff, in part material scoured from the cliff by waves and in part material brought down to the beach by rivers. The sediment is moved by currents set up by the waves—the back-and-forth motion as the waves move in, break, and retreat, and a less conspicuous net movement along the shore (*long-shore currents*) if the waves approach at an angle. The constant motion is a powerful agent for sorting and rounding of fragments, even more effective in general than the motion in a turbulent stream because it is longer-lasting. One can expect, then, that the sediment will be deposited in even, parallel layers, coarser material on the beach grading out into finer material in the quiet water at greater depths. Cross-bedding may form in the beach deposits and in offshore bars, but in general the bedding should be parallel. If conditions remain constant for a long time, individual beds should build up to a considerable thickness, perhaps punctuated at intervals by beds of coarser material deposited during storms. Many variations of this imaginative guess about what goes on under the sea surface are possible, depending on different assumptions about climate, underwater topography, and the nature of the material supplied to the waves.

It is pleasant to sit on a cliff and guess about what happens offshore, but for accurate information actual sampling of the offshore sediment is needed. A great deal of such information has been secured in recent years, both by sampling from ships and by diving for first-hand observations. As might be expected, the nature and distribution of sediments are in most places more complicated than simple guessing would indicate. In a general way the predictions about rounding and sorting of grains, uniform bedding, and transition seaward from coarse material to fine are supported, but exceptions are so numerous that the rules are not very useful. Exceptions can be traced to a number of causes: irregularities in the ocean floor, changes of sea level within the last few tens of thousands of years, currents other than those directly due to wave motion, changes of climate, and so on.

On a broader scale, recent exploration of the sea bottom has given us information about conditions of sedimentation for many miles offshore. As a general rule the sea floor immediately adjacent to a continent slopes very gently, forming a broad platform called the *continental shelf* on which most sediment from the land is deposited. At the seaward margin of the shelf the slope increases, forming the *continental slope;* at the base of the slope is another area of low gradient, the *continental rise,* which descends almost imperceptibly into the deep sea. As might be expected, most of the sediment that reaches the continental slope and continental rise is very fine-grained, forming impalpable blue-gray muds. But one of the startling discoveries of ocean exploration has been the finding of thick deposits of coarse sand on some parts of the slope and rise.

How could sand in huge quantities be carried so far from land, into regions where water movement is generally very feeble? A possible answer is through the agency of *turbidity currents,* sudden movements of sediment-water mixtures in which the sediment can remain suspended for long distances because of the extreme turbulence. Suppose, for example, that sand brought

Fig. 5-17. Continental shelf, continental slope, continental rise. The vertical scale is greatly exaggerated; actual slopes are 3-6° for the continental slope, and 1° or less for the shelf and rise.

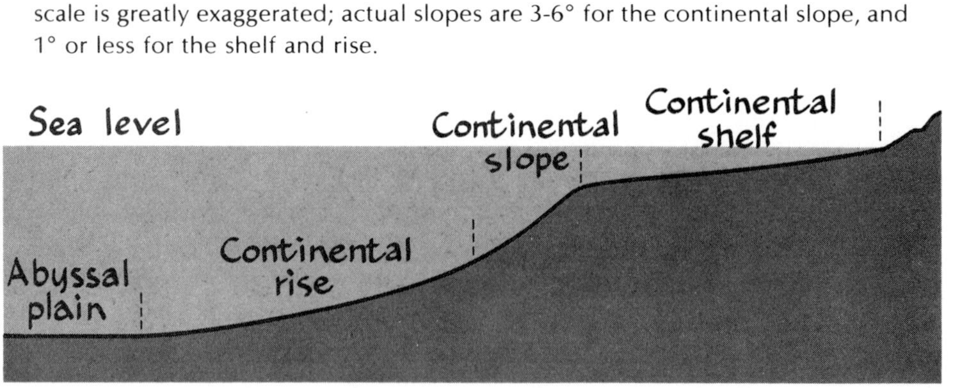

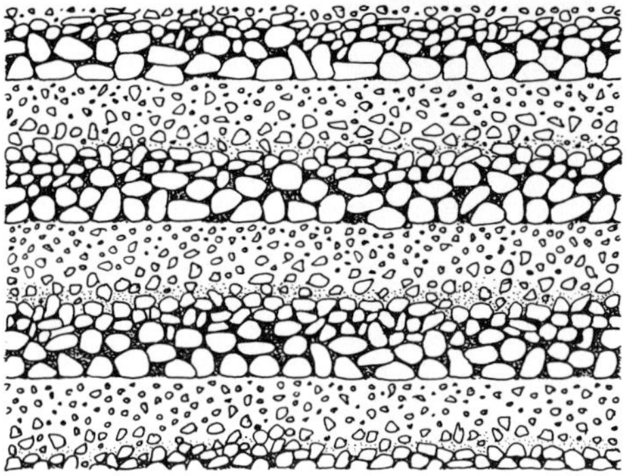

Fig. 5-18. Graded bedding.

from the land accumulates to considerable thickness near the seaward margin of the continental shelf. The sand, of course, is saturated with water and capable of flow if disturbed. A possible disturbance might be a minor earthquake, or perhaps the sand simply accumulates to such a thickness that the pile becomes unstable. If a sufficient mass is jarred loose, the sand-water mixture starts down the continental slope, gathering speed as it moves and becoming so turbulent that the sand cannot settle out. Such underwater currents can be produced in tanks in the laboratory and have been observed in lakes. Direct observation of large-scale currents in the ocean is probably impossible, but good evidence of their occurrence has come from the occasional breaking of transoceanic cables. Small flows of sand have been observed by divers in places where submarine canyons cut the continental shelf.

Although the evidence for turbidity flows in the ocean is largely indirect, it is so persuasive that these currents are commonly regarded as playing a major role in marine sedimentation. They are not limited to the huge flows that carry sand down the continental slope into deep water, but include many on a smaller scale that shift sediments in basins on the continental shelf. To such currents are ascribed several characteristics often noted in marine sediments and sedimentary rocks: *graded bedding,* referring to the gradual change within a bed from coarse material at the base to fine material at the top, produced as the material of a turbidity flow gradually settles out; *sole marks,* grooves and indentations on the surface of a bed produced by movement over it of a dense turbidity current; and thick sandstone layers with no sign of bedding, perhaps caused more by large-scale sliding than by active turbidity. If the history of an ancient layer of sedimentary rock is in doubt, these features will often help to make a marine origin probable.

So far we have talked only of the detrital sediments of the marine environment and the rocks derived from them. Chemical sediments as well are for the most part products of marine deposition. Calcium carbonate in the form of limy mud and coral reefs is accumulating today in warm,

Fig. 5-19. Sole marks. This is the under side of a sedimentary layer, with grooves produced by the rapid movement of this sediment over the surface of the layer beneath. The sediment in this specimen is the lower, coarse part of one cycle in a sequence of graded beds. (K.B. Ketner, U.S. Geological Survey)

shallow parts of the ocean, and such environments must have been more common in the geologic past to explain the huge amounts of limestone often found in sequences of marine sedimentary rocks. The origin of chert is less clear, since chert in any quantity is not known to be forming in the present ocean, but from the fossils it contains and the rocks associated with it we know that most chert was originally a marine sediment.

Suppose, then, you face the problem of deciding whether the sedimentary rock in an outcrop you are studying represents material once deposited in a lake, in the ocean, or on a river floodplain—a very common sort of problem in geologic work. What would you look for? The first and obvious answer, of course, is "fossils." If you find the tracks of reptiles, the bones of a horse, or impressions of plant leaves, then a floodplain or a shallow, temporary lake is indicated. If fish remains are abundant, you would seek expert opinion as to whether the kinds of fish you have found lived in fresh or salt water. If you can recognize the shells of barnacles or trilobites or brachiopods, a marine environment is established; for other shells you might again need advice from an expert. Thus fossil remains would make your problem easy. The difficulty is that fossils, *any* fossils, are rare in sedimentary rocks, and without extraordinary luck you would find none at all. Then what other evidence might you seek?

You should recognize, to begin with, that the problem may be insoluble. Even a trained geologist often cannot be sure, from a single outcrop, whether he is looking at a rock formed in the sea or on land. But some signs would give good indications. If the outcrop shows thick beds of sandstone or limestone or both, there is a strong presumption of marine origin. Shale beds might have formed in either fresh or salt water; thin, uniform beds suggest but do not prove origin in a lake. Lens-shaped beds showing cross-bedding, crude sorting of gravel, sand, and clay, and cross-sections of old channels would strongly indicate the work of streams. Piecing together evidence of this sort will often make possible a reasonable guess about origin, but the final decision may depend on a study of many outcrops besides the original one.

On a regional scale rather than the scale of a single outcrop the continuity of beds for distances of tens or hundreds of miles may give an indication of marine origin. In the northern Midwest, for example, thick beds of nearly flat-lying sandstone and limestone can be followed from outcrop to outcrop over enormous areas, often from one state to another. In the Appalachian Mountains the tilted beds that form the straight ridges and valleys also suggest by their continuity that their original sediments were in large part marine. In these areas long geologic study has brought to light many fossils, which not only confirm a marine origin for most of the beds but provide much information about details of the oceanic environments in which particular kinds of sediment were laid down.

One important fact about most sedimentary rocks of marine origin, revealed both by their fossils and by their physical characteristics, is that the original sediments were deposited in shallow water rather than in the deep sea. The presence of widespread marine sedimentary layers in the Appalachian region and across the Mississippi Valley proves that the ocean once covered a large part of North America, but it does not prove that the continent sank to the level of the deeper parts of the ocean. The seas that spread widely over the continental interior were shallow, like the present Hudson Bay or Baltic Sea. Sediments of the present-day deep sea are different in many respects from sediments of the continental shelf and continental slope, and similar differences have probably always existed.

Rocks formed from deep-sea sediments are scarce on the continents. This is one of the striking general facts that emerge from a study of sedimentary rocks: the characteristics of most ancient sediments can be matched with the characteristics of modern sediments formed on land or in shallow marine waters, but not with deep-sea sediments. The present continents, however much their rocks may have shifted during geologic time, have not bodily risen from or sunk down to oceanic depths.

Unconformities.

We have been talking as if sedimentation is a very common and widespread process. Every valley, every broad plain, the shallow water along every seacoast, according to the argument of the last few paragraphs, should be accumulating sediment. As the sedimentary layers build up, the lower ones should be in process of conversion to sedimentary rock. Just how accurate is this picture?

Consider the possible sedimentary environments near where you live. Is sediment actually accumulating to any great thickness? If you live in hilly country, mud and sand may pile up at the base of slopes or along stream courses during heavy storms; but such deposits last only until they are carried away by erosion during the next spell of rainy weather. If your home is in flatter country, creeks swollen by rain may occasionally overtop their channels and drop silt on their immediate surroundings, but again the sediment is easy prey to later erosion. If you live near the sea you may well imagine sand and mud being deposited just offshore, but storm waves that batter the coast may apparently in a few hours undo the work of months of slow deposition. Of course if you happen to live in or near an enclosed basin—on a lake shore or in a closed desert valley—sediment must be building up because there is no agent to remove it from the basin. But in most places, if you consider a long enough period, a state of rough balance is established so that sediment is removed about as fast as it is formed. Over the land surface generally, in fact, the long-term balance must be tipped in the direction of removal of sediment rather than accumulation, since the net effect of erosion is to wear down the land and carry the debris to the ocean.

Sedimentation, then, only exceptionally goes on long enough and continuously enough to build up the layers of sediment and sedimentary rock we have talked about in previous paragraphs. Special conditions are needed—say the slow sinking of one part of the earth's surface with respect to another, or the slow rise of a mountain range, or a gradual change of climate. Most sediments that have formed during the earth's long history have vanished without a trace, but the necessary special conditions for accumulation have existed often enough for sedimentary rocks to form in considerable quantity.

The conditions for accumulation are precarious, and can be destroyed or interrupted by any one of a number of geologic events. The site of deposition may be slowly elevated, or tilted; rivers supplying sediment may change their courses; mountains that furnish the debris may be worn down; the climate may grow drier or colder so that less sediment is produced. What evidence would be left in the sedimentary record of such an interruption?

Suppose, for example, that sedimentary layers on a river floodplain have been building up for many thousands of years. Every year, or every few years, the river floods its surroundings and drops mud and silt in a thin layer over part of the plain. The area of the floodplain is very gradually sinking, so that deposition of sediment serves to keep the elevation of its surface approximately constant. The sediment has piled up to a thickness of several hundred meters, and its lower layers are partly consolidated by compaction and cementation. Now conditions change: the slow sinking of the plain is reversed, and over many more thousands of years it gradually rises. The river no longer floods, but instead cuts a deep channel into the rising plain. Tributaries carve the plain into ridges and valleys, removing great quantities of the upper layers of sediment. The cutting continues until a broad, hummocky surface of erosion is worn into the rocky layers that once were near the bottom of the pile. Then again a change occurs: something in the deep interior of the earth causes the area of the plain to start subsiding once more, and the river becomes again a sluggish, meandering stream subject to occasional floods. Sedimentary layers again accumulate, on top of the eroded remnants of the older ones. And now we suppose that geologic eons pass by, that new movements of the earth's crust elevate the region so that erosion carves deeply, and that a new highway cut exposes both older and younger sediments to our view. From what characteristics would we be able to reconstruct this long history?

We would see in the lower part of the roadcut a stack of ordinary-looking sedimentary layers, perhaps horizontal or perhaps tilted, depending on details of the movements that have affected them. The layers would be thin-bedded, mostly shale but with occasional discontinuous layers and lens-shaped bodies of sandstone, perhaps (if we are lucky) interrupted by a lens-shaped channel cross-section filled with sand and gravel, perhaps containing charred fragments of the leaves and stems of plants; from these characteristics we infer that the original sediment was formed on a floodplain. Then we notice that at one point in the sedimentary pile the lower layers are cut by an undulating surface, that the layers above fill in the low points of the surface, and that the rock just below the surface is discolored and contains remains of plant roots. The surface tells us that for

Fig. 5-20. An angular unconformity near Town Creek, Georgia. The unconformity is the uneven surface separating nearly horizontal beds above from steeply dipping layers beneath. It represents a long period of erosion of the lower rocks before the upper beds were deposited. (W.B. Lang, U.S. Geological Survey)

a period sedimentation was interrupted, that the land was eroded and covered with deep soil before the next layers were laid down.

A buried surface of erosion of this sort is called an *unconformity.* To a geologist it is one of the most informative of all the structures he finds in sedimentary rocks. It gives him a glimpse into a major event of the geologic past, a time when movements of the earth's crust made large areas prey to erosion by streams.

Many kinds of unconformity can be recognized. The one we have just considered is an unspectacular but very common variety—an unconformity that means simply a temporary interruption in a long history of stream deposition. Such unconformities are often noted, for example, in the old

Fig. 5-21. Different kinds of unconformities. Unconformities may be developed on tilted beds, horizontal beds, or igneous rocks, but all represent prolonged erosion before deposition of the overlying beds.

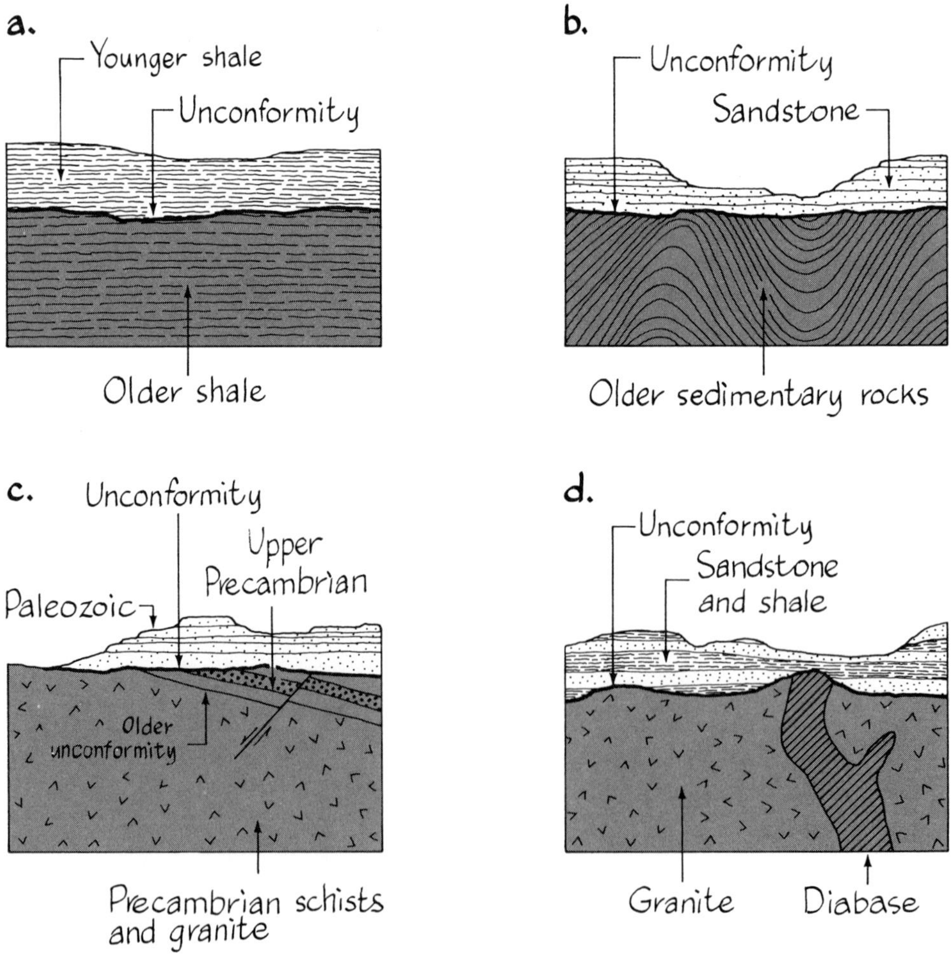

stream sediments that underlie much of the coastal plain of Georgia and the Carolinas. More profound changes in land elevation are indicated by unconformities that separate rock layers of marine origin: if rocks both above and below the erosion surface represent old marine sediments, the area must have been at one time beneath the sea, at a later time elevated above the sea and eroded by streams, then at a still later time once more submerged. The layers below many unconformities are tilted or folded, the more recent layers lying across the deeply eroded stumps of the old ones; this sort of unconformity indicates a time long enough for the older layers to have been consolidated, deformed by crustal movement, and worn down to a nearly level surface before the upper layers were formed. The rock below an unconformity need not be sedimentary, but may represent an erosion surface developed on granite or volcanic rocks. The only essential thing about an unconformity is that it must record a period of erosion before the upper strata were deposited.

The Grand Canyon again

The great gash that the Colorado River has cut through the plateau of northern Arizona is one of the finest geologic exhibits anywhere, furnishing classic examples of a wide variety of natural phenomena. We have used it before to illustrate landscapes carved by streams in flat-lying beds, and we now return for a closer look at its sedimentary rocks and structures.

This time we will have to do more than stand on the rim and look down. To study the rocks requires that we venture down one of the trails, a project that is deceptively easy to undertake but that means inevitably an arduous climb back to the rim later on. Until we get far down near the river all the rocks we encounter are sedimentary, beautifully exposed for examination in simple horizontal beds that we meet one after another. We are crawling down a deep cut in the earth's crust, and if we have the wit to interpret the rocks we will see a panorama unfolding before us that extends far back into the earth's history (Plate 9).

As we drop over the rim we find ourselves on a steep trail zigzagging down through a bed of limestone (called by geologists the Kaibab Limestone), whose thickness and continuity along the rim and across the canyon indicate its marine origin. Farther down the trail is another striking formation, a light-colored sandstone (Coconino Sandstone) whose grains are remarkably rounded and well sorted, and which shows cross-bedding on a large scale. These are characteristics associated with present-day dune sands, suggesting that the Coconino was formed by consolidation of the shifting sands of an ancient desert. Occasional bedding surfaces of the

Coconino represent sandy flats or water-holes on the desert. Below the Coconino is an abrupt change of color to a reddish formation (Hermit Shale), mostly shale but with some thin beds of sandstone, which forms one of the canyon's gently sloping shelves. Bedding in the shale is somewhat irregular, some layers have well preserved ripple marks and mud cracks, and some surfaces show imprints of fern leaves and tracks left by animals. These features indicate deposition on a floodplain, with temporary shallow ponds that slowly dried in the sun and preserved a record of their wind-rippled surfaces and of the small lizard-like creatures that walked across them; the red color suggests but does not prove a semi-arid climate. Still farther down is the most prominent of the cliff-forming beds in the canyon, the Redwall Limestone, which despite its name is a blue-gray rock that is stained only on its surface by red mud washed down from the formations above. Like the Kaibab, the Redwall Limestone is a rock formed from calcium-carbonate mud that once accumulated in a shallow sea, a conclusion that is verified by marine fossils visible in some layers.

Fig. 5-22. Animal tracks on a slab of Coconino Sandstone. (L.F. Noble, U.S. Geological Survey)

Fig. 5-23. The inner gorge (often called the "Granite Gorge") of the Grand Canyon. The steep cliffs just above the river consist of schist and quartzite, intruded by dikes of granite. On the eroded surface of these ancient rocks (unconformity 1) rest sedimentary beds tilted at a slight angle, and on the eroded surface of the tilted beds (unconformity 2) lies the Tapeats Sandstone, the lowermost of the horizontal cliff-forming beds of the canyon. (L.F. Noble, U.S. Geological Survey)

These are but four of the many beds we encounter along the trail down the canyon. From their simple characteristics we can put together a history of invasions by the sea, of times when this region was part of a broad floodplain, and of at least one time when it was covered by desert sands. Many of the formations are separated from adjacent ones by unconformities, indicating that during some long intervals no deposition was taking place.

At length we reach the brink of the inner gorge, the steep-sided trough at whose bottom flows the muddy turbulent river. Here we find a more spectacular unconformity. The erosion surfaces so far have been difficult to see, because they are nearly flat surfaces between parallel beds and hence seem little different from bedding planes. But now at the top of the gorge the lowermost of the horizontal beds (the Tapeats Sandstone, which contains marine fossils) rests on a surface cut across beds of sandstone and shale that are tilted at an angle of about 20°. Here there can be no question that the lower, older beds were deposited, consolidated, tilted, and subjected to long erosion by streams before the land sank under a shallow sea. The origin of the tilted strata is still uncertain: as alternating beds of sandstone and shale they might have formed either on a river plain or in shallow marine water, and no fossils have been found to establish one or the other.

This is still not the end of the story. If you have picked the right trail, you will be able to follow the tilted beds downward until they too rest on an unconformity, this time a surface eroded on metamorphic rocks intruded by granite dikes. These, the oldest rocks exposed in the canyon, are rocks formed by the alteration of still other sedimentary layers—an ancient series of sandstones and shales, changed by heat and pressure into the hard rock called quartzite and the spangled layered rock called schist.

And now the long climb out of the canyon will give you plenty of time to reflect on the history that the sedimentary rocks and their structures reveal. It is a history that goes back more than a thousand million years—a number that you could not read directly from the rocks, but that rests on straightforward evidence which we will consider later. The story of the canyon, you might reflect also, is but one example of the kind of geologic reconstruction that can be made in many places from a study of sedimentary rocks. You will seldom find elsewhere rocks so beautifully exposed, but a bit of geologic sleuthing is usually sufficient, even where outcrops are scarce, to put together the sedimentary story.

Fig. 6-1. Dinosaur bones in sandstone, Dinosaur National Monument, Utah. (Photograph by M.W. Williams, National Park Service)

Fig. 6-2. Restoration of one of the large flesh-eating dinosaurs (*Allosaurus fragilis*) of the middle Mesozoic Era, from Como Bluffs, Wyoming. (Courtesy American Museum of Natural History)

6

Fossils and Geologic Time

Of all the structures in sedimentary rocks, the most informative by far are fossils. When fossils are absent, as they usually are, we must depend on other characteristics like the kind of bedding and the composition, sorting, and rounding of grains for clues as to how the sediment was deposited. But if fossils are present and can be recognized, the origin of the sediment is immediately clear.

Take, for example, the famous fossils of giant reptiles in the Dinosaur National Monument of eastern Utah. A tilted bed of sandstone contains huge bones in great number, and a spot where they are particularly abundant has been shielded from the hot desert sun so that visitors can watch in comfort while workmen carefully remove some of the bones from the rock. In nearby exhibits the bones have been assembled into skeletons of the strange creatures who died here long ago. Even the untrained eye can recognize these at once as very different from any animals that now walk the earth. And it requires no special knowledge to see also that these were beasts that lived on the land rather than in the sea. The sandstone, if no fossils were present, might be difficult to recognize with certainty as a sediment deposited on land, but the scattered bones tell us at once that this must have been an ancient sandy flat on a river plain, perhaps near a water hole where many animals came and some were trapped and died. Once the sandstone here is identified as a floodplain deposit, we can follow the same bed by its physical characteristics for many miles, and guess that it represents land-deposited sediment even in places where no fossils can be found.

Thus fossils, despite their general scarcity, are indispensable for drawing firm conclusions about sedimentary origins. Not only do fossils give us a clear picture of the immediate environment where they were entombed; by noting the character of the sediment associated with fossils of different kinds, we can draw inferences about the history of similar beds elsewhere whether fossils are present or not. Both because they are powerful tools in interpreting sediments and because they are fascinating as a record of developing life, it is important to look carefully at these remains of plants and animals that lived before us on the earth.

What are fossils?

Fossils are the remains or other evidence of plants and animals found in rocks. They may consist of bones or shells or wood fragments that were covered with sediment and buried intact. More commonly the original material has been altered after burial: wood may be partly converted to carbon, shells and bones may have lost their original organic matter and become soft and powdery, the original substance may have been replaced by another material, or it may have been dissolved away to leave only an empty mold of the organism. Chemically the commonest substances

Fig. 6-3. Petrified logs in the Petrified Forest National Monument, Arizona. (N.H. Darton, U.S. Geological Survey)

preserved in fossils are the calcium carbonate of most shells, the calcium phosphate of bones and teeth, the silica of some one-celled organisms, and the carbon-rich material produced by partial decay of plants. The compound that most commonly replaces other substances is silica, generally in the form of chert or opal. One familiar and often spectacular result of silica replacement is petrified wood, in which the organic structures of the wood have been duplicated in faithful detail by silica deposited from solution (Plates 12 and 13). Replacement by silica is common also in shells of marine organisms that originally were calcium carbonate. For the most part fossils are limited to the hard parts of organisms, but very rarely impressions are preserved of such soft material as animal bodies, insect wings, and bird feathers. Fossils include also mere traces of past life such as animal tracks and worm borings. Any rock structure that gives evidence of the former existence of a specific organism is included in the definition.

It all seems so simple and straightforward: we take it for granted nowadays that a clam shell in limestone, a bone protruding from a sandstone outcrop, a delicate leaf revealed when thin shale layers are pried apart, all are remnants of organisms that lived when the sediments were being deposited. But it was not always so simple. Our modern assurance about the nature of fossils is the product of a long and tortuous development in intellectual history. The ancients recognized fossils as remains of animals,

Fig. 6-4. A fossil plant related to the modern horsetail, from shale beds of the Pennsylvanian Period (about 300 million years old). (John C. Frye, Illinois Geological Survey)

but accounted for them in the rocks of mountains by assuming great floods in the earth's more tumultuous past. In early Christian Europe the same easy explanation was adopted, with the addition of pinpointing a specific flood as the one which Noah survived. Difficulties arose when more careful observers noted that fossils were present deep inside of rocks, not scattered on the surface as they would be after a flood, and when theologians concluded that Noah's flood could be no more than a few thousand years in the past. Fanciful suggestions were made to resolve the dilemma: perhaps fossils were produced inorganically, by some occult radiation that could penetrate rocks and mimic the forms of living creatures at the surface; or perhaps Satan was playing games with gullible humanity, planting fossils in rocks to tempt the unwary into doubting the Biblical account. Only gradually the accumulating weight of observational evidence made the modern explanation possible.

How are fossils formed?

It is easy to say that dead organisms are buried by accumulating sediment, and that their hard parts are preserved because the sediment protects them from erosion and decay. This must indeed be the way fossils are formed. A moment's reflection, however, indicates that such a process cannot be very common.

Consider an animal that dies on land, say a dead cow on a hillside. How much chance is there for its bones to be preserved? Some will be broken, scattered, and eaten by predators. Some of those that remain may be partly buried in soil, but here they are subject to slow attack by air and water. The small amount of organic matter that bones contain will be oxidized; the inorganic material is largely calcium phosphate, a fairly insoluble compound but susceptible to gradual solution in water containing carbon dioxide. Over a period of years the bones will disappear—unless conditions just happen to be favorable for preservation. If, for example, some of the bones are washed into a valley where sediment is accumulating rapidly, if the climate is dry enough so that rainwater does not often penetrate the sediment once it is laid down, if the accumulating sediment includes impermeable clay layers, so that eventually the bones are effectively sealed from circulating water—then there is a slight chance that remains of the cow will one day become part of a sedimentary rock.

Or look at a seacoast. Where the coast is rocky, living things in great variety populate the tide pools and cling to the rocks. On sandy beaches shells of clams and snails and carcasses of dead fish are rolled to and fro by the waves. Muddy tide flats have colonies of thickly growing clams, and busy snails stir the mud of channel bottoms. Of all these living creatures and remains of living creatures, how many have a chance of preservation as fossils? The soft bodies, of course, have almost no chance at all, for they are eaten or decay before they can be buried. Shells and bones may fare better, but only if sedimentation is rapid; by far the greater number are pulverized by waves, eaten by other animals, or slowly dissolved as their protecting organic matter decays. This environment of shallow marine water is the most favorable of all for preserving organic traces, but even here only a tiny fraction of the abundant living forms can escape destruction before they are buried deeply enough to stand a chance of survival as fossils.

A review of these and other environments shows clearly how ineffective most sedimentary processes are for preserving a record of past life. The bodies of living things are highly destructible, even their hard parts; all except a very few of the creatures that have inhabited the earth have simply vanished without leaving a trace. It is no wonder that fossils are scarce, that many sedimentary rocks show no sign of organic remains. Perhaps the real cause for wonder is that fossils are preserved at all, and especially that in a few localities they are so very abundant.

Water that percolates through sediments is one of the agents by which fossils are destroyed. If it contains carbon dioxide and is unsaturated with calcium carbonate, it can slowly dissolve shell material; if it contains oxygen it can oxidize organic matter; and if the sediment is still unconsolidated, these processes may remove all trace of the organic remains that have survived burial. On the other hand, percolating water under the right circumstances may help to preserve fossils. If the sediment is compacted quickly into solid rock and if the movement of water is slow, the dissolving of shells or bones may leave cavities which retain the original shapes and hence are recognizable fossils. Or better, the water may contain dissolved material which very slowly replaces the original substance but retains the details of its structure, and which is less susceptible to later attack by other solutions. For example, as mentioned above, shells originally composed of calcium carbonate are often replaced by silica, and the silica is so resistant that these replaced shells may protrude from weathered limestone surfaces and even form talus piles beneath limestone outcrops.

The varying effects of percolating water are in part responsible for the differences in fossil content of different sedimentary rocks. As a general rule fossils are most abundant and most frequently found in limestone, partly because the original limy mud hardens quickly to solid rock after burial and thereafter is relatively impermeable. Another reason for the abundance is the fact that most limestone forms in shallow marine water where animals with preservable hard parts flourish in great numbers, but in water deep enough so that the shells are not pounded to pieces by waves. A second rock that often repays a search for fossils is shale, and again a part of the reason is the rapid compaction of the original mud and its later impermeability. The fact that mud is deposited in quiet water and is so resistant to the movement of fluids makes possible in some places the extraordinary preservation of delicate structures and even the imprint of soft parts of organisms. Sandstone and conglomerate, on the other hand, only exceptionally have good fossils. The original sediments are deposited in more turbulent environments, consolidation into rock is slow, and both sediment and rock are generally permeable. To all these rules there are conspicuous exceptions, but for would-be fossil hunters it is useful to keep in mind that chances of success are by-and-large considerably better in limestone and shale than in the coarser-grained sedimentary rocks.

Fig. 6-5. Bituminous coal with many fossil plant fragments. (John C. Frye, Illinois Geological Survey)

Fig. 6-6. Restoration of a coal forest of the Pennsylvanian Period, about 300 million years ago. (Courtesy American Museum of Natural History)

Why look for fossils?

Amateurs hunt fossils with much the same motives that they look for mineral specimens or old bottles or arrowheads: good specimens are handsome rarities, fine conversation pieces for a prominent place on the mantelpiece or in an exhibit case. Geologists are not immune to such motives, for they covet good fossil specimens to display in museums or to excite the envy of colleagues. But if you ask a geologist why the search for fossils is such an important part of his work, he will almost certainly suggest a number of more sophisticated reasons.

First and most obviously, fossils in a sedimentary rock are a record of the living things that flourished when the sediment was deposited. They are necessarily an incomplete record, because only a fraction of the creatures in a community will have hard parts that can be preserved. The dinosaur beds of the National Monument permit a vivid reconstruction of the vertebrates whose bones are so abundant, but they tell us little about the vegetation on which the reptiles lived or about the fragile insects that crawled on the ground beneath them or hovered around their heads. The fossil seashells that weather out in great numbers from limestone bluffs along the upper Mississippi River and its tributaries tell us much about the living forms that teemed in the shallow water of an ancient sea, but the soft-bodied things that must also have been present are not represented. Some of the coal beds of Illinois contain marvelously preserved plant remains that enable us to picture the strange forest that grew in a huge swamp, but only rarely do we get a glimpse of the animals that roamed through the forest. Despite the inadequacy of the record in any one place, by piecing together information from different localities we can reconstruct ancient communities and ancient environments in fair detail.

But fossils are important in other ways. If you are studying a series of sedimentary layers and you make a careful search for fossils, you will find that the fossil assemblages differ from one layer to the next. Each bed has a characteristic group of fossils, and the fossils are helpful in recognizing a bed as you follow it from one outcrop to the next. If you are mapping the rocks of a large area, you may find that individual beds change their character as you follow them for distances of several miles: they may grow thicker or thinner, or their color and grain size may change enough to make identification uncertain. But if you can find fossils, the persistence of particular kinds will enable you to recognize each bed for as long as you care to follow it.

This simple fact, that fossils may be used to identify sedimentary beds over long distances, was discovered early in the last century by a British surveyor, William Smith, in the course of his efforts to find rocks suitable for quarrying in various parts of England. It was the cardinal discovery that made modern geology possible. The sedimentary layers of any one area could be arranged in sequence according to their fossil content; then the same sequence, or parts of the sequence, could be recognized elsewhere, even at distances of hundreds of miles and even if the beds were concealed by soil and vegetation between one area and the next. The beds might be folded or broken by faults; they might change their character radically; in some areas one or more beds might be missing from the sequence; still, if fossils could be found, at least some of the beds could be matched, or *correlated,* from one end of the country to the other. British geologists were quick to follow the hint that Smith had given, and used fossils to decipher the complex record of sedimentation and crustal movement preserved in the rocks of the British Isles. Others used the same technique in Europe, and a little later in America. Fossils provided the key to unraveling sequences of sedimentary rocks the world over.

You may have wondered, in connection with our discussion of the Grand Canyon, how the geologic history so beautifully displayed in northern Arizona might be tied together with geologic events elsewhere. At the canyon we can work out the local history in great detail by looking at the characteristics of the sedimentary layers, but to grasp geology on a continental scale we need to know what was happening at the same time in

New York, or in Saskatchewan, or in southern Mexico. For answers to such questions fossils are indispensable. If we can match some of the fossils of a rock layer in New York with those in one of the strata of the Grand Canyon, we can guess that the two rocks were formed at about the same time, even though they differ greatly in composition and sedimentary structures. Our guess is on firmer ground if we can find a sequence of layers in New York that show the same succession of fossil assemblages we have seen in the canyon.

The matching or correlation of rocks between New York and Arizona is not quite this simple, as a moment's reflection will suggest. If a layer in the Grand Canyon consists of well sorted, cross-bedded sandstone that we recognize as an ancient dune sand, and a layer in New York is a thick-bedded limestone of marine origin, the two might have formed simultaneously and still contain very different fossils, simply because the animals that inhabit a desert are different from those that live in a shallow sea. Furthermore, many of the beds in the canyon are separated by unconformities, representing times when northern Arizona was above sea level and undergoing erosion; if New York at the same time was under the sea, sediments would be accumulating there whose fossils would not be represented at all in the canyon. Difficulties of this sort make correlation more complicated than simply matching fossils from one place to another, and require considerable ingenuity on the part of a geologist for their solution. But the problems are by no means insurmountable. The Grand Canyon sequence can be correlated very convincingly and in great detail with other sedimentary rocks in many parts of the world.

One more property of fossils makes them enormously important in geology. Where great thicknesses of sedimentary rocks are exposed, their fossil assemblages show a striking characteristic: fossils in strata near the top of the pile are more similar to organisms living at present than are fossils from near the bottom. For an example we go back once more to the Grand Canyon region. In the canyon itself, practically all fossils that have been found deviate widely from present-day forms. In the marine strata one can find a few fossils of clams and snails that show some resemblance to those of modern seas, but are different in detail; and the more abundant fossils are creatures unlike anything that lives today. Vertebrate fossils in these beds are limited to fish in the lower beds, amphibians in some of the upper layers that represent ancient river plains, and small reptiles in the uppermost beds, but all are different from modern relatives. North of the canyon in southern Utah are other sedimentary beds that lie on top of the canyon strata—the brilliantly colored red and white sandstones that form the cliffs of Zion Canyon. Rare fossils in these rocks are of a different sort, including bones of dinosaurs and their relatives, but still with little resemblance to animals in the world around us. Farther north, beautifully exposed in Bryce Canyon (Plate 8), are more sedimentary beds still higher in the sequence, and in these are found bones of animals we can recognize as mammals—not mammals of species that are living today, but creatures certainly closer to the present world than anything in the rocks below.

The conclusion seems inescapable that animal and plant populations have changed in the course of geologic time, and that the change has been in the direction of closer and closer resemblance to modern forms. The creatures whose fossils are found in older rocks were more primitive, in the sense that their organic structures were less complicated than some of those in the present world. Without trying to read any deeper significance into these facts, we can use them to arrange sedimentary rocks in a time sequence. Fossils enable us not only to correlate strata from one region to another, but also to assign each individual stratum to its place in the worldwide time sequence defined by the changes in life forms.

Is it any wonder that geologists spend so much time combing sedimentary rocks for traces of fossils? Their search is often unrewarded, but if fossils can be found they aid greatly in reconstructing past environments, in working out the relations of one sedimentary rock to another, in unraveling complex geologic structures, and in assigning relative ages. Even if only a few fossils can be found, and even if they are limited to one or two beds in a sequence, they make the geologist's conclusions far more definite than they could possibly be otherwise.

Telling time geologically

Fossils, we have noted, make it possible to arrange sedimentary beds in a time order, an order that is the same in all parts of the world. For a time order to be useful it must be divided into periods, so that a given rock can be pigeonholed in a block of time early in the sequence, half-way along, or late in the sequence. You will recall that human history is conveniently divided into ancient times, medieval times, and modern times, and then for detailed treatment into shorter intervals like the Fourth Dynasty, the Elizabethan Period, or the Nineteenth Century. For similar convenience, we need some method of sorting out the sedimentary record into periods of geologic time.

The progression of fossils is from primitive forms to organisms closely resembling modern creatures, so a reasonable basis for major time divisions is the degree of resemblance of fossils to their modern relatives. Rocks with fossils most unlike present-day forms are assigned to the *Paleozoic Era*

Fig. 6-7. Restoration of a seascape of the late Devonian Period, central New York State. Primitive fishes (*Dinichthys* and *Cheirolepis*) in a colony of several kinds of sponges. (Courtesy New York State Museum and Science Service)

Fig. 6-8. Early amphibians, animals related to the modern salamander who spent part of their lives in water and part on land. *Diplovertebron*, Pennsylvanian Period. Painting by Francis L. Jaques. (Courtesy American Museum of Natural History)

Fig. 6-9. One of the largest of the dinosaurs, *Brontosaurus*, a plant-eating creature who lived in swamps. Painting by Charles R. Knight. (Courtesy American Museum of Natural History)

Fig. 6-10. Pleistocene animals of southern California, including giant vultures, horses, and sabre-toothed tigers. Painting by Charles R. Knight. (Field Museum of Natural History)

Fig. 6-11. Miocene mammals of western Nebraska. Painted by Jay H. Matternes. (Smithsonian Institution Museum of Natural History)

(meaning era of ancient life), those somewhat closer to the present to the *Mesozoic Era* (era of medieval life), and those approaching the modern world to the *Cenozoic Era* (era of recent life). The divisions between eras are arbitrary, but are taken as times when living forms showed particularly great changes from one rock layer to the next.

The rocks of northern Arizona and southern Utah that we considered in the last section provide examples of all three eras. The horizontal rocks of the Grand Canyon—those above the inner gorge—all belong to the Paleozoic Era, those of Zion Canyon to the Mesozoic, and those of Bryce Canyon to the Cenozoic. Many kinds of fossils can be used to distinguish the three, but for the moment perhaps the vertebrates provide the easiest way to fix the eras in mind: primitive fishes in the early Paleozoic, amphibians and small reptiles in the late Paleozoic, giant reptiles in the Mesozoic, and abundant mammals in the Cenozoic. Of rocks that we have mentioned in other connections, the folded strata of the Appalachians and the widespreading horizontal layers of the northern Mississippi Valley are Paleozoic, the red sandstones of Connecticut and the dinosaur beds of Utah and western Colorado are Mesozoic, and the coastal plain sediments of the southeastern states and around the Gulf of Mexico are Cenozoic.

What lies under the Paleozoic? If we go back to the Grand Canyon, we recall that the lowermost of the horizontal beds lies on an unconformity, and that beneath the unconformity is another series of sedimentary rocks

Fig. 6-12. Restoration of a Pleistocene mammoth of southern California, sinking in a tar pit. The tar pits of La Brea (now a part of the city of Los Angeles) trapped and preserved the remains of many Pleistocene animals, including those shown in Fig. 6-10. (Los Angeles County Museum of Natural History)

tilted at a 20-degree angle to the horizontal (Fig. 5-23). How do we know that the lower series is not also Paleozoic? The answer is simply that it contains no fossils. More accurately, it contains structures that may represent primitive life forms like algae, but the remains of complex organisms so abundant in the lowermost Paleozoic beds are completely absent.

The reasoning here seems questionable, because one can imagine that someday more fossils will be found, and then the lower beds would have to be included in the Paleozoic. But it turns out that wherever Paleozoic beds rest on older sedimentary rocks—and they do in many places—the pre-Paleozoic beds contain no fossils except scant traces of the most primitive forms, while the earliest Paleozoic strata have many fossil species represented. In most places, as in the Grand Canyon, an unconformity separates Paleozoic from pre-Paleozoic, but in a few places the succession of beds appears to be continuous; in such spots the boundary may be difficult to fix, but in general it is marked by the sudden appearance of characteristic fossils in considerable number. One of the strangest facts about the whole geologic record is the contrast between abundant fossils at the base of many Paleozoic sequences and their near absence in the rocks below. Careful search of pre-Paleozoic rocks in recent years has brought to light occasional traces of primitive organisms in considerable variety, particularly algae, worms, and bacteria. But the abrupt change from these simple plants and animals to the relatively complex creatures of the lowermost Paleozoic is still most striking.

The rocks older than Paleozoic would most reasonably be called pre-Paleozoic, or perhaps given an era name of their own. Attempts have been made to divide these ancient rocks into eras, but the lack or scarcity of fossils makes it hard to define time divisions that can be applied over large areas. By long custom the rocks are called Precambrian rather than pre-Paleozoic, the name coming from the oldest time subdivision of the Paleozoic Era, the Cambrian Period.

The three major eras are subdivided into shorter time units called periods, again on the basis of the fossils contained in the rocks. The names of the periods are given in the table on pages 168 and 169. There is no point in memorizing the names at the moment, because they will become familiar as we look at some of the events in geologic history.

What does the fossil sequence mean?

Up to this point we have treated fossils simply as another kind of sedimentary structure, like cross-bedding or channeling or the pebbles in a conglomerate. We have used fossils to correlate strata from place to place, to arrange sedimentary layers in a sequence, to set up a universal time scale that holds for sedimentary rocks the world over. In a sense we have handled sedimentary rocks like pieces of a huge four-dimensional jigsaw puzzle, to be put together in three dimensions of space and a fourth dimension of time, and fossils are the markings that make assembly of the puzzle possible. Geologists find this a fascinating pastime, in part for the pure intellectual joy of solving a problem, in part because of the insight they gain into geologic processes, and in part for more practical purposes like deciding where to drill for oil or excavate for metallic ores. For all such activity fossils need be little more than mechanical devices to help us recognize different rock layers.

But surely fossils have more to tell us than this. What does it mean that communities of organisms have changed in the course of time, and in particular that the change has been in the direction of ever greater resemblance to modern forms? Observations of this sort cry out for a theory, for a set of assumptions that will provide a reasonable explanation.

The Geologic Time Scale

The geologic time scale, summing up what we know of the earth's past 4600 million years, developed from research in many parts of the world over the past two centuries. The research involved examination of rocks for fossil evidence, correlation of fossils from place to place and arrangement into developmental sequences, then the fitting of these sequences into the history of physical events recorded by minerals and rock structures. Radiometric dating has given ages in years, which confirm the relative ages deduced from study of rocks and fossils.

The terminology, as befits this kind of development, simply grew with time and added knowledge. Names of the three *eras* mean "ancient life," "middle life," and "recent life." The long time before the Paleozoic, for which no division into eras has been agreed upon, is called *Precambrian.*

Names of *periods* come from many sources. Some refer to areas where the characteristic fossils were first studied: *Cambria,* the Roman name for Wales; *Silures* and *Ordovices,* names of ancient tribes of western Great Britain; *Devon* in England, *Perm* in Russia, the *Jura* Mountains between France and Switzerland. The name *Triassic* refers to a conspicuous 3-fold division of these rocks in Germany; *Cretaceous* comes from the Latin for chalk: *Tertiary* and *Quaternary* are relics from the 18th century when it was thought that all rocks could be grouped into four divisions. The term *Carboniferous,* often used to include *Mississippian* and *Pennsylvanian,* refers to extensive coal beds in this part of the geologic record.

The *epochs* of the Cenozoic are named to indicate the increasing resemblance of fossil forms to living organisms of the present, from "ancient recent" (*Paleocene*) to "entirely recent" (*Holocene*).

Precambrian time, reaching back to the planet's formation some 4600 million years ago, is hard to subdivide because evidences of life are limited to scant remains of primitive soft-bodied forms. The composition of early Precambrian rocks suggests that the atmosphere had more carbon dioxide and less oxygen than at present, and that volcanic activity was more widespread. In the later Precambrian the climates and kinds of geologic changes were much like those of today.

The *Paleozoic Era* begins with rocks containing abundant fossils of marine creatures with hard, preservable shells. Vertebrates appeared early in the era, first fishes, then sluggish amphibians living on the border between land and water, then reptiles completely adapted to existence on land. Primitive land plants date from the Silurian, and later became abundant in huge "coal swamps." Shallow seas spread widely over the continents many times. Equable climates were the rule, but deserts existed at some times and places and far-spreading ice-caps at others.

The *Mesozoic Era* was the age of the great reptiles called dinosaurs. In the middle of the era appeared the first flowering plants, the first modern insects, the first birds. Tiny mammals played a minor role on the Mesozoic lands. Deserts were widespread for a time, and shallow seas again invaded the continents toward the end of the era.

The *Cenozoic Era* has seen the evolution of the tiny mammals surviving from the Mesozoic into huge creatures that dominated the lands. Many of the larger mammals are now extinct, and those that remain have lost their dominance to man—a newcomer who appeared a scant 3 million years ago. Much of the Cenozoic was a time of warm climate, but cooling toward the end culminated in the spreading glaciers of the Pleistocene.

History of Life by Geologic Eras, Periods, and Epochs

Precambrian—4600 to 600 m.y. ago . . . . Extent 4000 m.y.

Fossils limited to simple marine forms—algae, bacteria, worms.

Paleozoic Era—600 to 225 m.y. ago . . . . Extent 375 m.y.

Period (and extent)

Cambrian (120 m.y.): First abundant fossils, all marine.
Ordivician (45 m.y.): Trilobites and brachiopods abundant.
Silurian (35 m.y.): First simple land plants. Fishes became abundant.
Devonian (50 m.y.): First land vertebrates (amphibians).
Mississippian (40 m.y.): Coal swamps widespread. First reptiles.
Pennsylvanian (50 m.y.): Extensive coal swamps. (This period and Mississippian often called Carboniferous.)
Permian (35 m.y.): Primitive reptiles abundant.

Mesozoic Era—225 to 70 m.y. ago . . . . Extent 155 m.y.

Period (and extent)

Triassic (45 m.y.): First dinosaurs, first ammonites, first mammals.
Jurassic (50 m.y.): First flowering plants, first birds, first modern insects.
Cretaceous (60 m.y.): Large ammonites abundant. Dinosaurs and ammonites became extinct at end of period.

Cenozoic Era—70 m.y. ago to present

Tertiary Period—70 to 3 m.y. ago . . . . Extent 67 m.y.

Epochs (and extent)

PALEOCENE (10 m.y.) EOCENE (22 m.y.) OLIGOCENE (11 m.y.) MIOCENE (17 m.y.) PLIOCENE (7 m.y.)	Mammals evolved during these epochs, many showing a progression from small forms at the beginning to large forms in the Miocene and later.

Quaternary Period—3 m.y. ago to present . . . . Extent 3 m.y.

Epochs (and extent)

PLEISTOCENE (3 m.y.): The "Ice Age." Large mammals abundant. Man appeared early in this epoch or in latest Pliocene.

RECENT or HOLOCENE (11,000 years)

Fig. 6-13. Algae (stromatolites), primitive plants whose fossil remains are found in rocks of all ages from far back in the Precambrian to the present. The specimen on the left is from a Precambrian limestone in Glacier National Park, that on the right from a Cambrian limestone in New York State. (W.C. Alden, U.S. Geological Survey; R.W. Owen and G.M. Friedman, Renssalaer Polytechnic Institute)

What kind of theory is possible? We might propose, as a first guess, that organisms have been created and destroyed many times during earth history, that each fossil assemblage we find represents a special creation. This was a widespread notion when geology began to develop as a science in the early 19th century, a notion consistent with prevailing ideas that the earth's past was punctuated with cataclysmic events like worldwide volcanic activity, universal floods, and sudden large-scale upheavals of mountains and plateaus. Because such events, including the creation and destruction of life, were pictured as sudden and violent, assumptions of this sort are described in general as *catastrophic* theories.

The alternative general position is that events of the earth's history have been slower and more orderly, that rocks and landscapes and organisms have changed gradually from one form to another, and that past changes are of the same sort that we see taking place in the world around us today. Slow changes of this kind are described by the term *evolution,* in contrast to the sudden transformations embodied in catastrophism. Applied specifically to living things, evolution means the gradual change of one species of plant or animal into another. A succession of fossil assemblages, according to this theory, requires continual development of new species over a very long period of time; the close resemblance of later fossil assemblages to present-day organisms means that familiar plants and animals are simply the latest step in a continuous evolutionary sequence. The word evolution, if used alone, refers to this kind of biological development, but one can speak just as correctly of evolution in a wider sense—the evolution of landscapes and rocks, of planetary systems, even of stars and galaxies.

Evolution in a general sense, as an alternative to catastrophism, became increasingly popular among scientists and philosophers of the first half of the 19th century. The idea had been suggested earlier, but with the rapid development of geology after William Smith's discovery of the pragmatic usefulness of fossils the evidence for gradual evolutionary development of the inorganic part of the earth became overwhelming—particularly in the writings of Sir Charles Lyell. Biologic evolution, although strongly advocated by the Frenchman Jean Baptiste Lamarck near the beginning of the century, was much slower in gaining acceptance.

The objections to biologic evolution were partly theological, partly scientific. Transformation of one species into another ran counter to the story of Genesis, and would require far more time than a literal interpretation of Genesis allowed. On scientific grounds evolution was suspect because it could not be demonstrated in the present world: existing species seemed stable, showing no tendency to change either in the natural world or in the laboratory. Lamarck had hazarded a guess about a mechanism that would make evolutionary change possible, the inheritance of acquired characters, but available evidence seemed strongly against his suggestion. It was well after the middle of the century (1859) when Charles Darwin, on the basis of evidence from paleontology and from painstaking study of present-day organisms, succeeded in establishing the theory of organic evolution on firm grounds.

How did Darwin accomplish this feat? It would take us too far afield to discuss his evidence in detail, particularly the evidence from living plants and animals. Regarding the latter, let us note only that his meticulous observations of slight changes in species isolated in remote areas indicated that evolution is indeed taking place in the world around us, even though the changes are so slow in comparison with human life that they are difficult to demonstrate in the laboratory. The paleontological evidence needs more attention, since it rests on detailed study of fossil assemblages like those we have been discussing.

For one thing, Darwin could point out that changes in fossil assemblages are so numerous and often so slight that special creation of each assemblage would require an almost infinite number of creation-followed-by-destruction episodes. In any one area, where only a few assemblages are found in layers separated by unconformities, special creation may seem plausible; but for the whole sedimentary record, obtained by combining assemblages from many areas, the idea is patently unacceptable.

Furthermore, the development of a single kind of organism can sometimes be followed through a succession of sedimentary beds. Its numbers are small when its fossils first appear, presumably because of competition from established organisms. Then in higher strata some varieties become numerous, indicating that species were developing which were better adapted to their environment. If an organism is highly successful, species develop that can flourish in many kinds of environments, hence appear as fossils in different kinds of sedimentary rocks. In the uppermost beds of the sequence some fossils show extreme specialization, so much so that a change in environment was evidently disastrous, causing them to disappear from the later fossil record.

Thus fossils provide support for the principal points of Darwin's theory: new species arise from random variations in existing species; most variant forms disappear because they are less well equipped to survive than normal forms, but a few have modifications that make them better able to compete; the successful variants multiply, tending to spread into all parts of their environment; new variants appear, becoming species adapted to other environments; the spreading and variation continue until environments change or until other kinds of organisms arise that are even more successful.

The story of the reptiles is a good example. Fossils of small reptiles appear first in late Paleozoic rocks, presumably developing by a series of variations from an amphibian ancestor. The reptiles remain small and scarce in the Paleozoic, evidently dominated by the more numerous amphibians. Then in the early Mesozoic rocks amphibians become scarce, probably because of a drastic change in environment, and suddenly reptilian fossils are very numerous. In later Mesozoic rocks the number of reptilian species increases, some of the variants grow to large size, and some become adapted to life in the water and in the air. At the end of the Mesozoic fossils of the giant reptiles disappear completely, perhaps because they were too specialized to survive a change of climate, perhaps because they were unable to compete with the rapidly developing population of mammals.

A plausible mechanism (termed random variation and natural selection), the demonstration of the evolutionary history of many life forms from the fossil record, and evidence from present organisms that evolution is continuing today: these are the ingredients that made Darwin's theory so extraordinarily successful. Details of the theory have been refined in the century since Darwin—for example, we know much more today about the mechanism of the variations on which changes in species depend, and many gaps in the fossil record that worried Darwin have been filled. A number of details have been seriously questioned—for example, Darwin put too much emphasis on some kinds of competition in explaining natural selection. Some details are still troublesome—for example, a really satisfactory explanation still eludes us for the sudden dying out of large groups of apparently successful organisms, like the disappearance of the dinosaurs at the end of the Mesozoic Era. But the basic idea of evolution is so firmly supported by the record preserved in fossils that it can no longer be seriously doubted.

Fossils are not only a key for correlating sedimentary rocks, not only a means for arranging geologic events in time, but even more importantly a major piece of evidence for the conclusion that our planet, including all the living creatures on its surface, has developed gradually over immense reaches of time by processes that are still in operation.

Fossil hunting

We descend now from theoretical speculations to some much simpler questions. Suppose you want to hunt fossils yourself: where do you go, what do you look for, what can you expect to find?

To begin with, no specific answers of wide application are possible. Conditions for fossil hunting vary enormously from one place to another. It is generally true that limestone and shale are more promising rocks for a fossil hunter than sandstone and conglomerate, but the rule has many exceptions. If a shale or sandstone contains limy concretions—rounded masses cemented by calcium carbonate—it is often worthwhile to break open a few concretions to see whether they contain fossils at their centers. In some places fossils weather out of limestone conspicuously and accumulate as talus; in other places a deeply weathered surface conceals fossils that may be revealed on a fresh surface. But rules like this will be helpful only locally.

You will have better luck if you seek information for your particular locality. Good quarries or outcrops for fossil collection are generally well known; once you get a feel for the kind of rocks in which fossils are abundant, you can extend your search from known outcrops to others. If your state has an active geological survey, you can almost certainly get detailed information about fossil localities. Standard textbooks on paleontology (the science of fossils) and publications of the U.S. Geological Survey are other possible sources, but their information may be too generalized for your purposes.

If you simply start out on your own, with no background information, you may well be disappointed. Fossils in most rocks are scarce, and a beginner often misses the poorly preserved remains that may be present. But of course there is always a chance that you will stumble on a good specimen of a rare organism, enough of a chance to make the hunt worthwhile.

Suppose you find a fossil assemblage: what can you do with it? Unfortunately the interpretation of fossils is an exceedingly complicated subject, one that requires years of study to master. Even most geologists, in fact, receive so little training in paleontology that they are not competent to make more than general guesses about the meaning of fossil assemblages. When a geologist finds fossils, he collects the best specimens he can obtain of each variety, and submits the collection to a paleontologist for identification. You would do well to follow his example. Your collection may give you some exciting hints about the kinds of creatures that lived long ago, but only an expert can fill in the details.

The following paragraphs are brief descriptions of some of the commonest fossils that an amateur may reasonably expect to encounter. A few of these are distinctive enough so that he can make guesses about their geologic age and about the environment in which the organisms lived.

Trilobites are common and conspicuous fossils in some sedimentary layers of the early Paleozoic Era, especially in the Appalachian region and the Middle West. Distant relatives of modern crabs and lobsters, these creatures lived exclusively in the sea. A few trilobites are found in later Paleozoic rocks, but Mesozoic and Cenozoic rocks contain none at all. The history of the trilobites is a typical evolutionary saga: first appearance of primitive forms in the earliest Paleozoic, rapid development of great numbers of species and increase in size of individuals so that they were the dominant animals of early Paleozoic seas, then the growth of specialized

Fig. 6-14. A trilobite (*Paedumias yorkensis*) from the lower Cambrian of Pennsylvania. (Smithsonian Institution Museum of Natural History)

forms and gradual reduction in numbers until they became extinct at the end of the era.

Trilobite fossils range in length from a centimeter or so to more than ten centimeters. As the name implies, they are conspicuously divided into three parts or lobes by grooves down the back. All three lobes are segmented. At one end is a broad head, commonly with large eye sockets, and in many species a tail protrudes at the other end. Some fossil specimens are tightly coiled, indicating that trilobites could protect themselves by curling up like modern sowbugs.

The interpretation of a trilobite find is clearer than for most fossils. You know at once that the sediment containing them is marine and that its age is Paleozoic; if the fossils are abundant, the early part of the Paleozoic is almost certain.

Brachiopods are another kind of marine animal whose fossils are especially abundant in Paleozoic rocks. Unlike trilobites, they remained numerous until the end of the Paleozoic; a few species survived into later times, and some are even present in the modern ocean. But if you find a limestone studded with many brachiopod shells, as you can at several places in the northeastern part of the United States, you may be sure that the rock dates from the Paleozoic Era.

Fig. 6-15. Brachiopods (*Sowerbyella compacta*) from Alabama. (Smithsonian Institution Museum of Natural History)

Fig. 6-16. Restoration of an early Paleozoic sea floor, showing trilobites, cephalopods, brachiopods, corals, and sponges. (Smithsonian Institution Museum of Natural History)

Brachiopods superficially resemble some kinds of clams. A complete fossil has two shells, but differs from a clam in that the two shells are not mirror images. The shapes of the shells differ greatly from one species to another; an abundant variety, especially in late Paleozoic rocks, has a form reminiscent of an old-fashioned oil lamp, hence is aptly described by the common name "lamp shells."

Pelecypod is the general term for a group of animals that includes such familiar creatures as clams, mussels, oysters, and scallops. These are common fossils in rocks of all ages from near the beginning of the Paleozoic down to the present, and in rocks formed from lake and river sediments as well as marine sediments. If you recognize a fossil simply as a pelecypod, you gain little information about either age or environment of deposition of the enclosing sediment. But if you take the fossil to an expert he may be able to tell you a great deal, because the species of pelecypods changed

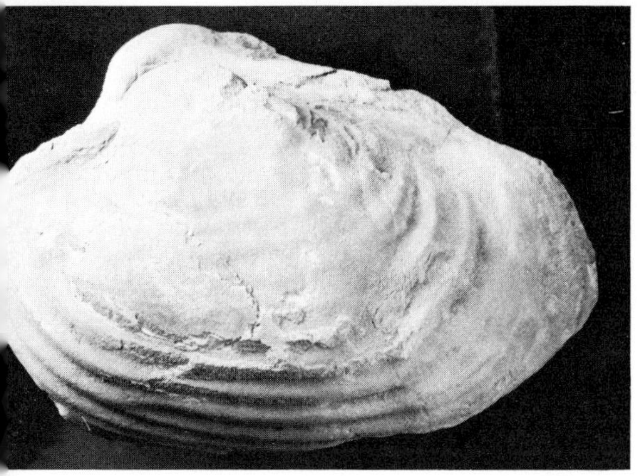

Fig. 6-17. Two fossil pelecypods, the one on the left (*Grammysioides*) from Devonian beds in Ontario and the one on the right (*Orthonota undulata*) from Devonian beds in New York. (Smithsonian Institution Museum of Natural History)

greatly during geologic time and particular varieties are characteristic of certain ages and certain kinds of sediment.

Like brachiopods, pelecypods build two shells, but now the shells are mirror images. When the organism dies the shells may be separated, so that fossil shells are frequently single. The shells are most commonly a few centimeters in diameter, but range up to a meter or more. Some are decorated with grooves and ridges, others are very plain.

Gastropods, commonly called snails, have an even wider diversity of habitat than pelecypods, for some species are adapted to an air-breathing existence on land. Like pelecypods their fossils are found in rocks of all

Fig. 6-18. Two varieties of fossil gastropod, the one on the right (*Bambexia*) from Devonian beds in Michigan and the ones embedded in rock (*Turritella mortoni*) from Eocene beds in Virginia. (Smithsonian Institution Museum of Natural History)

geologic ages from the early Paleozoic, and unless you can recognize particular species the finding of a fossil snail is not helpful in identifying either age or environment of sedimentation.

Cephalopods in modern seas include squids, octopuses, and the pearly nautilus. The fossil record shows that they were more abundant, in both numbers and variety of species, in seas of the past. The common cephalopods of the early and middle Paleozoic grew long, straight, cigar-shaped shells, some attaining lengths of more than five meters; in the late Paleozoic some species became coiled, and the coiled varieties (of the order called *ammonites*) proliferated in the Mesozoic Era until they became as dominant in the sea as the dinosaurs were on Mesozoic lands. If you live in the western states you may be lucky enough to find one of these handsome fossils, but if your find belongs to one of the larger species you will have a hard time moving it—big specimens are several feet in diameter and weigh a few hundred pounds. The modern pearly nautilus is a distant relative of these Mesozoic monsters, which, like the dinosaurs, died out before the Cenozoic Era.

Corals are small animals that lead a sedentary existence in shallow sea water, attaching themselves to rock by means of a horn-shaped or tube-shaped skeleton made of calcium carbonate. Some varieties are solitary, but the ones you are most likely to find as fossils are those that have

Fig. 6-19. A cephalopod of the subclass called *ammonites* (*Harpoceras*) from Jurassic beds in Yorkshire, England. (Smithsonian Institution Museum of Natural History)

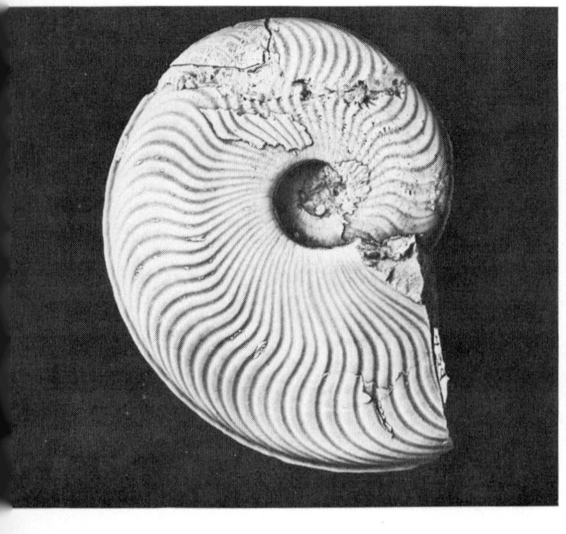

Fig. 6-20. A colonial coral (*Hexagonaria*) from Devonian beds in Michigan. (Smithsonian Institution Museum of Natural History)

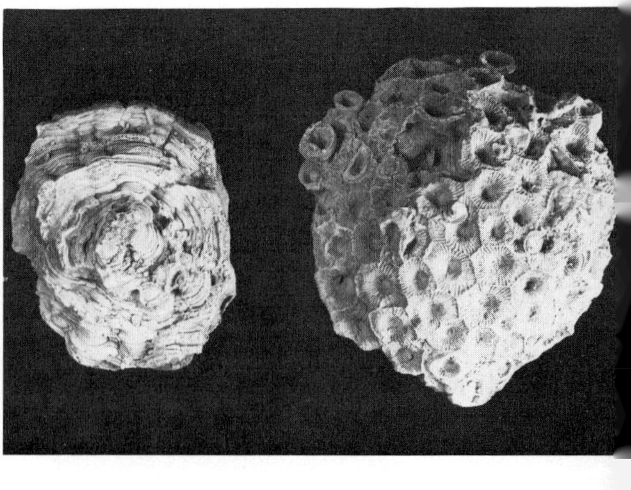

built huge colonies like the present-day coral reefs fringing islands in the Caribbean Sea and the South Pacific. Corals are found in rocks of all ages, and can be used for dating only if separate species can be recognized. Parts of some limestones consist of coral reefs marvelously preserved; spectacular examples dating from the late Paleozoic form outcrops in western Texas. Such rocks are often polished for decorative stone, the intricate coral structures forming beautiful colored patterns. Ancient reefs are important also as reservoirs for petroleum, open cavities in the original reef or in spaces between broken blocks from the reef providing ample space for oil to accumulate.

Vertebrate fossils, represented by bones and teeth, are not common except in a few places. A vertebrate find is exciting, but unless you have had some training in anatomy it will not be very informative. If you do happen to stumble on well preserved skeletal remains, it would be a good idea to report your find to a paleontologist, because vertebrate fossils are sufficiently rare to make each one of possible value in providing new details of the geologic record.

How do we guess times in years?

Sequences of fossils enable us to arrange rocks in order from early to late, no matter how discontinuous the outcrops may be or how complex the folding and faulting of the rock layers. The overall fossil sequence in a large area can be grouped into three major divisions (eras) and many smaller subdivisions (periods and epochs), divisions which we describe as intervals of time and which we recognize over the entire earth. The fossils themselves provide overwhelming evidence for a change in species through time, a change in a general direction from primitive forms to the complex organisms of the present world. Our whole discussion of fossils has been saturated with the idea of time—yet not once has a time been specified in hours or days or years. How do we go about assigning numbers to the times we have been considering?

From fossils alone the effort to derive numbers is frustrating. In a general way we know that evolution is slow, and geologic times must therefore be long. Species of organisms in the world around us seem stable; in the course of a human life we are not aware of any change occurring from one species to another. Even if we try to produce change artificially, speeding up natural selection by segregating individuals with particular characteristics, the development of organisms that will not interbreed with the parent forms is difficult. From reports of accidents in which organisms have become separated from their relatives and isolated on remote islands,

Fig. 6-21. Vertebrate fossils: Eocene fishes (*Priscacara liops* and *Diplomystus humilis*) from the Green River shale in Wyoming. (Smithsonian Institution Museum of Natural History)

we know that different—but not very different—species have originated in times on the order of a few thousand years. Multiply this by the enormous number of radical changes that have occurred in many lines of organisms in the fossil record, and we know for certain that we are talking about times at least on the order of millions of years, more probably hundreds of millions. But beyond this vague guess the study of fossils cannot help us.

Of the many ways that have been suggested to get more quantitative estimates of geologic time, the only one that has proved generally useful is measurement of the extent of decay of radioactive elements contained in rocks. The figures for ages of geologic events quoted in previous chapters have all been obtained by this method. We will look into the problems of radiometric age measurement later on.

Helpful as radiometric measurements are, they do not solve all problems of determining geologic times. For one thing they are expensive, requiring much time and much laboratory equipment, hence cannot be used directly in the course of geologic field study. Another difficulty is their limitation to certain kinds of rock: obviously a radiometric determination requires that a rock contain a measurable quantity of a radioactive element, and many common rocks do not meet this requirement. Because of these difficulties, radiometric ages must be supplemented by the qualitative

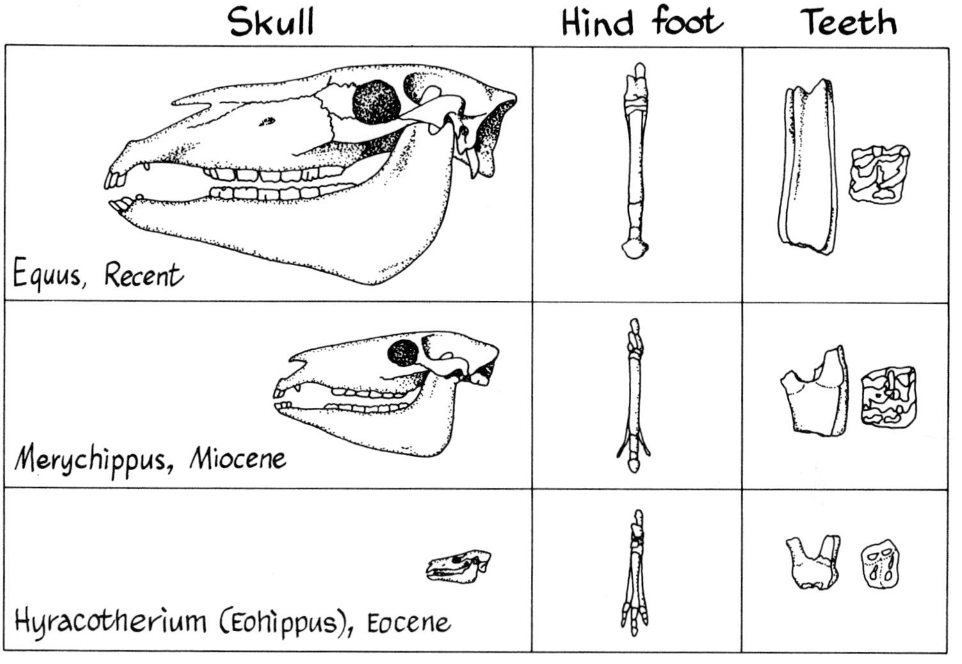

Fig. 6-22. Three stages in the evolution of the horse during the Cenozoic Era. The size of the skull (and the whole organism) has increased greatly, the number of toes has decreased from three to one (four to one on the forefoot), and the tooth has grown longer and more complex. With these changes the horse has become adapted to living on grassy plains, where swiftness of foot and ability to eat coarse vegetation are essential for survival.

ages given by fossils. If a few rocks can be very accurately dated, and if their position in the geologic time scale can be determined by their fossil content, then other rocks can be arranged between these guideposts by their fossils, and their ages can be fairly accurately guessed. As more and more guideposts are set up by radiometric determinations, the accuracy of interpolated ages improves. Thus age determination in general depends on a combination of fossil study and radiometric analyses. Fossils are the common tool for age determination in field work, but to translate the relative ages given by fossils to actual ages in years requires that at least some of the rocks containing fossil assemblages be dated radiometrically.

Fig. 7-1. Half Dome in Yosemite National Park. The dome, the steep cliffs beneath it, and the high mountains beyond are all carved from the huge mass of granitic rock that forms the backbone of the Sierra Nevada. (U.S. Department of the Interior, National Park Service)

7

Rocks That Come from Below

For two chapters now we have concentrated on sedimentary rocks. From one point of view these are the most interesting of geologic materials, because they contain a record of conditions at the earth's surface and particularly of the plants and animals that have lived on or near the surface. But while sediments were accumulating at the surface other processes were going on below, giving rise to the rocks we call igneous and metamorphic—and to some geologists these rocks are more fascinating than the sedimentary variety.

One reason for the fascination is the simple fact that greater mystery surrounds the origin of these rocks. We don't see them in process of formation, in the same sense that we can watch the raw materials of sedimentary rocks accumulate. Volcanic rocks, to be sure, we can watch as they freeze from molten lava, but the source of the lava remains hidden. All that we know about the history of igneous and metamorphic rocks we gain from indirect and fragmentary evidence, not from direct observation.

Another reason for interest in these rocks from the depths is just the commonness of their occurrence. Not only are they abundant, but they often form the most prominent features of a landscape. The great volcanoes of the Pacific Northwest and southern Alaska are good examples, but many others come to mind. The White Mountains of New Hampshire, Stone Mountain in Georgia, the steep cliffs of Yosemite Valley are carved in masses of granite formed long ago and exposed by deep erosion. The rugged coast of Maine and much of eastern Canada are underlain by metamorphic rocks. Despite their obscure origin, igneous and metamorphic rocks are responsible for much of the world's scenery.

An understanding of these rocks requires acquaintance with a considerable vocabulary. We have looked at igneous rocks in two previous chapters, and learned some of the words needed to describe them. We will meet a few additional terms in this chapter, but emphasis now will be less on description and more on the structures that igneous rocks form and the information they give about conditions under the earth's surface. Metamorphic rocks we have not yet encountered. We will need to enlarge our vocabulary a bit to describe them, but again we will focus on their meaning with respect to internal conditions rather than on descriptive details.

Igneous and metamorphic rocks are easily confused, and the distinction between them needs preliminary emphasis. Igneous rocks are those that have formed by the cooling of molten material (*magma*); metamorphic rocks have formed by the alteration of other rocks under the influence of heat and pressure, *the material remaining largely solid during the alteration.* The history of an igneous rock always involves melting; the history of a metamorphic rock is a transformation of its minerals while the rock remains in the solid state. A typical specimen of each bears the marks of high temperature during its formation, often in the form of intergrowing crystals of silicate minerals, but the metamorphic specimen must show in its composition or structure that it has not undergone melting.

How hot does the earth get inside?

Since both igneous and metamorphic rocks have been at higher temperatures than sedimentary rocks, we start this inquiry by asking what sort of temperatures we might expect to find under the earth's surface.

That temperatures increase within the earth needs no more proof than the simple observation that mines become warmer as they are made deeper. Miners know from long experience that working conditions in deep mines would be intolerable without efficient ventilation. Similarly, thermometers lowered down deep drillholes record a steady rise in temperature as depth increases. From many measurements made in different parts of the world, the average increase is about 30° Celsius (or Centigrade) for every kilometer of depth (or 20° Fahrenheit for every 1000 feet). In places the gradient may be as low as 10°C/km or as much as 50° (or even higher in volcanic areas), but a reading near 30° is common. Actual measurement can go no deeper than the bottom of the deepest drill-holes, 8 or 9 km from the surface, but if we assume that the gradient is constant downward we can calculate the temperature at any depth.

Such a calculation indicates that to reach a temperature of 1100°C, high enough to melt basalt in the laboratory, requires a depth of only about 35 km (the dotted line in Fig. 7-2). More siliceous lavas like andesite or rhyolite melt at lower temperatures, hence could be molten at even shallower depths. The melting points are increased by pressure, so that the required depths are actually somewhat greater, but even so the common igneous rocks should be capable of existing as liquids at fairly shallow levels in the earth.

Can we infer that the whole earth is largely molten below depths of 20 or 30 miles? This would be a necessary conclusion if the temperature continues to increase downward at a steady rate of 30°/km. At the earth's center, some 6400 km from the surface, the calculated temperature would be of the order of 200,000°C, far hotter than the surface of the sun—hot enough not only to melt but to vaporize all ordinary rocks. Is there any evidence for such extreme conditions, or is something wrong with our basic assumption about the constancy of the temperature gradient?

Speculations about the structure of the earth up until a couple of centuries ago were based on a model much like the one given by our simple temperature extrapolation: the earth was a molten globe surrounded by a thin solid crust, and volcanoes were places where cracks in the crust permitted molten material to escape. A vestige of this ancient idea persists

Fig. 7-2. The estimated increase of temperature with depth in the outer part of the earth. The position of the line is not entirely certain, especially at greater depths, but it differs markedly from the line (dotted) that would be projected from the 30°/km measured in deep mines and drillholes.

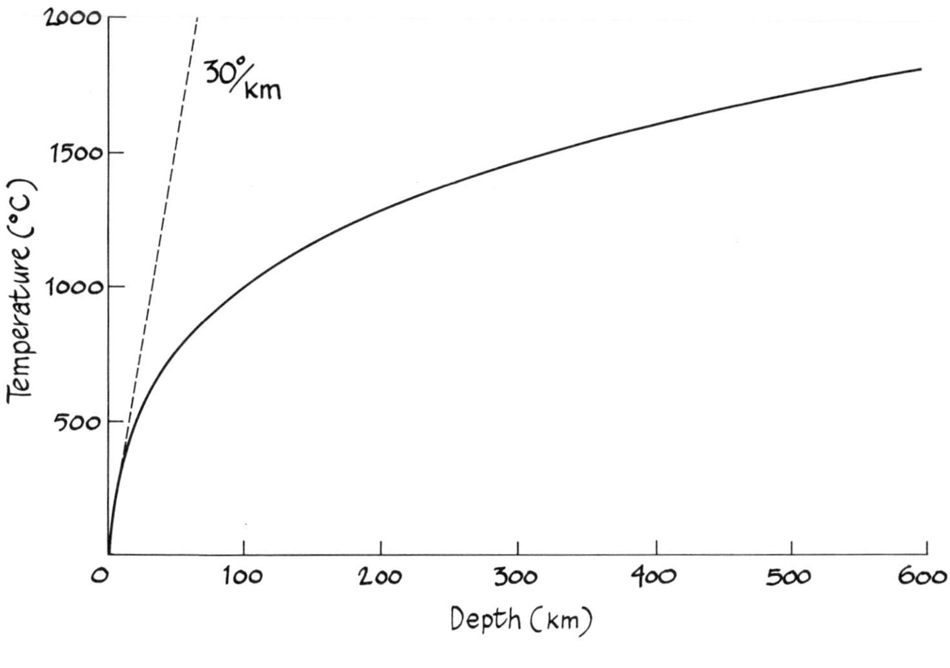

in the common expression "crust of the earth" for our planet's outermost layer. But today we can demonstrate convincingly that such a model is far from correct.

Part of the demonstration depends on analysis of the vibrations from large earthquakes that travel through the earth, and part depends on the earth's reaction to the tide-generating forces produced by the gravitational attraction of the moon and sun. Both kinds of evidence show that the earth behaves as a rigid body, hence that its interior must be largely solid. Evidently our model is wrong: temperatures in the deep interior cannot increase downward at the steady rate of 30°/km which we measure near the surface. The temperature at modest depths of 30 or 40 km may indeed be high enough for some molten rock to exist, but below this the increase in temperature must be so gradual that rock is kept largely in the solid state by the high pressures.

It would be interesting to speculate further about temperatures at great depth, but this subject is best deferred until later. For the moment our primary concern is the origin of igneous and metamorphic rocks; and the important conclusion is simply that temperatures are high enough for some local melting to occur at depths of a few tens of kilometers, but not high enough to make the whole interior of the planet molten.

The question of pressure

High pressures, we noted in the last section, raise the melting point of rocks and hence permit rock material to remain solid even when temperatures are far above the point where melting would occur under normal conditions. What magnitude of pressures are we talking about, and how do rocks respond to them?

Pressures in a fluid, as you well know if you have had any diving experience, increase rapidly as you go to greater depths. The pressure results simply from gravity, from the weight of overlying layers of fluid. It is easy to calculate: the weight of water resting on a square centimeter of surface, for example, increases by one gram for every centimeter of height (since a cubic centimeter of water weighs a gram), or by a kilogram for every 10 meters. It happens that a kilogram per square centimeter is approximately the pressure of the atmosphere at sea level, which is often taken as a unit of pressure. Thus in water the pressure increases by approximately one *atmosphere* for every 10 meters of depth below the surface, or by 100 atmospheres per kilometer. Pressures in rock due to weight of overlying material are greater than in water, simply because rock is heavier. Most ordinary rocks have densities roughly three times that of water, so that pressures in rock increase by about 300 atmospheres per kilometer of depth.

Unlike temperature, pressure can be assumed to increase fairly steadily downward. Some variation will result from changes in density of rocks and from the variation of gravity with distance to the earth's center, but for distances of a few hundred kilometers down from the surface pressure is roughly proportional to depth. Hence at 30 km the pressure is about 9000 atm and at 100 km about 30,000 atm.

Do such pressures have any effect on ordinary rock, besides increasing the melting point? The question can be made more pointed by asking it in a different way. Suppose that we drill a large hole into a mass of flawless granite which extends from the surface to a depth of a few tens of kilometers. How deep can we drill before the granite at the bottom of the hole is crushed by its own weight and starts to flow into the hole?

The crushing strength of granite, a figure we can obtain from standard tables, is about 30,000 pounds per square inch or 2100 kg/cm^2. Thus a pressure of 2100 atmospheres is sufficient to crush granite, and this kind of pressure would be reached at a depth of less than 10 km. In other words, we would have difficulty in drilling to such a depth, and in keeping the hole open after we drilled it, because the granite would tend to flow and fill in the bottom of the hole.

This calculation is not very realistic because granite near the bottom of the hole is supported by rock on all sides, and unless our hole is very large the support would help to prevent flow. Nevertheless the strength of rock clearly puts a limit on the depth to which we can drill and on the ability of rock to maintain open cavities at levels more than a few kilometers below the surface. Rocks with lower strength than granite would start to fail at shallower levels—and this is by no means of theoretical interest only, because the slow collapse of mine tunnels is a major problem in the maintenance of deep mines. Caverns extending far down toward the center of the earth cannot exist except in the imagination of science-fiction writers.

Even if openings are not present, pressures at depths of a few tens of kilometers are sufficient to make solid rock move in response to pressure differences. At first encounter this seems a strange statement. Ordinarily we think of a rock like granite as indefinitely strong, unmoved by any force short of a powerful explosion. We often refer to granite, in fact, as a symbol of great strength. Yet pressures in the earth not far beneath the surface are so high that even granite can be deformed and flow like putty if the pressure in one place is less than that in another. It is important to keep this in mind when looking at the structures of igneous and metamorphic rocks.

Note that here we are talking about *solid* rock, not liquid. At depths of a few tens of kilometers there may be local pockets of liquid rock also, which would flow even more readily in response to pressure differences. But in this discussion of pressure we are emphasizing that all rock material, no matter how strong it may appear at the earth's surface, is capable of movement *as a solid* if pressures are sufficiently high.

Metamorphic rocks

In discussing sedimentary rocks we noted that the conversion of sediment to rock usually requires burial of sediment beneath other layers. If

Fig. 7-3. A contact of granite (light color, on the left) with metamorphic rock. In the left foreground is a chunk of the metamorphic rock that was picked up by granite when it was molten. (C.D. Rinehart, U.S. Geological Survey)

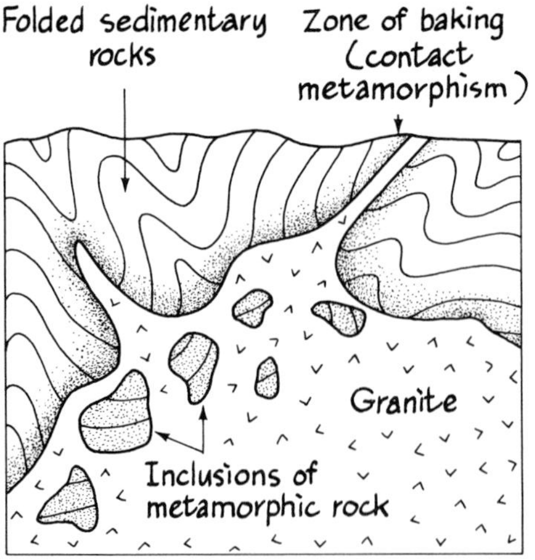

Fig. 7-4. Sedimentary rocks metamorphosed at a granite contact.

sediment is being deposited in a basin that is slowly subsiding because of movements in the crust, the early layers may be buried to depths of thousands of feet or even miles. If subsidence goes on long enough, some of the sediment may be pushed down into the regions of high temperature and pressure we have considered in the last two sections. This result is even more probable if subsidence is supplemented by more active crustal movements that result in folding and breaking of the sedimentary layers, pushing them high above sea level in some places and far below the surface in others. What happens to sedimentary rocks subjected to this kind of rough treatment?

If burial is carried to extremes, the rocks would simply melt. Many sedimentary rocks have chemical compositions much like igneous rocks, and at depths where igneous rocks would be molten the sedimentary varieties must also melt.

Fig. 7-5. Photomicrograph of marble. The crystals of calcite (light gray) have grown large and interlocking. Impurities in the original limestone have recrystallized to other minerals (darker gray), of which the most abundant is the ferromagnesian mineral hornblende. Magnification about 20 times. (W.R. Hansen, U.S. Geological Survey)

Fig. 7-6. Photomicrograph of a thin section of quartzite. The grains are intergrown, not separate as they are in a sandstone. The grains are almost entirely quartz; the different shades of gray are not actual grain colors, but are produced by prisms in the microscope that polarize light. Magnification about 10 times. (D. Carroll, U.S. Geological Survey)

But the melting is a complicated process. Any rock consisting of several different minerals does not become liquid at a single temperature but melts gradually over a temperature range, some of the minerals or mineral combinations melting before others. A sandstone consisting largely of quartz and feldspar, for example, on heating produces at first a liquid containing much of the quartz and the part of the feldspar that is rich in sodium and potassium; only if the temperature is raised by another 100° or more will the calcium-rich feldspar and the minor amounts of ferromagnesian minerals go into the melt. In a previous paragraph we mentioned the "melting point" of basalt as about 1100°C; this is a gross oversimplification, because basalt has no real melting point. If you measure the temperature of a moving basalt flow near its source you will get a reading in the neighborhood of 1100°, so that this is a rough estimate of the temperature at which the rock becomes molten. But the moving flow contains solid crystals, and it will continue to flow more and more sluggishly as it cools and other solids appear. Fluid movement stops completely only when the temperature has dropped to about 800°. Sandstone, basalt, and all other heterogeneous rocks have a melting *range* rather than a melting *point.*

Now we inquire, does anything happen to a sedimentary rock *before* it starts to melt? To make the question specific, suppose that a sequence of shale, sandstone, and limestone has been buried and carried down to a level 10–20 km below the surface, and suppose it has been intruded by a large dike of granitic magma. The rocks will have a temperature of a few hundred degrees just because of the depth, and near the dike the temperature will be a little higher but not high enough to cause melting. Will the rock show any effect of this new environment?

A theoretical prediction would be hard to make, but we can approach

Fig. 7-7. A metamorphosed conglomerate. The rock has been subjected to high temperature, and to pressures strong enough to squeeze the pebbles into elongated spindle-shapes. The distortion of quartz pebbles in a rock like this is convincing evidence of the enormous forces at work in the earth's crust. (John Shelton)

Fig. 7-8. Slate. In this outcrop the *foliation* resulting from metamorphism (flat surfaces sloping gently to the right) is in a direction different from that of the original *bedding* of the shale (the planes shown by faint lines sloping left). (A.A. Drake, Jr., U.S. Geological Survey)

an answer by looking at outcrops of sedimentary material which has been exposed to this kind of situation. Sedimentary rocks near a granite contact generally show considerable alteration. Limestone beds have become coarse-grained, the large calcite crystals showing complex intergrowths. Pure quartz sandstones have become very hard, and a handlens shows that their quartz grains have enlarged and grown together. The shale has become a hard, brick-like material, too fine-grained for a handlens to pick out separate crystals; a more powerful microscope would show that the original clay minerals have disappeared, and a variety of harder minerals in interlocking grains have taken their place. Evidently the heat has altered, or *metamorphosed,* the original sedimentary materials. We give new names to the products: *marble* for the metamorphosed limestone, *quartzite* for the metamorphosed sandstone, *hornfels* for the metamorphosed shale.

We change the scene a bit, and suppose now that the same sedimentary layers are not just passively heated by a small body of granite but are actively squeezed and pushed while they are subjected to the same fairly high temperatures. Again the temperature is not high enough for melting to start, but the rock is under enough pressure so that it flows readily in response to pressure differences. The flow means that the rock is intimately deformed: its grains are rotated, broken. and moved past one

Fig. 7-9. Schist. The left-hand specimen shows the rough foliation and spangled surfaces produced by orientation of mica flakes in planes, and the right-hand specimen shows the intricate crumpling of foliation surfaces found in many schist outcrops. (Joachim Hampel)

Fig. 7-10. A schist outcrop near New Haven, Connecticut. (H.T. Stearns, U.S. Geological Survey)

Fig. 7-11. Gneiss.

another along innumerable planes of shearing. Might we expect that the rocks would respond to this treatment just as they did to simple heating?

Again we look to natural outcrops for an answer. It is less easy to find places where we can be sure of seeing rocks that have undergone the imagined conditions, but in some areas we do encounter rocks of altered structure that are similar in sequence and in chemical composition to sedimentary rocks of adjacent areas. The change is most apparent in rocks with the composition of shale: the altered rock is fine-grained and layered like shale, but it is harder and its layers are very thin and have shiny surfaces. This is the familiar rock called *slate.* Under strong magnification the slate appears to consist chiefly of tiny quartz grains and mica flakes, with the flakes conspicuously oriented in parallel planes. The planes account for the layering of the rock, and the sheen of the surfaces is the reflection of light from myriads of mica flakes lying parallel to the layers. Evidently the principal changes in the rock are the conversion of original clay minerals to mica and the growing together of new-formed minerals to give a harder and more coherent structure. It is easy to imagine that the shale has been heated and squeezed by differential pressure, the new mica flakes growing in response to the pressure along planes of intimate movement within the rock.

The contrast between slate and shale is important. Both are fine-grained, both are layered, but the layering in slate results from growth of oriented mineral grains during movement rather than from sedimentation. Layering produced by oriented grains is called *foliation,* and we describe slate as a *foliated* rock. The foliation is often parallel to the original sedimentary layering, but may be at any angle to it. If movement within the rock has been fairly slight, the original bedding may be visible in the altered rock as bands of different color cutting across the foliation.

In some areas the alteration of shale to slate is only the first step in a series of progressive changes. Slate layers followed from outcrop to outcrop may become coarser-grained, giving a rock with visible mica flakes and with quartz grains big enough to be seen easily with a handlens. The mica is still arranged in roughly parallel planes, so that rock surfaces split along these planes have a spangled appearance. This coarse-grained foliated rock goes by the name of *schist* (Plate 24). If followed farther the schist layers may change to still another kind of rock distinguished by banding of dark and light material, the light-colored bands consisting chiefly of quartz and feldspar and the dark bands having abundant mica oriented parallel to the bands; this rock is called *gneiss* (pronounced as if it were written "nice").

The progression from shale to slate to schist to gneiss results from increasing temperature and pressure and increasing length of exposure to high temperature-pressure conditions. The change from one rock to another is a continuation and intensification of the same process—deformation and intimate movement within the rock resulting from differential pressure, at temperatures high enough to cause some minerals to change to others. The principal mineral changes are clay minerals to mica at relatively low temperatures, and mica to feldspar at higher temperatures. Perhaps it has occurred to you that the composition of gneiss is much the same as that of granite, chiefly quartz plus feldspar plus mica. In other words, the end product of this metamorphic sequence in solid rock is a set of minerals similar to those that are stable during crystallization of a magma. If the temperature of metamorphism becomes any greater the rock must start to melt—and in fact gneiss and schist often have lenses and pockets of coarse quartz-feldspar mixtures that probably represent places where a little melting did actually occur.

The limestone and quartz sandstone that accompanied the shale in the original sedimentary pile are altered by the metamorphism, but their changes are less dramatic. Each rock consists largely of a single mineral (calcite or quartz), and the minerals cannot change to others on heating. The mineral grains may be crushed and broken by movement in the rock, but at the prevailing high temperatures new crystals will grow from the crushed material. Neither calcite nor quartz has any tendency to form plate-like or needle-like crystals, so the resulting recrystallized rock cannot show conspicuous foliation. The limestone and sandstone, therefore, change to marble and quartzite, just as they would if they were simply heated at a granite contact.

Thus limestone becomes marble and sandstone becomes quartzite under any conditions of metamorphism, while shale is transformed to hornfels by simple heating and to a foliated rock (slate, schist, or gneiss) if the heating is accompanied by differential pressure.

Where does metamorphism start and stop?

It should be apparent that this brief discussion of metamorphic changes only skims the surface of a fearsomely complicated subject. We have considered a drastically idealized situation: a sedimentary assemblage consisting of pure quartz sandstone, pure calcite limestone, and a shale with no minerals other than quartz and clay, and metamorphism produced only by simple heating or by heating plus differential pressure. If the sedimentary rocks are less pure or if they include other varieties like tuff or coal or rock salt, and if metamorphism involves the action of fluids as well as heat and pressure, the variety of possible products is enormously multiplied. This is one of the most complicated subjects in all geology, and we will do well to leave it with this simple introduction.

To recognize the commoner metamorphic rocks you need only keep in mind that the greater number are foliated, meaning that they are layered and have shiny or spangled surfaces because of orientation of mineral grains; the foliated varieties are grouped according to grain size into slate, schist, and gneiss, and the gneiss is distinguished also by its banding; the unfoliated rocks marble and quartzite are recognizable by their mineral composition, their grain size, and their intergrowing crystals; the unfoliated rock hornfels may be hard to distinguish from lava, and is easily recognizable only if it retains sedimentary structures.

The essential thing about metamorphic rocks is their formation under the earth's surface, by the action of heat and pressure on pre-existing rocks, under conditions such that the rock remains solid during the transformation. To apply a definition of this sort obviously involves a good deal of hair-splitting. How far under the surface does the metamorphism occur? What temperatures and pressures are required to make a change truly metamorphic? If a rock gets hot enough for melting to start, is the change still considered metamorphic? In other words, what are the limits of metamorphic processes that distinguish them from sedimentary processes on the one hand and igneous processes on the other?

The change from sediments to sedimentary rocks involves compaction, cementation, and recrystallization, all of which take place at least a short distance under the earth's surface after the sediments have been buried. Temperatures and pressures depend on depth of burial. Somewhere there must be a fine line between these changes and these temperature-pressure conditions and the ones we would describe as metamorphic. Where is the line to be drawn?

The answer is that there can be no line which is satisfactory in all cases and to all investigators. If the temperature has risen above 100 or 150°C and if the formation of new minerals is evident, most geologists would call the product a metamorphic rock rather than sedimentary. But defining the limit of metamorphism leaves much room for argument.

The limit at the high-temperature end is at least as hazy, probably more so because the interpretation of rock structures formed at high temperatures is more difficult. At some contacts between "metamorphic" gneiss and "igneous" granite, for example, the two rocks become interlayered or intimately mixed; some of the gneiss may have melted to form granite, and some of the granite may resemble gneiss because its dark minerals were oriented by flow in the liquid as the magma cooled. Such areas defy attempts to locate a sharp contact between igneous and metamorphic rocks, and the precise nature of the processes recorded in the rock structures has occasioned long debates.

Trying to establish the limits of metamorphism is one more example of the futility of haggling over precise definitions in geology. We are trying to set limits where none exist in nature. Obviously there is a smooth gradation from sedimentary to metamorphic processes on the one hand, and from metamorphic to igneous processes on the other. Any attempt to fix rigid boundaries is foredoomed to failure. Between sedimentary processes that take place near the surface and igneous processes that take place at depth there is a wide range of temperature-pressure conditions under which rocks are altered in the solid state to other rocks with distinctive characteristics. It is convenient to describe this kind of alteration as metamorphism and the

Fig. 7-12. Mixed igneous and metamorphic material at a granite contact. Is the granite pushing its way into cracks in the metamorphic rock, or is the metamorphic rock partially melting to form granite? Questions like this occasion long arguments among geologists. (Paul Bailly)

products as metamorphic rocks, but efforts to make the definition more precise seem hardly worthwhile.

Igneous rocks

To look back a moment, we have begun to form a picture of the strange conditions that must exist at depths of a few miles or tens of miles anywhere beneath the earth's surface. Simple measurements of temperature indicate that rocks get rapidly hotter downward, reaching temperatures at depths of 20 or 30 miles at which many would start to melt. Very large masses of molten rock cannot be present, however, because the earth responds to tide-raising forces and to earthquake shocks as if it is largely a rigid solid. Evidently the downward increase in temperature is low enough, and the increase in pressure is high enough, to keep the material largely solid, although it is close to its melting range and probably at all times has pockets and discontinuous layers of molten material. Pressures can be estimated from weights of overlying layers, and prove to be so high at depths below about 10 miles that solid rock must flow wherever differences in pressure exist. Rock that is formed near the surface and that is carried down by movements of the crust to depths of a few miles is unstable under the new conditions of high temperature and pressure, and is slowly transformed to a metamorphic variety. If the rock descends far enough the temperature may become so high that it begins to melt. Our picture of internal conditions, then, is a pretty active one, involving slow movement of both liquid and solid rock, local melting, and mineral growth and transformation.

We turn our attention now to the liquid material, or magma, that probably exists at all times in small masses at depths between 40 and 200 km, and that may be produced in larger amounts when surface rocks are carried downward by crustal movement. If pressure is lower in one place than another, the magma is more mobile than its solid surroundings and hence moves to relieve the pressure difference. In particular, if the adjacent rock is cracked or broken so as to provide a path for upward movement, the magma may climb high into the relatively cool rocks near the surface. If it actually breaks out on the surface, we witness a volcanic eruption. Much of it never reaches the surface, but cools and slowly crystallizes while still enclosed in solid rock.

What sort of substance is this magma? In a previous discussion we have likened its consistency to that of toothpaste, or more accurately to that of molten glass. We need now to look at its composition more closely.

If you ask a geochemist to analyze a piece of congealed magma—say a chunk of basalt or granite—he will report that the principal element present is oxygen and the next most abundant is silicon. Besides these two he will find important amounts of a half-dozen others, all metallic elements: aluminum, iron, magnesium, calcium, sodium, and potassium. These are present in widely differing amounts in different kinds of igneous rocks. In particular, a dark-colored rock like basalt will have several percent of iron and magnesium; a light-colored rock like granite will have somewhat more silicon and much more sodium and potassium. In these chemical details magmas show their great resemblance to ordinary glass, for glass also has silicon and oxygen as major constituents, and colored varieties generally contain abundant iron.

Chemical analyses of typical granite and basalt are given in the table. The analyses show the proportions of the elements both as the elements themselves and in the form of oxides. The statement in terms of oxides is the common way of expressing rock compositions, although the oxides are not present as such in the rocks. Note the much higher amounts of Si in the granite, and of Fe and Mg in the basalt.

The chemical make-up of a rock is reflected in its mineral composition, and of course it is the mineral composition that you can recognize in hand specimens or in the field. Granite, we recall from Chapter 4, consists chiefly of quartz and feldspar with only minor dark minerals—a composition that accounts nicely for the high content of Si, Al, Na, and K that the chemist

Chemical Composition of Typical Granite and Basalt

	Percent by weight of oxides			*Percent by weight of elements*	
Oxide	Granite	Basalt	Element	Granite	Basalt
SiO_2	73.86	50.83	Oxygen(O)	49.01	44.58
TiO_2	0.20	2.03	Silicon(Si)	34.48	23.73
Al_2O_3	13.75	14.07	Titanium(Ti)	0.12	1.23
Fe_2O_3	0.78	2.88	Aluminum(Al)	7.28	7.45
FeO	1.13	9.06	Iron(Fe)	1.43	9.06
MnO	0.05	0.18	Manganese(Mn)	0.04	0.14
MgO	0.26	6.34	Magnesium(Mg)	0.16	3.82
CaO	0.72	10.42	Calcium(Ca)	0.51	7.45
Na_2O	3.51	2.23	Sodium(Na)	2.60	1.66
K_2O	5.13	0.82	Potassium(K)	4.26	0.68
H_2O	0.47	0.91	Hydrogen(H)	0.05	0.10
P_2O_5	0.14	0.23	Phosphorus (P)	0.06	0.10

reports. Basalt is made up chiefly of plagioclase feldspar and ferromagnesian minerals, as would be expected from its lower content of Si and higher Fe and Mg. These mineralogical differences provide a convenient classification of igneous rocks, as noted in Chapter 4 (page 97). Recall that the major grouping is into *extrusive* and *intrusive* rocks—those that cooled rapidly in lava flows at the earth's surface, hence are fine-grained, and those that are coarse-grained because they crystallized slowly beneath the surface. Within each group the rocks are divided according to mineral content: those that contain quartz (granite and rhyolite), those with no quartz and with plagioclase in excess over ferromagnesian minerals (diorite and andesite), and those with no quartz and with ferromagnesian minerals in excess (gabbro and basalt). These six are the commonest igneous rocks, but they are only a small fraction of the number of varieties that can be distinguished.

Three other names should be added to the list, just because they appear so frequently in both popular and scientific literature. One is *peridotite,* a coarse-grained, dark-colored rock consisting entirely of ferromagnesian minerals—not common in outcrops, but important because it is probably the principal material of the earth below depths of a few tens of kilometers. Chemically, as one would guess from the mineral composition, peridotite has even less Si and more Fe and Mg than basalt. Then there are two common names for varieties of rhyolite, or perhaps better for rocks with the chemical composition of rhyolite but very different appearance: the glassy black rock called *obsidian,* and the light frothy rock called *pumice.* Obsidian is a lava that cooled so fast, or was so viscous, that crystals did not have time to form; the dark color results from myriads of very tiny iron-oxide grains (magnetite) scattered through the glass. Pumice is glassy

Fig. 7-13. Banded obsidian. The rock is chiefly black glass, with the chemical composition of rhyolite. Parts of the glass that were full of tiny gas bubbles have been drawn out into gray streaks by movement of the material when it was a viscous melt. (California Division of Mines and Geology)

rhyolite expanded by escaping gas in the last stages of cooling. Both obsidian and pumice are often intimately associated with normal rhyolite in volcanic areas.

Finally, two words in common use to express the general chemical relations of igneous rocks are the adjectives *mafic* and *felsic.* A mafic rock is one with abundant ferromagnesian minerals, hence rich in magnesium and iron; a felsic rock is one rich in silicon and aluminum. Thus basalt and gabbro are mafic rocks, granite and rhyolite are felsic, diorite and andesite are intermediate. Some geologists prefer to use "basic" rather than mafic, and "acid" rather than felsic, but these words may lead to confusion because they have a different meaning in chemistry.

The forms of igneous and metamorphic rocks

After this dose of theory and nomenclature, it is time for a look at igneous and metamorphic rocks in the field.

We should keep in mind that most of these rocks have their origin out of sight, at depths of at least several miles in the earth. Volcanic rocks, of course, we can see forming at the surface, but their material comes directly from depth. Intrusive igneous rocks and nearly all metamorphic rocks attain their present composition and their present textures far under the surface; we would not see them at all unless deep erosion had removed the rocks that once covered them. The forms of intrusive and metamorphic rocks that we can reconstruct from outcrops are forms that were shaped by forces whose action we cannot directly observe.

Another important general fact about these rocks is that simply by the circumstances of their origin they must consist of interlocking crystals of minerals that are mostly fairly hard and dense. We can expect, then, that they will be harder and more resistant than the general run of sedimentary rocks, hence often exposed in ridges, cliffs, and hilltops. There are local exceptions, of course, and metamorphic rocks differ widely among themselves in resistance to erosion, but in a region of marked topographic relief and a variety of rock types the igneous and metamorphic varieties tend to be more abundant at higher elevations than sedimentary rocks.

Volcanic forms we have looked at in some detail previously (Chapter 2). There we considered the typical landscape features produced by a mafic lava, basalt: a steep-sided conical hill with a crater at its top, and lava flows extending down from the crater or heading in fissures near the base of the cone. These are the commonest of all volcanic structures, the forms that the word "volcano" calls to mind if it is not otherwise qualified. Not all mafic lava behaves in this way, however. If you go to Hawaii you will find that

Fig. 7-14. The island of Lanai, one of the many shield volcanoes of the Hawaiian Islands. This one has long been extinct, but it preserves the typical shape of a low dome with very gentle slopes.

the islands are built almost entirely of basalt—but the famous volcanoes have very gentle, rounded slopes, not at all like the conical hills you expect. If you are lucky enough to be in Hawaii while an eruption is in progress you will quickly find a reason for the gentle slopes: the flowing lava is extremely fluid, resembling water in its movement more closely than toothpaste or molten glass (Plate 5). It races down even gradual slopes at several meters per second, spreading out as it cools into thin flows with gently sloping tops. If you take a piece of the rock to a chemist he will tell you that it is unusually low in silicon and rich in iron and magnesium, and that these peculiarities of composition account for the fluidity. Volcanoes

Fig. 7-15. Typical Hawaiian lava. The extreme fluidity of this mafic lava is indicated by the smoothly rounded, ropy surfaces of the flows.

Fig. 7-16. A volcanic dome in southern California. The lava has welled up to form a large mound, and has moved only a short distance as a thick, viscous flow. A smaller dome in process of formation is shown in Fig. 2-9. (California Division of Mines and Geology)

like those of Hawaii, often called *shield volcanoes,* are commonly produced by very mafic lavas of low viscosity.

The relation of viscosity to silica content becomes even more obvious if we look at the other extreme of compositions, felsic lavas so rich in silicon that quartz forms prominent crystals. Volcanic shapes produced by such lavas generally consist of round, steep-sided domes of solid rock with jagged surfaces, and if lava flows are present at all they are short, blocky, and several tens of feet thick. Evidently the molten rock was so sticky that most of it simply rose slowly into the volcanic orifice and stayed there; if any flowed out it was too viscous to move far. Such volcanoes are called *volcanic domes.* The most famous American example is Lassen Peak in northern California.

The correlation of volcanic forms with silica content of the lava is by no means perfect. The intermediate lava andesite can either build typical steep-sided cones consisting of lava flows and beds of ash and cinders, like Mt. Rainier in the Cascade Range of Washington, or it can behave as a very viscous lava and form domes. The viscosity of lava evidently depends on other factors in addition to silica content, perhaps on temperature and amount of dissolved gas. Another complication that makes any simple relation impossible is the fact that long-lived volcanoes generally erupt lavas of different compositions at different times in their careers.

Fig. 7-17. Lassen Peak, an eroded dacite dome in northern California. The hills in the foreground are underlain by thick lava flows and deposits of pyroclastic material. (National Park Service)

Fig. 7-18. A diabase dike cutting gently dipping flows of basalt.

Fig. 7-19. West Spanish Peak, central Colorado. The ridges in the foreground are vertical dikes, a few of the very numerous dikes forming a radial pattern around the old volcanic structure, made conspicuous by erosion of softer material from their sides. (G.W. Stose, U.S. Geological Survey)

What about forms assumed by magma that never reaches the surface? The simplest to envision are the sheet-like or tabular bodies that consist of magma intruded along cracks. The general term for such bodies is *dikes* (Chapter 4); if the crack was parallel to the layering of sedimentary rocks, the special term *sill* is often used. Dikes and sills may consist of any kind of igneous rocks, coarse-grained or fine-grained. Generally, as might be expected, thin sheets have a smaller average grain-size than thick ones. A mafic rock with a grain-size intermediate between that of basalt and gabbro is so common in sills and dikes that it is given a special name, *diabase.* Dikes are particularly conspicuous in volcanic areas that have been partly eroded, the dikes representing the filling of cracks in the old volcanic structures. A good example is the set of conspicuous dikes that radiate out from the central part of Spanish Peaks south of Denver, Colorado. The best known sill in the United States is the Palisades Sill in New Jersey, which forms the cliffs (the "palisades") across the Hudson River from New York City; this is a sill of diabase once intruded into Mesozoic sandstone and now partly uncovered by erosion.

Larger intrusive structures have a variety of forms and have been given a corresponding variety of names. For present purposes the only one we need consider is the largest form of all, the kind of intrusive body called a *batholith.* Batholiths are so big that the concept is hard to grasp. Suppose, for example, that you make several trips across the Sierra Nevada in California, choosing a different highway each time. As you cross the central part of the range on any one of your routes you will see broad areas of granite, especially well exposed because they were scraped clean by glaciers not many thousands of years ago. In imagination—or in actuality if you hike along some of the High Sierra trails—you can piece together your observations to reconstruct an enormous body of granite making up most of the backbone of the range, something like 500 miles long and averaging about 100 miles wide. Where rivers and glaciers have carved deeply into the rock, as in the famous valley of Yosemite, you see no bottom to the granite; it extends downward indefinitely, as far as exposures permit you to see. This is a typical batholith: a body of intrusive rock at least several tens of square miles in exposed area, with no visible floor. The principal rock of such bodies is granite. The more mafic intrusive rocks, gabbro and diorite, by themselves are seldom found in masses this large, but they often make up minor parts of a granite batholith.

There is a seeming contradiction here. We have said that the earth is largely a rigid solid, that there can be no large bodies of molten material at least in its outer parts, yet here we are talking about enormous masses of rock that show in their composition and structure that they have crystallized from a liquid. If you examine the Sierra Nevada batholith carefully you will find a possible way to resolve the difficulty: the rock is not

Fig. 7-20. Common intrusive forms.

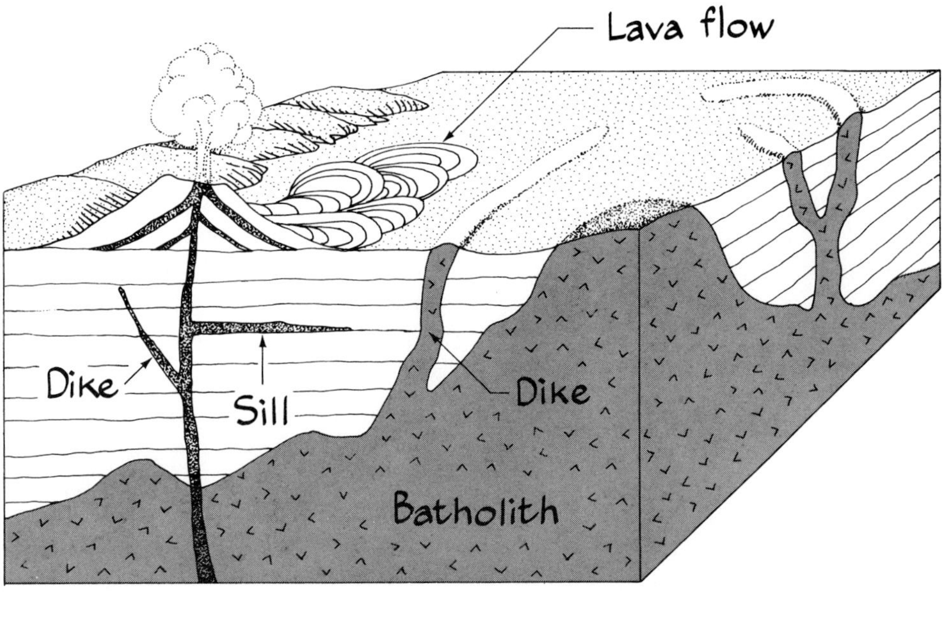

Fig. 7-21. Part of the roof of the Sierra Nevada batholith. Erosion has cut down through the metamorphic rocks (dark color, at top) to expose the granite (light color) that once intruded them. Only a thin remnant of the roof is preserved, probably a part of the roof that extended down into the granite as a prong or "pendant." (John Shelton)

uniform over the batholith, but consists of different varieties of granite distinguishable by their mineral composition, grain size, and textures. The batholith is not a single mass of material that was all molten at once, but consists of many smaller bodies that were liquid and crystallized at different times. At any one time only a small part of the entire granitic mass was molten. Other large batholiths on close examination prove to have similar composite structures.

For igneous rocks, then, we can recognize and give names to a variety of forms. Metamorphic rocks cannot have "forms" in the same sense, because their geometric patterns are determined largely by the forms of the original rocks that have undergone metamorphism. Our question about metamorphic rocks would be better phrased by asking in what kinds of situations they occur.

The Sierra Nevada provides a beautiful exhibit not only of a typical batholith but also of many kinds of metamorphic rocks. In your travels across the range, if you look carefully, you will spot places where the granite makes contact with other rocks. Such contacts are of special interest to a geologist, because they represent places where hot, molten material has moved against relatively cool, solid rock. It is in these places that we find the varieties of metamorphic rock produced by simple baking, without much movement due to differential pressure: hornfels, marble, and quartzite. Locally in the Sierra Nevada the baking has been accompanied by injection of hot solutions from the crystallizing magma, as is shown by the spectacular development of coarse brown garnet crystals and a number of green minerals in the metamorphic rock. Farther from granite contacts, particularly on the west slope of the range, are widespread outcrops of other metamorphic rocks, particularly slates, produced by metamorphism that involved both heating and differential pressure. Evidently the granitic magma that formed the batholith was intruded into rocks that had already undergone some metamorphism, and then heat from the magma caused further baking at its contacts. This kind of rock assemblage—granites intruded into foliated metamorphic rocks, with unfoliated baked rocks at their contacts—is found so often in many parts of the world that we can generalize with a statement that intrusion of granitic magma commonly accompanies metamorphism.

The generalization cannot be pressed too far, because some broad areas of metamorphic rock include only few and minor exposures of granite. Much of New England, for example, is underlain by schist, representing the highly metamorphosed equivalent of early Paleozoic sedimentary and volcanic layers. The schists occur in great variety, making a fascinating but difficult subject for geologic study. Granite can be found in small, isolated areas, but it is a minor part of the geologic structure rather than a major component as in the Sierra Nevada.

Huge areas of eastern Canada are also underlain by schists and gneisses, in places accompanied by much granite and in places by very little. These are ancient rocks, for the most part Precambrian, and the relation of granite to metamorphism is less clear than in California and other parts of the west.

More questions

A brief discussion of rocks produced in the earth's interior necessarily opens up a hornet's nest of troublesome questions. From simple considerations of temperature and pressure we can get a rough idea of conditions at depths of a few tens of kilometers and of the ways rocks might be expected to behave, but we have nothing like the solid basis of firsthand knowledge that gives us confidence in talking about sedimentary rocks.

Take the question of geologic age, for example. What do we mean by the "age" of an igneous rock? The time when the rock solidified, or the time when the magma was formed? By general agreement we mean the

Fig. 7-22. Dikes of granite (a coarse-grained variety called *pegmatite*) intruding metamorphosed volcanic rocks. Is the igneous rock older or younger than the metamorphic rock?

former—the date when crystallization occurred—but of course with the understanding that the rock material as a coherent body of magma may have existed for a long time previously. How do we determine such ages? Generally, in ordinary field work, by the relation of igneous rocks to adjacent sedimentary rocks. A lava flow sandwiched between a layer of sandstone and a layer of limestone, for example, must be younger than the underlying rock and older than the rock above. A gabbro dike is younger than the youngest rock it intrudes and older than any rock that lies on top of it. If we need ages more accurate than can be given by this elementary geometry, many igneous rocks contain enough radioactive material so that a radiometric date can be determined.

For metamorphic rocks the question of age is more complicated. Any metamorphic rock obviously records two dates—the date when the original rock was formed, and the date when it was metamorphosed. The age is taken by convention as the time of original rock formation. It is sometimes determinable from fossils, if metamorphism has not altered the rock drastically, and otherwise must be guessed from relations to rocks whose age is known. Radiometric methods permit determination for some metamorphic rocks of both the original age and the time of metamorphism.

Then there are deeper questions about the origin of magmas, the movement of magmas, the changes in magmas during crystallization, and the mechanism by which rocks formed near the earth's surface are carried down to depths where metamorphism can take place. The granite magma that forms a batholith—has it always existed as pockets of molten material in the earth, or does it form by melting of other rocks? By what means does it rise from depth into space that is presumably already occupied by other rocks? Is the mafic magma that ultimately forms basalt and gabbro generated in the same place as granite magma, or does it come from greater depths? The crystallization of magma is a slow and complex process, involving the formation of different minerals in succession over a large temperature range; does this process offer a possibility for one kind of magma to change into another? Or could a magma change its composition if chunks of rock fall into it and are melted? Intrusive igneous rocks and metamorphic rocks are often found closely associated; does this mean that the formation of magmas has some relation to metamorphism? The list of questions could go on and on. Some of them take us to the frontiers of geologic thinking, where no firm answers are yet possible. We will come back to such questions later on, when we take a longer and more sophisticated look at conditions in the earth's interior.

And finally, a host of more practical questions arises from the common association of ore deposits with igneous and metamorphic rocks. Every

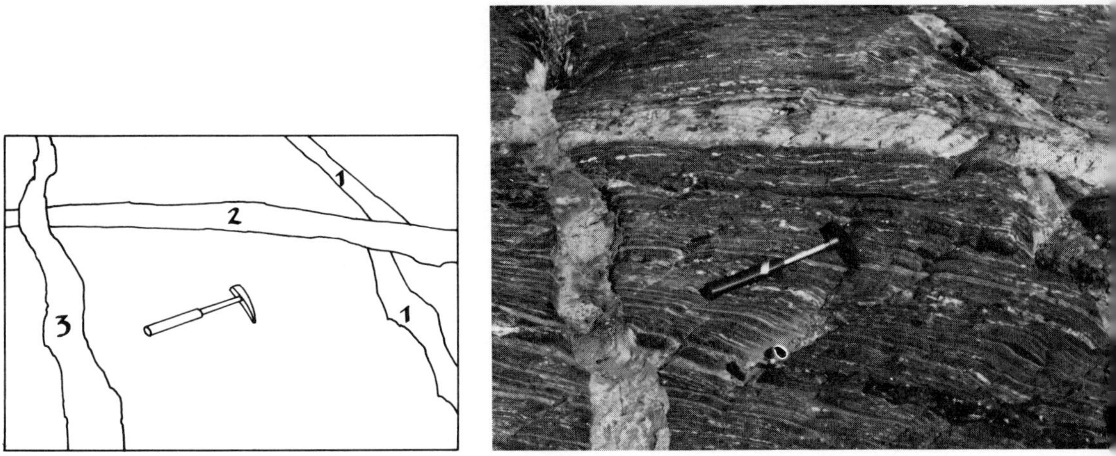

Fig. 7-23. Three diabase dikes cutting metamorphosed sedimentary rocks. Are the dikes older or younger than the metamorphic rocks? Which of the dikes is the oldest? Which the youngest?

prospector knows that the contact of a granitic mass with adjacent metamorphic rocks is a good place to explore for concentrations of gold, silver, copper, and many less common metals. The famous copper deposits of Butte, Montana, Bingham Canyon in Utah, and many places in southern Arizona are located in and near granite bodies. Many of the "bonanza" deposits of gold and silver in the western states were located at old centers of volcanic activity. The Homestake mine in South Dakota, largest gold producer in the country, consists of a network of quartz veins cutting schist. Not only metallic ores but deposits of precious stones occur with igneous and metamorphic rocks: beryl and zircon in dikes associated with granites, garnet in schists, diamond in a variety of peridotite. About all these occurrences one can ask why the metal or the precious stone is found in or near the deep-seated rock. Was the valuable substance somehow concentrated by the same conditions of high temperature and high pressure that formed the adjacent rock? If so, why do we not find metallic ores near the contacts of all granites, or near all volcanic centers, instead of only a few? Such questions have plagued geologists since their subject became a science, and there are still no satisfactory answers.

Fig. 8-1. The top of Mt. Rainier, Washington. This old volcanic cone is being re-shaped by weathering and erosion. What kinds of erosive activity are evident in the picture? (U.S. Department of the Interior, National Park Service)

Fig. 8-2. Monument Valley, a stream-eroded landscape in northeastern Arizona. The steep cliffs and isolated pinnacles are remnants of once-continuous horizontal sandstone beds, left because of minor differences in hardness by the streams that eroded the rest of the landscape. (John Shelton)

8

The Not-So-Eternal Hills

Rocks exposed in outcrops are subject to attack by air and water. Some parts of the attack we have talked about before, particularly the formation of sediment by the erosive action of streams and waves. But the subject is a complicated one that needs a closer look. With a background of information on the make-up of many kinds of ordinary rock, we are now in a position to see just how the attack takes place and what geologic results it can produce.

Rocks are attacked both passively, by slow decay in contact with water and the gases of the atmosphere, and actively by the abrasive action of wind and of moving water and ice. The passive attack is described as *weathering,* the active attack as *erosion.* The two words have somewhat elastic definitions, as is hardly surprising since weathering and erosion often take place together. Erosion is often used as a general term for all kinds of attack on rocks, weathering then being a particular kind of erosion that involves little transport of the weathered fragments.

If there were no atmosphere . . .

The results of weathering and erosion become more vivid if we try to imagine a world without air, and hence without water—since water on continental surfaces all has its origin in rain. Actually not much imagination is needed, because astronauts have brought back from the moon detailed photographs of just such a world. The moon has been without air and water for at least the greater part of its history. Its surface, as might be expected, consists entirely of igneous rocks, if we except the fragmental material produced when the igneous rocks are shattered by meteor impacts.

Fig. 8-3. A landscape on the moon. What are the most important differences between these landscape forms and those in comparably hilly country on the earth? For another lunar landscape, see Plate 25. (NASA)

Samples of rock brought back to the earth for study are amazingly fresh, showing none of the discoloration and cloudiness generally shown by mineral grains in terrestrial rock samples. Hills on the moon are commonly smooth-sided because of the constant rain of meteors that have struck them over long reaches of time; there are no patterns of connecting valleys to indicate the work of streams, and no rounded valley bottoms to show the former presence of glaciers. Shapes of hills and mountains suggest their formation by impact of huge meteors, volcanic eruptions, and movement of solid rock along faults, but without modification by the canyon-cutting, slope-carving, sediment-heaping activity of running water. In places we recognize lunar mountains of different ages, because the older ones have slopes somewhat more softened by the long meteor bombardment, but the amount of erosion is negligible compared with erosive activity on earth. The moon is a dead world, made up of rocks and landscapes little altered from the time of their formation.

Our earth is larger than the moon and has a more colorful history of volcanic eruptions and movements of solid rock continuing down to the present day. But if the atmosphere were lacking the planet's surface would be much the same as what we see on the moon—broad plains of lava, mountains of fresh igneous rock, slopes softened only by occasional landslides and by the breaking of rocks in meteor impacts. Most of the time we are hardly conscious of the weathering and erosion going on around us, but to these processes we owe the vast difference between familiar landscapes and the stark scenery of our satellite.

How do rocks decay?

That rock surfaces slowly change on exposure to wind and weather needs no proving. We think of gravestones in old cemeteries with their barely decipherable inscriptions, or of the carved marble figures on European cathedrals that have lost much of their detail, or perhaps of a recent highway cut through fresh rock that is stained and darkened by only a few years' exposure. Nor is there any mystery about some of the factors on which rate of weathering depends: a somewhat soluble rock like marble or limestone shows the effect of weathering faster than granite does; weathering in a dry climate is slower than weathering in a humid climate; a rock containing abundant iron minerals, like gabbro or peridotite, develops a stained surface faster than a quartz sandstone. Some of the important agents of weathering are equally obvious: a surface on which water stands frequently weathers faster than a dry surface; in cold climates rock surfaces are broken up by water freezing in cracks; iron minerals produce staining because they react with oxygen in the air to form ferric oxide. If the factors of weathering are so evident and so commonplace, why study the subject at all?

The difficulty with weathering as a subject for investigation is in part its slowness and in part its very commonplace character. Rocks are weathering all around us, obviously, but so slowly that we seldom notice changes until they are pointed out. When the subject is brought up it brings to mind rock structures just sitting and moldering through long ages, like the temple of Angkor Wat in Cambodia or the ruined Mayan cities of Yucatán. It is a gentle process, an unobtrusive process, lacking the immediate appeal (both to laymen and to most geologists) of the more dramatic geologic happenings like earthquakes and floods and volcanic eruptions. Yet a good case can be made for the proposition that of all geologic processes weathering is by a wide margin the most important, for the simple reason that its end product is the soil on which our existence depends. A little acquaintance with weathering proves that it is both more complicated and more interesting than it first appears.

What happens when a rock weathers? It may be attacked by physical processes or chemical processes or both. Of the strictly physical processes, the most important is the breaking up of rock by water that seeps into cracks and then expands on freezing. If you have done any mountain climbing, you have perhaps been struck by the abundance of small angular fragments of fresh rock littering the surface of many high peaks—fragments

Fig. 8-4. Differential erosion in southern Utah. The rocks are three horizontal beds of sandstone, a white bed sandwiched between two darker ones. Evidently some parts of the rock are more easily eroded than others, and the differences in resistance account for the odd shapes. Remembering that this is a region of arid climate and fairly high elevation, what erosional processes would you think are most active in carving these rocks? (C.B. Hunt, U.S. Geological Survey)

Fig. 8-5. The top of a 12,000-foot ridge in eastern California, covered with angular blocks pried out of the rock by water freezing in cracks. Similar frost-wedged blocks cover much of the surface in Fig. 8-1.

produced by the very frequent freezing and thawing in this environment. The chemical processes of decay are more complicated, and are different for different kinds of rock.

Some rock varieties simply dissolve when they stand in contact with water. Rock salt is a familiar example; gypsum rock is another, although it is considerably less soluble than salt. A quartzite or quartz sandstone also weathers chemically by simple solution, but the solubility is so very small that these rocks are attacked more slowly than most and hence often stand up as prominent cliffs and ridges.

Limestone also is a somewhat soluble rock, but the process by which it dissolves is a little more complicated. Water that comes in contact with rock surfaces is generally very slightly acid, because it contains carbon dioxide dissolved from the air and the carbon dioxide reacts to form carbonic acid:

$$CO_2 + H_2O \rightarrow H_2CO_3$$

This symbolic equation means simply that the compounds CO_2 and H_2O unite to form a compound containing all three elements, which is recognizable as an acid because its formula starts with "H." The amount of CO_2 in air is very small (0.03 percent by volume), and carbonic acid is very weak, so that the amount of acidity in ordinary water is detectable only with sensitive instruments. The acidity plays no role in the dissolving of rock salt or quartz, but it aids in dissolving limestone. This is because the chief mineral of the rock, calcium carbonate, is readily attacked by acids:

$$2H^+ + CaCO_3 \rightarrow Ca^{++} + H_2O + CO_2$$

(In this equation H^+ represents any acid, because it is set free when acids dissolve in water; Ca^{++} is calcium ion which is released from the solid compound and set free in the solution.) This reaction is the one on which the familiar "acid test" for limestone depends: when you place a drop of strong acid on a limestone surface, the gas that bubbles off violently is the CO_2 represented in the equation. The process of dissolving in nature produces no visible evolution of gas because the acid is very much weaker, but the chemistry of the change is similar.

Silicate minerals behave somewhat like carbonate minerals during weathering, but with added details that make the reactions seem complex. Take a simple feldspar as an example: the overall change is an alteration of the original compound, $KAlSi_3O_8$, to a clay mineral which we can symbolize $H_4Al_2Si_2O_9$. How may such a change be brought about? We again start with a water solution containing a little carbonic acid, let the acid be symbolized by H^+, and guess that the first step will be the replacing of K by H, just as the dissolving of $CaCO_3$ involves replacing Ca by H:

$$H^+ + KAlSi_3O_8 \rightarrow K^+ + HAlSi_3O_8$$

Now there is no known compound with the formula $HAlSi_3O_8$, so we guess that it breaks down quickly:

$$2HAlSi_3O_8 + 9H_2O \rightarrow H_4Al_2Si_2O_9 + 4H_4SiO_4$$

The compound H_4SiO_4 is called silicic acid, and represents the form that silica (SiO_2) assumes when it dissolves in water. Our two equations, then, have the net effect of setting free potassium and silica in solution and producing clay as a solid. This reaction is so slow that it cannot be duplicated at low temperature in the laboratory, but an indication of its essential correctness is the fact that both K^+ and H_4SiO_4 are readily detectable in small amounts in solutions that have stood in contact with powdered orthoclase.

If a silicate or carbonate mineral contains iron, one other important reaction is the oxidation of the iron by atmospheric oxygen. The product of oxidation is one of the brightly colored ferric oxide minerals, the red compound *hematite* (Fe_2O_3) or the yellow-brown hydrated oxide called *limonite* (roughly $Fe_2O_3 \cdot H_2O$). The reaction for iron carbonate (a mineral called *siderite*) may be symbolized

$$2FeCO_3 + \tfrac{1}{2}O_2 \rightarrow Fe_2O_3 + 2CO_2$$

The chemical equations are a convenient shorthand, but they are not necessary to an understanding of the important chemical changes that occur during weathering. By way of summary, without using equations: The specific substances responsible for chemical weathering are water, carbon dioxide, and free oxygen. Water acts in part by simply dissolving minerals, but its solvent effect for many minerals is aided by the weak acid formed by its combination with carbon dioxide. Of common rock-forming minerals the only ones affected by free oxygen are those containing iron. The products of complete weathering are metal ions and a little silica in solution, clay minerals and iron oxides as solids, and carbon dioxide gas released to the atmosphere.

Decayed rocks

We have looked at the processes and products of decay in some detail. Now how do we recognize weathered rocks when we meet them in the field?

The ultimate solid products of weathering, chiefly iron oxide and clay

Fig. 8-6. A glaciated granite surface in the Sierra Nevada. Much of the grooved and polished surface left by glacial erosion is still preserved, despite several thousand years of exposure since the glacier disappeared. In the foreground is an area where weathering has penetrated the surface. (John Shelton)

minerals, form very slowly and at different rates from different minerals. This means that a rock surface exposed to weathering will in general be only partly decayed, the iron oxide and clay being mixed with fragments of the original minerals in various states of disintegration. The exposed surface of the rock is commonly soft and crumbly, and if you scrape off some of the surface layer for examination with a handlens you will see a dirty assemblage of still recognizable mineral grains plus a fine-grained powder whose yellow or brownish color indicates that iron oxide is present. This surface layer may be only a millimeter or two thick, or it may be so thick that you will not be able to get a fresh sample of rock even by strenuous use of your hammer. It may be absent altogether, if the rock surface has been exposed to weathering only recently or if it is kept clean by waves or swiftly flowing water.

Take granite as an example. The ferromagnesian minerals are generally the most vulnerable to attack, so that around the grains of black mica and hornblende you see brown stains or pockets of brownish powder. The feldspar weathers more slowly, but on a weathered surface the feldspar grains are softer and more cloudy than on fresh surfaces. The quartz is not altered chemically, and its dissolving is so exceedingly slow that quartz grains persist without apparent change in the weathered material. The depth of weathering will depend on where the granite is. If you are looking at granite in high mountains, where glaciers scraped the surface clean a few thousand years ago, weathering may be so slight that only careful comparison of specimens from the surface and from a few inches down will show any difference at all. If you are studying a granite cliff at lower elevations you will find most of the surface discolored and softened, especially where cracks have permitted water and air to penetrate deeply, and below the cliff you will find piles of debris consisting largely of unaltered quartz and partly weathered feldspar. And if you are examining granite in a roadcut in Georgia, where the warm climate makes weathering reactions fast, you may find weathered material even at the base of the cliff,

Fig. 8-7. Weathered granite in the Black Hills of South Dakota. Weathering is especially marked where air and water have penetrated deeply along cracks. (S. Paige, U.S. Geological Survey)

many tens of feet below the original surface. The rock may still look like granite, but you will find that you can crush it between your fingers because the feldspar crystals have been replaced by white powder and the ferromagnesian minerals by yellow or red powder, and only the quartz grains are left intact.

A word about color. Rocks that contain iron minerals generally have weathered surfaces stained in various shades of red, yellow, and brown, because the iron has been oxidized ("rusted") and the tiny particles of ferric oxide have been distributed through the weathered material by percolating water. Ferric oxide is the great coloring agent in nature. It shows a remarkable variety of hues, depending on its grain size and its degree of hydration (amount of chemically combined water). Whenever you see an outcrop that is red, brown, yellow, purple, or any intermediate color, you can say confidently that the rock contains iron oxide, at least in its surface layers. You cannot infer, however, that the amount of iron is very large. Ferric oxide is so powerful a coloring agent that only a few percent disseminated through a rock will give it a strong color. Rocks that do not contain iron may retain their original color during weathering, or may be bleached or darkened as their organic matter is altered.

The softness of weathered material results in large part from the formation of clay minerals. These substances are produced by the weathering of any silicate minerals that contain aluminum, but in largest quantity from feldspars. The change from feldspar to clay is an extremely slow reaction. Generally in a weathered rock various stages in the process can be detected under a microscope, from perfectly clear, fresh feldspar crystals to those that are filled with tiny specks of clay and finally to some so completely converted to clay that only vague outlines of the original crystals remain. In hand specimens the same changes are apparent in the increasing cloudiness and chalkiness of the feldspar grains. The process is so slow that the grains retain their integrity even when much of their substance is converted to clay, and the partly altered grains are an abundant constituent of sands produced as a weathering rock disintegrates.

Surfaces of weathered rocks are generally rough, as might be expected from a process in which some constituents decay faster than others and the products of decay are removed by gravity or by rainwater. A special sort of roughness is commonly developed on rocks that are somewhat soluble, like limestone or gypsum. These rocks, consisting largely of a single mineral, might be expected to dissolve uniformly and hence to give smooth surfaces. But water during rains tends to concentrate in some areas more than others, so that grooves and fluting with widths on the order of a few centimeters are commonly formed. Deeper grooves develop where water is concentrated along cracks, and on a larger scale the weathered surface may be conspicuously hummocky (Fig. 4-13). The typical grooved, fluted, and hummocky surface of limestone outcrops is one of the characteristics we have mentioned earlier (Chapter 4) which aid in identifying this rock.

The end-product of decay

Weathering is so slow that we cannot watch it take place, but weathered outcrops permit us to see different stages and to arrange specimens in a tentative order of progressive decay. It is reasonable to ask where the process leads, or what the rock would look like if weathering should go to completion.

Fig. 8-8. Deeply weathered granite in northern Georgia. Weathering solutions have moved along cracks and percolated into the solid rock, leaving rounded masses of partly decayed granite at the centers of blocks outlined by the cracks. In this warm, humid climate the weathered material between the residual masses is bright red from disseminated hematite. The white material in the cracks is the clay mineral kaolinite, formed by complete decay of feldspar and carried into the cracks by percolating water. (R.R. Compton)

A weathered granite should ultimately become just a mass of fine-grained clay and iron oxide with residual grains of quartz. Material of this description we have noted in the deeply weathered granites of Georgia and other southeastern states. If the decaying granite has not been disturbed, it may retain an apparent igneous structure even when it is so far altered that you can crush it between your fingers (Fig. 8-8). The top part of such a granite outcrop, however, will differ in appearance from the rest: it has lost all trace of igneous structure, and instead shows a rough horizontal layering. The top layer where plant roots penetrate is darker than the material below, and a second layer is distinctly reddish. The layered material we call *soil*.

If you examine a pinch of the soil with your handlens, or better if you sort out the constituents of the soil by dispersing it in water before you examine it, you will find that its composition is not much different from that of the decayed granite below it. The clay, iron oxide, and quartz will show somewhat different proportions in the different soil layers; the upper layer lacks ferric oxide completely and contains a good deal of organic matter, while the second layer is obviously enriched in ferric oxide. The soil has evidently been formed by some sort of modification of the weathered granite. What changes have taken place?

One change is the addition of organic matter, representing dead plant fragments from the vegetation growing on the surface. You could guess that the organic matter would provide nutrients for abundant bacteria, and your guess is easily confirmed if you are familiar with microbiological techniques. Then you might reflect that organic compounds, both the original ones and the simpler ones produced by bacterial action, are good reducing agents (substances capable of removing oxygen from other materials); hence ferric oxide could not persist in the upper layers, but the iron would be reduced and converted to a more soluble form. Perhaps it would occur to you also that the decay of organic matter gives carbon dioxide as a major product, so that water in the soil would have abundant CO_2 to dissolve, and might be more acid than water at the surface. If you put these thoughts together with the reflection that water from rain soaks into the soil during storms and then rises back toward the surface during dry spells, you might guess that the up-and-down movement of water would carry material from one layer to the next, especially material made more soluble by the acidity and the reducing conditions, and thus that water movement over a long period of time is responsible for the soil layers.

One need not travel to Georgia, of course, to find soils in process of formation. Wherever rock—*any* rock—has undergone deep weathering and has not been disturbed by erosion or movements of the earth's crust, the

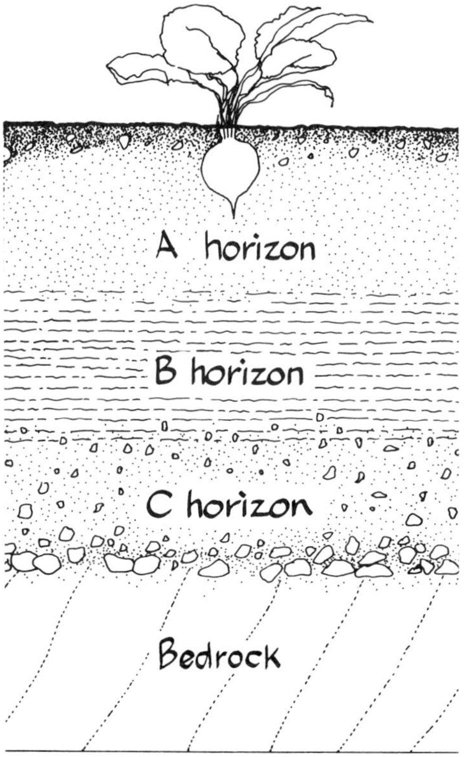

Fig. 8-9. An idealized *soil profile,* or cross-section through a soil. Three principal layers or *horizons* can commonly be distinguished: the upper or A horizon, in which organic matter accumulates and from which mineral matter is leached; the B horizon, in which some of the leached material is deposited, especially clay and ferric oxide; and the C horizon, in which fragments of bedrock have only started to decay. Fig. 8-8 shows the transition between B and C horizons in a Georgia soil.

same upward transition from decayed rock to layered soil can be observed. Soil formation from any kind of weathered rock involves the same two processes as for Georgian granite: the addition of organic debris accompanied by bacterial activity, and the transfer of material from one layer to another by vertical movements of the soil water. The condition of lack of disturbance is important; soil is easily moved by erosion, hence may be stripped from one place and piled up in another. If you try to find in nearby outcrops a typical textbook example of the transition from fresh rock to weathered rock to soil, you will probably be disappointed because of this tendency of soil to move. You may find thin soil resting directly on apparently fresh bedrock, or soil obviously different in composition from the material underneath, or deep soil with no apparent layering. To find an exposure where you can be sure that the soil you see is actually forming in place may take a good deal of hunting.

Soil is the end stage of weathering, and in human terms by far its most important result. Without soil there could be no plants on the land, and without plants there could be no animals. To preserve the soil, to improve the soil, to understand the soil are activities essential to human existence. They constitute a special branch of science, called *soil science* (or *pedology*). Seemingly this should be a part of geology, inasmuch as it deals with an inorganic part of the earth. But because soil science is oriented so heavily toward agriculture and because it requires techniques different from geological techniques, the two subjects have grown far apart. The study of soils is a science in its own right, and for our purposes here we can touch on only a few simple generalizations that have pertinence for geology.

If you have thought over our discussion of soil formation from granite, you may have wondered why the same language could not be applied to any other rock. Had the weathering rock been basalt or gabbro rather than granite, its ferromagnesian minerals would again have gone principally into solid ferric oxide plus metal ions and silica in solution, and its plagioclase feldspar would have decayed to clay minerals; the completely weathered material would not have had much quartz, but otherwise the differences would be minor. Even if the original material had been limestone the resulting soil might be similar, because the soil would consist chiefly of minor impurities from the original rock after the calcium carbonate had dissolved—and the commonest impurities in limestone are clay and iron oxide. Can we suggest, then, that soils derived from all rocks will be much alike, containing as inorganic constituents clay and iron oxide and differing chiefly in amounts of quartz and other very insoluble minerals?

For soils that result from *complete* weathering, this statement is fairly accurate. Regardless of the original rock, complete decay gives material consisting largely of clay minerals and iron oxide. Differences among soils, then, are to a considerable degree determined by differences in climate, because the nature of a soil depends on how much organic matter it contains and on how effectively water movement can transfer materials from one soil layer to another. Some indication of the original rock may persist, even when soil-forming processes have gone on for a long time. Soils derived from granite or sandstone generally retain grains of quartz, a soil formed from schist may have flakes of mica, a soil developed on limestone may contain rock fragments and may be somewhat alkaline. Climate is the chief determiner of soil type, parent material a secondary influence. The effect of climate is particularly marked on the soils of plains and gently rolling hills, where erosion is slow and soil formation can go on undisturbed for a long time. Parent material has its greatest effect on steep slopes, where topsoil is removed by erosion before soil formation has greatly modified the weathered rock.

The influence of climate stands out in some of the major soil types of the United States. The famous red soils of northern Georgia and Alabama form from a variety of rocks in a warm, humid climate. The high temperature speeds up organic decay, preventing the accumulation of much organic matter and favoring complete oxidation of iron; the high rainfall means that frequently water percolates completely through the soil, carrying dissolved material in solution to nearby springs and rivers. Thus the soil in many places consists of little but red iron oxide and clay, and is poor for growing crops because it lacks organic matter to give it easy "workability" and is impoverished in the soluble mineral nutrients that plants require. Farther north along the Atlantic coast, where the climate is humid but colder, soils lose their red color and take on shades of gray and brown because organic matter is preserved in greater quantity and iron is less completely oxidized. Percolating water again removes much soluble mineral matter, so that these soils also require much fertilizer for successful crop production.

The best soils for agricultural purposes are those that develop in regions of moderate rainfall and moderate temperatures, a combination that gives the highly prized "black-earth" soils of Iowa and adjacent states. The coolness permits accumulation of much organic debris, which accounts for the black color, and the lower rainfall insures retention of some soluble material. In soils of the desert areas of the western United States the retention of soluble material becomes even more complete, too complete for agricultural purposes. Percolating water seldom or never moves entirely through the soil, but simply redistributes salts by its up-and-down movement, often forming a layer of precipitated material, chiefly calcium carbonate (*caliche*), at or near the ground surface. With sufficient irrigation water such soils can be made suitable for crops by flushing out the very soluble materials, but the tendency of salts to accumulate is a continuing problem in many irrigated areas, for example Arizona and southern California.

This description of regional soil types is painted with a very broad brush. In any one climatic region many different soils can be found, depending on local topography and kinds of bedrock. The classification of soils and the relating of soil types to their environment is an exceedingly complicated subject, hardly suitable for further exploration here.

How long does soil take to form?

As always when we are discussing a geologic process, the question of time looms in the background. The formation of soil from weathered rock is a slow process involving complex chemical and bacteriological reactions. How long does it take to convert weathered rock into material of the right composition and consistency to support abundant plant growth?

As might be expected, there is no easy answer. The simple observation that areas stripped of topsoil (areas subjected to strip mining, for example)

Fig. 8-10. Stream erosion and removal of topsoil accelerated by overgrazing in eastern California. (U.S. Department of Agriculture, Soil Conservation Service)

support only sparse vegetation for decades means that the process is slow in comparison with human lifetimes. Soil formation must clearly be faster in some places than in others, for it is largely a chemical process, and the rates of most chemical reactions are sensitive to temperature and to the presence of moisture. Estimates of the time required for soil to form are generally in the range of a few thousand years, but such a figure can be no more than a very rough average.

Even though estimates are highly uncertain, the rate in most places is slow enough so that loss of topsoil is a serious matter. The upper layers of soil are a precious material, in that we depend on them for our food supply and for maintaining the trees and grassland that make landscapes pleasing to the eye. Too often carelessness in construction projects and over-intensive use of agricultural lands lead to accelerated erosion of topsoil, hence to loss of an important resource that will take many generations to replace.

Slides

From a geologic point of view, quite apart from human welfare, weathering serves the function of preparing material for easy removal by the more active agents of erosion. We turn now to these more active agents, and look first at one that doesn't really seem active at all: the force of gravity. Gravity can be an agent, of course, only if material is ready to fall—a block dislodged from the face of a cliff, melting snow in an overhanging cornice—and it is active only in the sense that loose material can fall or slide in enormous quantities, sometimes very rapidly.

If you live in hilly country you can scarcely be unaware of some of the effects produced by gravity. Most obvious is the talus at the base of cliffs, piles of angular blocks of all sizes that have clearly fallen from the cliff above as weathering has weakened their moorings. Perhaps you have also seen landslide scars on slopes—semicircular gouges on hillsides, with tongues of hummocky, fissured material extending downslope. Or you may have spotted cleared tracks through the forest on steep hillsides, where snow

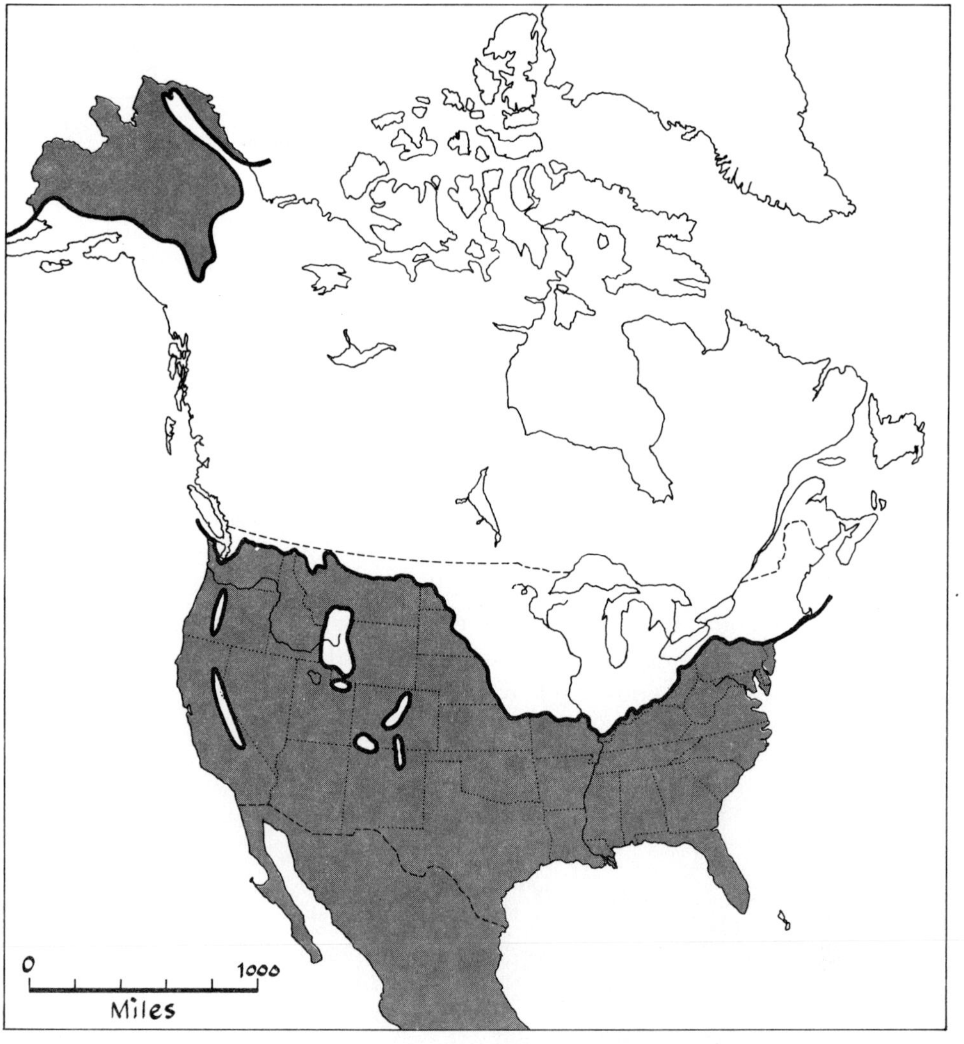

Fig. 8-18. The maximum extent of glaciation in North America during the Ice Age. (From a compilation by the U.S. Geological Survey)

they have been under a giant rasp. Isolated boulders of large size, some rounded and some angular, are common landscape features, and many are different in composition from bedrock in their vicinity. The orientation of scratches (called *striae*) is revealing: if their directions are plotted on a map, they seem to radiate from an area in northern Quebec east of Hudson Bay. Such observations are convincing evidence for the former existence of an ice cap in eastern Canada, much like the present-day ice caps of Greenland and Antarctica.

The eastern Canadian ice cap spilled across the border into the northeastern United States, as is clear from similar evidence—the lakes and swamps, the peculiar drainage, the rounded hilltops, the scattered boulders of rock types that are found in outcrop only far to the north, the occasional exposures of grooved and polished bedrock surfaces. Also conspicuous in some places in the northern states are low ridges of hummocky topography, made up of unsorted, unstratified sediment with all grain sizes from very fine clay to huge angular boulders; this is the kind of material, called *till*, that you might expect a glacier to push ahead of itself and to leave as a ridge (*moraine*) when the ice melts (Fig. 5-5). Detailed study of such evidence has revealed many details about the old ice cap. It spread not from a single center but at least three centers in Canada; it advanced and retreated many times, even disappearing altogether for long intervals; it first formed more than two million years ago, and its last remnants can be dated at only a few thousand years in the past. Geologists speak of this period of widespread glaciation as the *Pleistocene Epoch* or the *Ice Age*.

Glaciers of another kind lie in valleys in some of the high western mountains, and are especially numerous in the coastal ranges of Alaska (Plate 6). These are tongues of ice confined to valleys, fed by heavy snowfall at their high upper ends and moving slowly down-valley under the influence of gravity. Measured rates of movement are a few meters or tens

Fig. 8-19. The toe of a small glacier in the Wind River Mountains, Wyoming. A low moraine marks the end of the glacier, and melt-water is strewing outwash gravel over the valley floor in the foreground. Other small valley glaciers are shown in Fig. 8-1. (Mark F. Meier, U.S. Geological Survey)

Fig. 8-20. Part of the huge Malaspina Glacier in southeastern Alaska. The differential movement of the ice is spectacularly shown by the trains of debris carried on the ice surface from the mountain spurs. (Austin Post, U.S. Geological Survey)

of meters per year. The position of the lower end of a valley glacier represents a balance between motion and melting; some glaciers are moving so fast that their snouts are maintained far down where the average annual temperature is well above freezing. Although the movement cannot be seen directly, its major erosional effects are obvious from even brief study: the ice cuts into the sides of its valley as well as its floor, giving the valley a U shape with precipitous sides; at the head of the glacier the pulling of ice and rock away from the walls produces a round, steep-walled amphitheater called a *cirque*; at the foot of the glacier, and often along its sides as well, the ice pushes debris into morainal ridges.

Valleys showing these features are common in high mountains, even where glaciers are now completely absent. The suggestion is obvious that in mountainous areas as well as in eastern Canada glaciers once existed but have now vanished. For additional proof one can cite the grooved and polished rock surfaces and the many small lake basins in valleys that were

Fig. 8-21. A glaciated valley in the Sierra Nevada. What characteristics of this valley suggest that it was occupied by ice not many thousands of years ago? Note the delta being built by a small stream into the lake in the right foreground. (Eliot Blackwelder)

once glaciated. Note that the effects of valley glaciers and ice caps are in many ways similar, but in one important respect very different: ice caps, riding over hills as well as valleys, tend to leave smoothly rounded topography; valley glaciers, by carving steep-walled cirques and valleys, make topography more rugged. The sharp pinnacles and knife-edge ridges of the world's most spectacular mountain scenery are a legacy of the valley glaciers of the Pleistocene.

Sitting on a peak in the Rocky Mountains of Colorado, or exploring the northern Minnesota lake country by canoe, one is struck by the manifest evidence of deep erosion by moving ice. But in the larger picture of the erosion of the continents, glaciers play only a minor role. Valley glaciers change the shapes of the valleys in which they lie, but the original formation of the valleys goes back to erosion by streams before the glacial ice accumulated. Ice caps are capable of rounding hills and gouging out lake basins and piling up morainal ridges, but from a long perspective these are only superficial changes in the stream-carved landscape over which the ice moved. Pleistocene glaciers in the two million-odd years of their existence produced many striking landscape changes, but their long-time effect on North America is small compared with the work of the streams that preceded and followed the age of ice.

Comparatively superficial also is the erosional activity of wind. This is hard to believe during a dust storm on the desert, when howling gales drive sand along the ground and fill the air with clouds of yellow dust. If you look for evidence of the work of wind, you will find rock surfaces on desert ridges, and sometimes boulders on the desert floor, which show unmistakable polishing and grooving by driven sand. Locally in desert valleys and on the lee sides of ridges you will find heaps of sand, sometimes in the picturesque dunes which are popularly supposed to cover all deserts; the sand has clearly been picked up from nearby plains and dropped in places where the wind slackens. When you reflect that the piled-up sand must have been winnowed of the finer-grained silt and clay that once accompanied it (Fig.

5-4), and that the silt and clay have been carried completely out of the desert by strong winds, you may wonder if moving air does not have an important role in maintaining and even deepening the broad desert basins of the western states. Just how effective winds may be in this activity remains uncertain. But the tell-tale patterns of branching V-shaped valleys in desert mountains are mute evidence that even in very dry climates the erosional work of running water is dominant.

Again the problem of rate

Once more, as we must do so often, we change perspective and look at the continent as a whole. Its rocks are slowly weathering, and some of the weathered material is being converted to soil. Weathered rock and soil, together with some material freshly carved out of bedrock, are being moved downhill by streams. The ultimate goal of the movement is the sea,

Fig. 8-22. Glaciated valleys at the crest of the Uinta Mountains, Utah. The steep-walled amphitheatres at the heads of these valleys, called *cirques,* are common features of glaciated mountains. They are formed at the heads of glaciers, where the ice periodically freezes to the base of the slope and then, as the glacier moves, carries with it parts of the rock to which it is frozen. Note the small lakes dammed by moraines. (John Shelton)

although some of the debris may remain for long periods as deposits of lakes and floodplains. The wearing down of the continent by streams is aided along shorelines by waves, in areas of loose debris by wind, and in high mountains by glaciers. These are important geologic facts, but they are all fairly obvious qualitative generalizations. Can we make them more precise? In particular, can we say anything about how fast the overall process of erosion is wearing away the surface of the continent?

The question sounds preposterous, because obviously some parts of the continent are being eroded far more rapidly than others, and to measure the rates in different parts and then to add them all up seems a hopelessly complex undertaking. We can approach an answer indirectly, however, by measuring the amounts of debris carried by major rivers near their mouths. Even this is no mean task, because it requires measuring the load of sediment in each river at intervals over a period of at least a year and preferably much longer. The result cannot be strictly accurate, but at least it gives an order-of-magnitude estimate of the rate at which North America is being worn down. Measurements of this sort suggest an average lowering of the land surface by about a foot in every 5000 years.

A figure like this is an invitation to speculation. Has the continent always been eroded at this rate? If so, the lowering of the surface would amount to a thousand feet in 5 million years, and less than 20 million years would suffice to reduce the continent to a plain approximately at sea level. Such a result is impossible because erosion is continually opposed by volcanic activity and movements of solid rock, so we cannot answer our question about constancy of rate with such a naive approach. We might have more luck by asking, on what does rate depend? Two factors must clearly have a dominant influence: climate and topography. If mountains were higher and more numerous in the past, erosion would have been faster; if the average elevation was less, erosion would have been slower. The influence of climate is less obvious, for greater rainfall would have two results that would work at cross-purposes. Streams would have more water, so could carve the land more effectively; but at the same time more vegetation would flourish, and erosion would be slowed. The existence of widespread glaciers in the Pleistocene means that the climate during the past million years was wetter or colder than at present, so that in parts of the continent erosion was probably faster. The indications from fossils and the structures of sedimentary rocks that shallow seas once spread widely over the continent suggest that at times in the geologic past erosion was much slower. Thus an answer to the question is not simple, for it depends on our ability to reconstruct from geologic evidence the climates and topography of the distant past.

By weighing many kinds of evidence, geologists have concluded that the average rate of erosion at present is probably faster than it has been during most of geologic time. Mountains are higher and more numerous, and shallow seas more restricted in area, than they have been at most periods in the past. In the age-long struggle between erosion on the one hand and the processes of uplift on the other, the latter at present have gained the upper hand.

Fig. 9-1. A room in a limestone cavern. The cavern has formed by the slow dissolving of limestone by groundwater. The strangely shaped sheets, stalactites, and stalagmites are the result of deposition of calcium carbonate from water as it drips from the ceiling. (Courtesy of Luray Caverns, Virginia)

Fig. 9-2. Jupiter Terrace at Mammoth Hot Springs, Yellowstone National Park. The terrace has been built of calcium carbonate (a variety called *travertine*) carried to the surface by hot groundwater and deposited as the water cools and evaporates. (U.S. Department of the Interior, National Park Service)

9

The Ways of Water Underground

Springs have excited the wonder and curiosity of observers in all ages. Water flows from soil or rocks in seemingly inexhaustible supply, with no obvious source and showing little effect from weather changes or from human attempts to interfere. It is hardly surprising that people of the ancient Mediterranean civilizations, living in a region where water was far from abundant, often associated deities with particular springs.

If we look at springs with concepts of erosion from the last chapter still in mind, we note at once that here is another manifestation of water in motion, hence an agent of erosion that we have so far neglected. In most springs the movement of water is slow, so that we would not expect much active physical wear, but the water has obviously been long in contact with rocks beneath the surface and may have accomplished there a good deal of erosion by dissolving the more soluble rock constituents. We think then of limestone caverns, many of which are associated with springs, and wonder if these open spaces might be a result of such solution of rocks by water. Or we think of hot springs and the mounds of travertine or silica often built up around their mouths, and conjecture that the mounds represent material which the hot water has dissolved from rocks at depth. Or we let our minds wander and think of more general questions, for example the source of water that seeps into wells almost as fast as we draw it out. Clearly the movements and activities of water underground merit attention.

Where does spring water come from?

We no longer see nymphs in sacred groves around springs, but superstitions still cling to the behavior of underground water. Ask a resident of a mountain area far from the sea about the source of water in his spring, and the chances are good he will tell you it comes from the ocean, passing through fissures in the intervening rock and losing its salt by filtration through sandy layers—all this in defiance of the laws of gravity and elementary chemistry. Or if the spring is in granite he may tell you that the water is "juvenile" water, associated with granite in the deep interior since the earth was formed and now coming to the surface for the first time. When you suggest that the spring may consist only of rainwater that has fallen on the hills above and is percolating slowly downhill through weathered material and small cracks in bedrock, he will probably shake his head in disbelief. The amount of rain is too small, the spring water is too pure, the spring keeps running with hardly diminished flow even through long dry spells—with arguments like these he will try to show how silly your hypothesis is.

You may not be able to convince him, but if you choose to try you can bring out a good many cogent points. Springs never occur on peaks or ridges, but always on slopes or in valleys where there is a considerable drainage area above. The composition of the water can be reasonably explained as rainwater with a trace of material dissolved from rocks in the vicinity. Actual measurements of flow would demonstrate that rainfall over the drainage area is more than sufficient to account for the water, and that the flow does diminish somewhat during dry spells. Observations of this sort in many parts of the world have convinced scientists that springs can be explained with no other source of water than rain.

This does not mean that the mystery of springs is entirely solved. For some springs the visible drainage area may seem inadequate, and the actual drainage area that feeds them may be hard to delimit, because structures in the underlying rocks may force water to move in unexpected paths. The precise location of a spring may be unexplainable by anything in the geology exposed at the surface. And for hot springs there is a continuing argument among geologists as to whether at least a small part of the water may be supplied by vapor escaping from a body of cooling magma at depth. Springs are largely rainwater that has moved slowly through soil and rock from higher elevations, but for any individual spring detailed questions may remain unanswered.

How does water move underground?

In the last section and in several previous discussions we have talked about water "percolating" through soil and sediment and rock. It is time that we make this concept more precise.

If you spill water on the ground it will linger a moment and then disappear as it soaks in. If you continue to add water in the same spot the wet area will grow in diameter but the water continues to disappear from the surface. If you dig down into the wet spot you will find that the water is moving downward as well as sideward, the boundary between wet soil and dry being a sharply defined surface that expands in all directions. How is the water moving? Evidently by working its way through the small openings between soil particles, or "percolating" through the soil. The rate of percolation is determined largely by the size and connectedness of the openings, and this in turn by the size of the particles. In a sandy soil the water moves fast, but movement through a clay may be so slow that a large part of the water will run off the surface rather than soak in.

Now in imagination extend this experiment to a heavy rain falling on the same soil. Water at the surface is continually renewed, and the downward penetration can go on indefinitely. Some of the surface water will run off in sheetwash or rivulets, the proportion of runoff to the amount that soaks in depending on the *permeability* (ability to transmit fluids) of the soil and the heaviness of the rain. If the rain continues, the percolating water may move completely through the soil and weathered rock into fresh rock below, penetrating the rock along cracks and perhaps into open spaces

Fig. 9-3. Percolation of water through the interstices of soil.

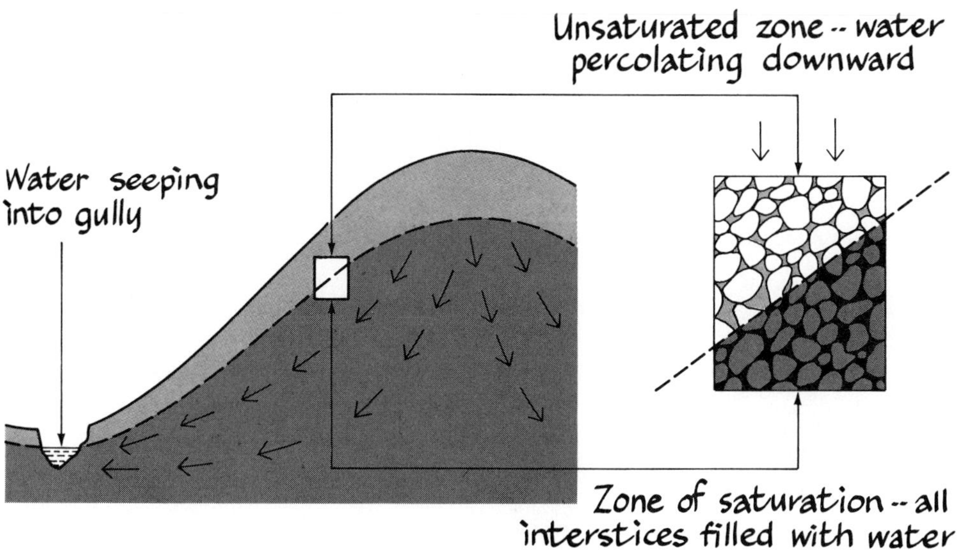

between its grains. Eventually a considerable thickness will be *saturated* with water—meaning that all available open spaces in the rock and soil are filled. Now if the area is cut by a gully, the walls of the gully provide a way for some of the water in the saturated zone to escape. Water will move sideways, feeding a stream in the gully, and a continuous circulation will be set up of rainwater falling on the surface, percolating water moving downward and sideward through the soil, and water escaping into the stream in the gully. When the rain stops the movement into the gully will continue for some time, the soil and weathered rock acting like a huge saturated sponge that slowly drains.

Most movement of water underground is this sort of slow percolation. In contrast, another popular superstition about groundwater pictures it as moving in underground rivers, which come to the surface as springs and which need only be tapped to form flowing wells. Such streams do exist in regions underlain by limestone, but they are rare. Almost everywhere the motion is much less spectacular, just the slow crawl of myriad bits of water from one tiny open space to another in soil and rocks.

It is the slowness of movement that makes springs possible. If all rainwater found its way into open channels underground, it would drain quickly into the nearest valley and springs would be active scarcely longer than the rain was falling. Just because soil and rock soak up water like a sponge and then release it slowly, springs can persist from one rain to the next.

The rate of percolation is obviously very different in different materials. Through sand and gravel it is fast, through clay very slow. In sandstone and conglomerate it may be rapid if the rocks are incompletely cemented, or slow if most openings have been filled with calcite or quartz. Through other rocks the motion depends largely on the number and size of cracks. Solid limestone is very impermeable, but the rock commonly has cracks and cavities enlarged by solution and hence may permit rapid movement. Unfractured lava has little permeability, but in a pile of old

Fig. 9-4. Big Springs, Idaho. The bedrock here is basalt, in horizontal flows one on top of another. The blocky, fragmental surfaces of the flows provide channels for groundwater flow, so that springs emerge at various levels in this basalt cliff. (O.E. Meinzer, U.S. Geological Survey)

Fig. 9-5. Groundwater movement along a fault. The San Andreas fault cuts across this arid landscape in southeastern California from upper right to lower left, and the clump of trees on the line of the fault shows where groundwater comes to the surface. (John Shelton)

flows the surfaces of some flows may be made up of fragmental blocks with enough space between to permit much fluid movement. Intrusive igneous rocks like granite and diorite, and metamorphic rocks like schist and gneiss, are ordinarily very impermeable except along fracture zones.

Differences in rate of percolation account for the location of springs. Ideally a spring appears at a place where permeable material overlies an impermeable bed: groundwater drains down through the permeable rock but is blocked by the impermeable layer, then flows along the top of this layer to the nearest point on the surface. If flow everywhere along the top of the layer is possible, a zone of seepage may result instead of a flowing spring, but generally the flow is concentrated by irregularities of the bedding or by fractures of high permeability. Springs often are conspicuous along a fault line, the ground-up material caused by fault movement evidently providing the necessary permeability for groundwater flow. The theory is straightforward, but in many places it is hard to determine for a particular spring just what relation between permeable and impermeable material is responsible for its location.

Where drill a well?

Suppose that you have bought land with the intention of building a house, and that you are far from an urban water system. You know that there is water in the ground beneath you that can be reached by drilling. Then the problem arises: Where should the well be located so as to ensure an ample water supply in all seasons and at as shallow a depth as possible?

This is a problem that has faced individuals and communities from time immemorial. Groundwater obtained from wells is the obvious source of supply at any site away from streams or lakes or springs, and it is one of the prime natural resources on which civilized life depends. But the ways of groundwater are mysterious. In some places a well can be drilled to great depth without encountering more than a trickle of water, and in others a shallow well may yield an abundant supply. How can one tell, standing on the surface, where the most favorable spot for drilling will be?

The usual answer, even in this scientific age, is to employ a "water witch" or "dowser," an individual who claims the occult power of locating water by traversing the ground with a forked stick held horizontally in front of him, and watching where the end of the stick is "pulled" downward by a "force" emanating from water at shallow levels beneath. The witch may be perfectly sincere, and he is often successful. In many parts of the country groundwater is so abundant that a good supply is available from a well drilled almost anywhere, so that some measure of success is guaranteed; and if the witch has had much local experience he can use that experience, consciously or unconsciously, to guide the motion of his delicately poised stick to the most favorable spot. This is the aspect of groundwater study where superstition seems most durable. The reasons for its persistence are obvious: a good water supply is all-important, to an individual or a community; the movements of groundwater are understood in general, but remain mysterious in detail; the location of successful wells is chancy, subject to no immediately obvious rules; and the witch can claim knowledge that is "scientific" although not yet recognized by the "scientific establishment." A situation like this is ideal for those who would exploit a superstition, either in good faith or for the fat profits obtainable by bilking the credulous.

If, in trying to locate your well, you appeal for aid to a geologist rather than a water witch, you may be disappointed in his apparent lack of assurance. He will talk to you learnedly about the behavior of groundwater in general, he will indicate some parts of your property that are more favorable for drilling than others, but he will hesitate to recommend a particular spot and will probably not guarantee success. Contrast these tentative answers with the calm certainty of the witch that he can find the one place where drilling will give you the most abundant water, and the force of the old superstition becomes still clearer.

The geologist's story will go something like this. During heavy rains in a humid region the open places in soil and bedrock are filled with water, and some of the water drains out as it moves downward and sideward into valleys. When the rains stop the water continues to drain, so that soon the upper part of the soil lacks water except for that which is held in surface films on soil particles (and is thus available to plants between rains). The still-saturated lower part of the soil and rock has a well defined upper surface which moves downward gradually as drainage continues. The surface of the saturated zone, called the *water table*, may drop many feet or tens of feet below ground level if a dry spell follows the rain, and then is again raised to the ground surface during the next storm. Thus the water table moves up and down with the weather, but in regions of well distributed rain it often stays much of the time at a fairly definite level that is spoken of as the "depth of the water table." For a well to be successful it must be drilled at least to this depth, and if it is to furnish a dependable supply throughout the year it must reach the lowest level to which the water table sinks during droughts.

The range of up-and-down movement of the water table depends in part on topography and in part on the permeability of soil and rock. Since movement of fluid below the water table is downward and sideward, it is fastest under hills with steep slopes, and faster in permeable material like sand or gravel than in clay. Thus the water table, so to speak, duplicates the surface hills and valleys in subdued form, and is closer to the surface topography in fine-grained material than in coarse. It follows that a well in clay will give you a permanent water supply at shallower levels than a well in sand, but the rate of flow of water into the well is so much smaller that the amount obtainable at any one time may not be adequate for your needs. If a geologist can secure any information about the permeability of

Fig. 9-6. Cross-section to show the relation of the water table to streams and wells.

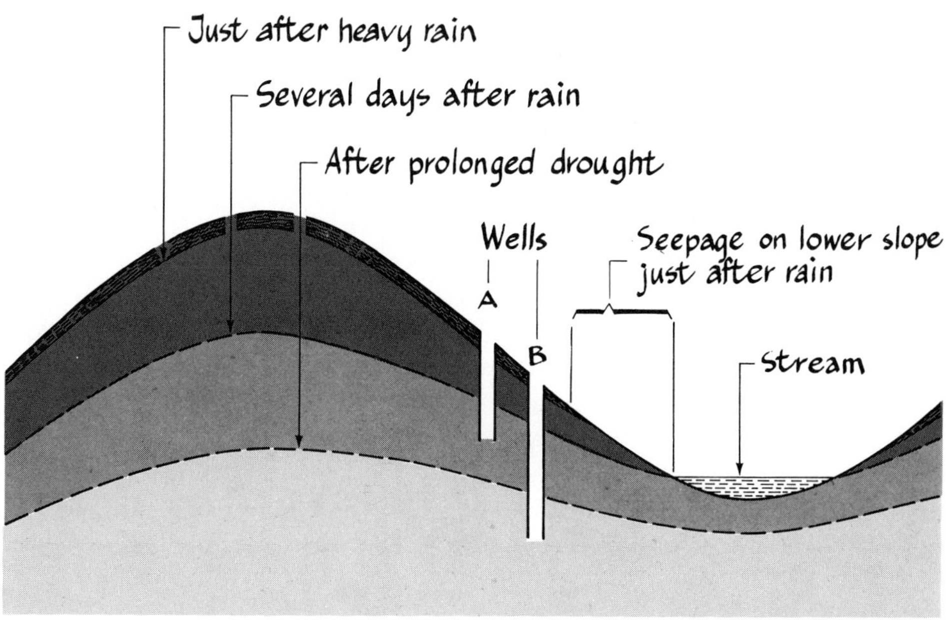

different materials underlying your property—and he usually can—he will give you information of this sort about the behavior of the local water table. Thus he can point out general areas where conditions are favorable or unfavorable for drilling, much more dependably than the dowser can, although there is always the possibility that the dowser will pick by accident a better specific spot.

The water table is a useful concept not only in predicting and explaining the behavior of wells, but also in accounting for changes in springs and streams. Water can flow to the ground surface only at or below the water table, so in a humid region underlain by reasonably permeable rock the level of streams and lakes must represent the local position of the water table. If a drought causes the water table to fall unusually low, streams and lakes may dry up temporarily because no water can flow in from saturated material above their level. Permanent springs also must have saturated rock and soil at higher levels nearby, and grow feeble when the water table drops unduly.

Many kinds of water table

We have been talking as if the material through which groundwater moves is fairly uniform, perhaps different from place to place in permeability but in any one spot remaining homogeneous to great depths. In places where this description is even approximately accurate the problems of groundwater are simple, and either a geologist or a water witch has an excellent chance of locating good wells. But nature is seldom this kind.

More commonly the materials under an area where wells are to be drilled are a complex mixture of permeable and impermeable soil, sediment, and rock. Take, for example, an alluvial plain along a mountain front, a gently sloping area made up of old alluvial fans spreading out from the major valleys and coalescing to form the plain. The fans are built up of a heterogeneous assemblage of sand, gravel, and clay, roughly sorted and stratified, the sand and gravel being deposits in old channels where storm waters rushed down the fans and the clays representing deposits from quieter waters which moved during lesser storms or spread out from the main channels during major storms. We assume that at present stream flow is controlled so that flooding seldom occurs, but rain from the hills soaks into the upper ends of the fans and fills much of their permeable material with water. Where would the water table be in such a situation?

As water drains out after the fan material has been saturated during a heavy rain, drainage will be fastest through the old buried channels and slowest in the parts that are largely clay. The water table will thus be an exceedingly irregular surface, high where clay is abundant and low where coarser materials are concentrated, near the ground surface at the lower ends of the fans and far from the surface at the upper ends. Wells drilled into the fans almost anywhere would encounter water, but those that happen to strike clay would have inadequate flow and those high up on the fans would need to be very deep to ensure a continuous supply in dry weather. The problem of finding adequate water is the problem of locating the old sand-and-gravel-filled channels, and the position of the channels is not indicated by any feature on the surface.

A somewhat similar situation is often encountered in glaciated areas of the northeastern states. Sediments left by the Pleistocene ice cap include the boulder-plus-clay material called till that was heaped up by the glacier itself, and also abundant sand and gravel dropped by copious melt-water. The till is unstratified, made up largely of fine clay-sized material ("rock flour"), thus highly impermeable; the sand and gravel have been roughly sorted and stratified by running water, hence are very permeable. Wells located in the coarse melt-water deposits are successful, those in till generally poor. Study of topography can often aid in recommending well sites, but the problem is difficult where the sediments are deeply buried.

In many places water is obtainable not only from unconsolidated sediments but from layers of sedimentary rock beneath. Sandstone is the most common material to supply water, but in places cavernous limestone or the tops of buried lava flows can serve the same purpose. Any layer through which water flows readily, be it a widespreading bed of sandstone or a narrow channel deposit in an alluvial fan, goes by the name of an *aquifer*. If an aquifer is confined between layers of impermeable material the concept of water table may be difficult to apply. In effect there are often two water tables, one in the soil and weathered rock above the upper impermeable layer and one in the buried aquifer; the latter may be determined by water entering the aquifer at a great distance and not by the local rainfall.

A particularly interesting situation arises if an aquifer confined between impermeable beds is slightly tilted. Water enters the aquifer where its upper end has been exposed by erosion, and moves slowly down the tilt. If a well is drilled into the permeable layer at a point far down-tilt from the outcrop, the ground surface at the well site may have a lower elevation than the place where the water enters. The weight of water through the length of the aquifer must then force water to rise in the well and flow out at the surface, forming an *artesian* well. Such wells, of course, are highly prized because they need no pumping. If much water is withdrawn from the well, it must after a time lose its ability to flow, because the rate of replenishment

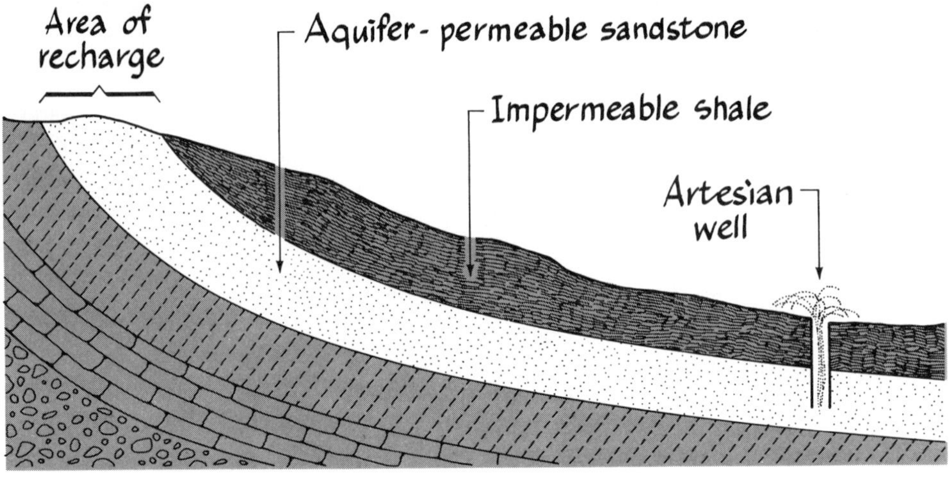

Fig. 9-7. Cross-section of an inclined aquifer and an artesian well. If the level of water at the upper end of the aquifer is above the well site, water will flow out onto the surface. If water is used in large amounts so that the level drops considerably, active flow may cease but the level in the well is still higher than the water table in its vicinity. In either case the well is an *artesian well,* but *artesian flow* can occur only if the intake is maintained higher than the well site.

at the outcrop is insufficient to keep the level of water in the aquifer above that of the well.

Thus the simple statement that a well must penetrate below the water table suffices for areas of homogeneous material, but where rock compositions and structures are variable the statement is not useful without many qualifications.

Groundwater in arid regions

Developing a good water supply is not always easy even where rainfall is high, but problems multiply in desert areas.

Rains on the desert are characteristically heavy and brief, producing much runoff that sweeps sand and gravel down mountain slopes and finer-grained material across wide basins into ephemeral playa lakes at their centers. The small fraction of the rainfall that does soak into the ground percolates downward only a few centimeters or a few meters; then when the rain stops and intense evaporation resumes at the surface, much of the water moves back upward. Only in very unusual storms does the entire mass of soil and weathered rock become saturated, and in the drier areas (for example southern Arizona and the Los Angeles basin in California) this probably never happens at all. Hence the water table, if it exists, lies commonly hundreds of meters below the surface, and its level is generally determined by conditions at a distance rather than by local rainfall.

Alluvial fans along the margins of desert basins and stream deposits in major valleys leading back into the adjacent ranges may collect a good deal of water in confined aquifers or in pockets of sand and gravel underlain by impermeable clay. Such aquifers may hold enough water to maintain local springs and seepages through the long periods of dry weather. The search for groundwater in arid regions is commonly a search for these confined layers or *perched* water tables, maintained high above the regional water table by their floors of impermeable material. Another possible source is water trapped in the sediments of large desert basins that have no drainage to the outside, even for underground water; the water is generally in part salty, but enough of it may be fresh to serve as an adequate supply for sizable communities. The difficulty with exploiting water from alluvial fans or basins in arid regions is that the supply is necessarily limited and is not replenished rapidly. Although the supply may seem inexhaustible when wells are first drilled, over-use poses a constant threat. The problem of water in arid regions generally involves a choice between adjusting use to the rate of replenishment of the local supply, or bringing in water from more humid regions that have water to spare.

Streams and large fresh-water lakes in arid regions are obviously not maintained by inflow from the sides, as they are in humid regions, because the water table lies far below them. Permanent streams that cross deserts

Fig. 9-8. A perched water table.

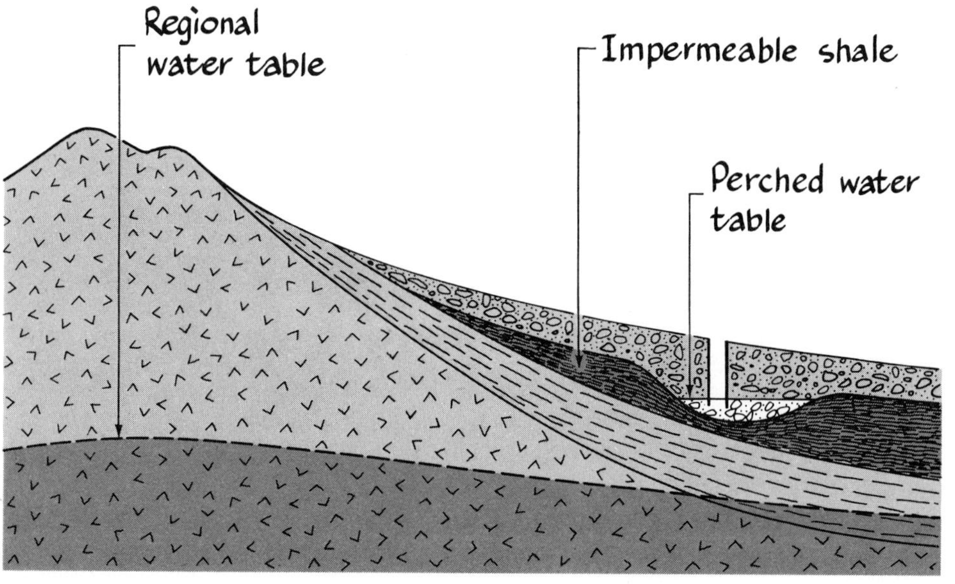

get their water from the higher rainfall of adjacent areas, and in their path across a desert feed water into the ground rather than receiving it. A good example is the Colorado River, which is fed by rains and melting snow in the high Rockies of Colorado and then for most of its thousand-mile course across the canyon country of Utah and Arizona and the desert of eastern California loses more water than it picks up.

Erosion by groundwater

At first thought the erosional activity of groundwater would seem to be negligible, simply because it moves so slowly. The percolation of tiny volumes of water along cracks and from one small cavity to another would surely be ineffective in dislodging or transporting any but the tiniest solid particles. The crystal clearness of most spring waters is good evidence that underground water carries little material in suspension.

On the other hand, the prolonged and intimate contact between groundwater and the surfaces of grains in soils and sediment provides ample opportunity for solution. Water that enters the ground contains abundant dissolved gases from the atmosphere, particularly carbon dioxide and oxygen, which enhance its ability to corrode the materials through which it moves and to dissolve any soluble constituents. In effect the zone of

Fig. 9-9. A large crack in limestone widened by solution. (Tennessee Valley Authority)

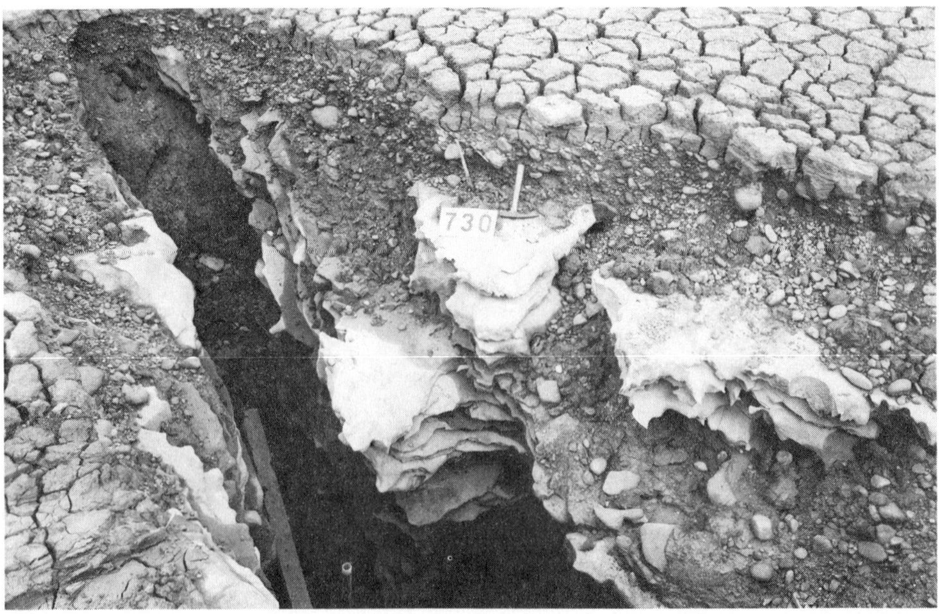

weathering extends through the part of the soil and rock above the water table, where interstices are filled with air except when the water table rises during storms. The reactions of weathering go on all through this zone—the dissolving of calcium carbonate by the slightly acid water, the decay of organic matter, the slow breakdown of silicate minerals to form clay and to put metal ions and silica into solution, and the oxidation of iron compounds to form ferric-oxide minerals. As the water table rises and falls with changes of weather the products of weathering are repeatedly flushed out, the soluble materials becoming part of the groundwater and moving slowly with it into nearby streams and springs and lakes. Below the water table similar processes can go on, but to a much smaller extent because air is largely excluded and the dissolved gases from above are rapidly depleted by reaction within a short distance from the water table.

We have spoken of the dissolving and leaching action of groundwater before, in connection with the development of soil from weathered rock. Soil formation takes place generally well above the water table, in the first few feet or tens of feet below the ground surface. As rainwater moves through this uppermost zone its dissolved oxygen is diminished and its carbon dioxide is increased by contact with decaying organic matter. Thus the amounts of the two gases that it carries downward are variable, depending on rate of flow and quantity of organic matter available at different times. The effect of the water on mineral grains is correspondingly variable, and the complex process of soil formation depends on an overall average of reactions that differ in intensity from time to time. The topmost soil layer in humid climates, as we have seen before, is generally leached of its metallic ions, and iron oxide is often concentrated in a layer just below the top one. The metal ions in large part are carried down to become part of the dissolved material in the groundwater below the water table.

The effectiveness of removal of the metallic ions depends on the amount of water moving through the soil, and therefore chiefly on climate. In regions of high rainfall leaching may be so thorough that not much soluble material is left, and groundwater that finds its way into streams and springs may have a composition little different from rainwater. In semiarid regions the amount of unleached soluble material may be substantial, particularly salts of calcium and magnesium, so that groundwater tapped by wells or issuing in springs leaves deposits when evaporated ("hard" water). Under still more arid conditions the very soluble salts of sodium and potassium may be left in some layers of the soil, and groundwater that has been in contact with them may be unusable because of its salt or alkali content.

Groundwater has figured in other earlier discussions with reference to the cementing of sediments into sedimentary rocks. Cementation, the opposite process to leaching, means the deposition of material out of solution in groundwater. Under what conditions would this happen? A little thought will suggest several possibilities. Perhaps groundwater percolates into a sediment whose pores are open to the atmosphere, so that the water simply evaporates and leaves its dissolved material behind. If the water is saturated with calcium carbonate, the amount it can hold is delicately dependent on the amount of dissolved carbon dioxide and on the acidity; if the groundwater moves to a place where CO_2 can escape or where reaction with adjacent grains makes it more alkaline, some of the calcium carbonate must precipitate. Perhaps the temperature of the water changes, or perhaps groundwater from different sources becomes mixed. Perhaps the dissolving and precipitation take place locally and almost simultaneously; it is a general rule that very small grains are more soluble than large grains, because they have more edges and corners exposed, so groundwater at any point in a sediment may dissolve the smaller particles and deposit their substance as cement around the larger ones. The possibilities are numerous, and it is small wonder that cementation of sediments is a widespread geologic phenomenon.

The kinds of cement commonly found are limited in number, and represent materials whose presence in groundwater is readily understandable. Calcium carbonate is probably the most abundant. It is a compound that groundwater can dissolve directly from limestone or marble, or indirectly by reaction of carbonic acid with silicate minerals containing calcium. Silica is another familiar cementing agent, generally in the form of quartz but sometimes as opal. Its presence in groundwater is explainable by the breakdown of silicate minerals during weathering, the dissolving of fossil shells of silica-secreting organisms, and to a very slight extent by the dissolving of quartz and chert. Ferric oxide may form as a cement if groundwater containing a dissolved ferrous compound moves to a place where organic matter is absent and oxygen can gain access. Clay as a cementing material is harder to explain; it may have been carried mechanically as very tiny particles, or the clay may have been deposited with the original sediment and somewhat recrystallized by long contact with percolating groundwater.

Thus groundwater unobtrusively accomplishes a good deal of erosion and deposition, simply by virtue of its ability to dissolve material slowly from ordinary rocks, to transport its dissolved load for long distances by slow percolation, and to deposit some of it when conditions change slightly. The amount dissolved greatly exceeds the amount deposited, as is evident from the fact that streams and springs fed by groundwater contain far more dissolved material than ordinary rainwater.

How fast does groundwater erode?

If streams consist largely of water that has worked its way through the ground, and if the groundwater supplies them with dissolved material which they then carry to the ocean, erosion by groundwater must be part of the overall process by which the continents are worn down. How important a role does it play? Is the amount of dissolved substance a large part or a negligible fraction of the total quantity of material that streams carry away from the continents? Here once again we are dealing with a question of rates of slow but long-continuing processes.

The procedure for getting a rough answer is obvious. To estimate the amount of suspended solid material that streams carry or roll or drag into the sea we looked at measurements of the suspended loads of major rivers averaged over a year's time. We follow a similar procedure here, in imagination collecting samples of water, filtering off the solids, and evaporating to dryness, then estimating total flow and adding up figures for the major drainage basins. The result is startlingly high: of the total load carried by streams, approximately one-third is in solution. The figure for average rate of lowering of the continental surface quoted in the last chapter, one foot per 5000 years, actually includes both dissolved and suspended material; thus solution alone would wear down North America at a rate of about a foot in 15,000 years. Even if streams carried no fragmental material at all, the continent could be reduced to a plain by solution in groundwater in a time that is short compared with the age of the earth.

A corollary has probably occurred to you. The dissolved material carried down from the continents is added to the salt of the sea. If the rate of removal is surprisingly fast, the concentration of ocean salt must be increasing at the same rapid rate. Can we conclude that the present salt content is the result of solution, largely by groundwater, through all of geologic time, and that the ocean is therefore becoming steadily more salty? This turns out to be one of the more profound geologic questions. The answer is probably negative, because dissolved material is being constantly removed from sea water, but whether addition and removal completely balance is not certain.

Speleology

This Greek word is the way cave explorers describe the subject about which they wax so enthusiastic. The huge caverns that in places can be followed for miles underground have always been objects of fascination, not only for speleologists but for many who prefer to learn about caves in less

Fig. 9-10. A lava cave. In contrast to limestone caves, a lava cave has no relation to groundwater. It forms when the surface of a thick basalt flow freezes, and the still-fluid lava beneath finds an exit and drains out. Such caves are common in areas of abundant lava, but they lack the deposits of calcium carbonate that make limestone caverns so spectacular. (California Division of Mines and Geology)

hazardous ways than actual exploration. Even if we no longer assume that caves are entrances to Hades, as many of our ancestors did, they retain an air of mystery that is often reflected in literature.

Openings in rocks can be traced to many causes. Lava caves are common in areas of extensive basalt flows, representing places where a solid crust formed over still-liquid rock which subsequently flowed out. Sea caves form where wave action penetrates into weak parts of a cliff, usually along zones where the rock has been weakened by closely spaced fractures or faulting. Shallow caves may be hollowed out by wind driving sand grains against soft layers in cliffs, or by undercutting of cliffs where stream energy is concentrated on the convex side of bends. But the big caves, the ones that reward speleologists with miles of winding passageways and occasional rooms decorated with spires and columns and rock curtains, are the ones developed in beds of limestone. These include the famous caves of Virginia and Kentucky (Fig. 9-1), Carlsbad Cavern in New Mexico (Plate 7), and smaller ones of local repute that are found in almost any region where limestone is a common rock.

Laboratory evidence for appreciable solubility of calcium carbonate, plus field evidence for the slow dissolving of exposed limestone surfaces and the widening of cracks along which water can penetrate, leaves little doubt about the origin of limestone caves. They form where groundwater has moved for long ages through fissures in a somewhat soluble rock, gradually enlarging its channels by dissolving the walls, carving out rooms and irregular chambers in places where the rock was particularly soft or highly fractured. The groundwater makes enough room for itself so that it can actually flow as a river, and such rivers are sometimes found in caves. More commonly the river has disappeared because of a change in climate or drainage or a general uplift of the area, and the original presence of the river must be inferred from signs of erosion on the cavern walls and floor.

The rock structures that lend such strange beauty to many limestone caverns are a fine illustration of the chemistry that governs the dissolving and precipitation of calcium carbonate. This compound, you will recall from previous discussions, is slightly soluble in pure water but much more so in acid. Water exposed to air contains the very weak carbonic acid, formed when carbon dioxide dissolves. When water containing carbon dioxide stands in contact with calcium carbonate, the compound dissolves until the solution is saturated. Thereafter any change in the solution may cause more of the solid to dissolve, or some of the dissolved material to precipitate out. In particular, if the solution loses carbon dioxide (and therefore carbonic acid), the calcium carbonate becomes less soluble and some must precipitate; if the solution acquires more carbon dioxide it is capable of dissolving more of the solid. Temperature, pressure, and other salts in solution have slight effects on the solubility, but the major influence is dissolved carbon dioxide.

Chemical symbols, if you have any acquaintance with them, will help in keeping track of the solubility changes. In Chapter 7 we wrote for the dissolving of calcium carbonate in any acid:

$$CaCO_3 + 2H^+ \rightarrow Ca^{++} + H_2O + CO_2 \qquad 9\text{-}1$$

For a weak acid like carbonic acid a more accurate equation would be:

$$CaCO_3 + H_2CO_3 \rightleftarrows Ca^{++} + 2HCO_3^- \qquad 9\text{-}2$$

Or H_2CO_3 can be written in the equivalent form $H_2O + CO_2$, to show more clearly the influence of carbon dioxide gas:

$$CaCO_3 + H_2O + CO_2 \rightleftarrows Ca^{++} + 2HCO_3^- \qquad 9\text{-}3$$

The formula HCO_3^- represents an ion, bicarbonate ion, and the combination $Ca^{++} + 2HCO_3^-$ stands for dissolved calcium carbonate. The double arrow in both of the last two equations means that the reaction can go either way. If CO_2 is increased the reaction goes to the right and the amount of dissolved calcium carbonate increases. If CO_2 is decreased the reaction goes to the left and calcium carbonate precipitates. The equations serve merely to express the essential chemistry more conveniently and more precisely than the ordinary language of the preceding paragraph.

Now in a limestone cavern water dripping from the ceiling contains carbon dioxide which it has absorbed from air in the soil above. It should therefore be saturated with calcium carbonate dissolved from the cavern's roof (the third equation going to the right). While a drop is suspended at the ceiling it may lose some of its carbon dioxide, because the amount of this gas in the cave atmosphere is probably less than that furnished by decaying vegetation in the soil. Hence some of the calcium carbonate in the drop should precipitate (the third equation going to the left). This sequence

Fig. 9-11. A sinkhole formed by sudden collapse of the roof of a limestone cavern in Alabama. (T.V. Stone, Geological Survey of Alabama)

Fig. 9-12. Karst topography near Bowling Green, Kentucky. (John Shelton)

of reactions, repeated over and over again as other drops form and fall, leaves a deposit of calcium carbonate attached to the roof, a deposit that eventually grows downward to become one of the "rock icicles" (*stalactites*) that are familiar features of caves. If water from the fallen drop lingers on the floor of the cavern, further loss of carbon dioxide (aided by evaporation of the water) leads to another deposit, this time a pile of calcium carbonate growing upward from the floor (*stalagmite*). Variations of this process which may lead to pillars and thin screens of calcium carbonate can readily be imagined.

The ability of groundwater to dissolve large amounts of limestone and to form caves and channels underground produces conspicuous landscape features in many limestone regions. If a flat-lying limestone bed is near the ground surface, the presence of caves or enlarged fissures beneath is often shown by sinkholes—rounded or elongated depressions formed by the dissolving away of a cave's roof or by collapse of surface material into an opening beneath. Open channels under the surface provide places for concentrated stream flow, and some surface streams simply disappear because their water goes underground into a limestone channel. If this process of solution and collapse and diverted stream flow goes on long enough, a limestone bed may be eroded away except for remnants in the form of isolated steep-walled hills and pillars. A landscape of this sort, with sinkholes and disappearing streams and sometimes with isolated limestone hills, goes by the name of *karst,* after a region in Yugoslavia where it is well developed. The best example in this country is the landscape of western Kentucky.

Hot groundwater

Still more mysterious than ordinary springs are those that issue from the ground at temperatures higher than normal. For these we face the problem of explaining not only the source of water but the source of heat.

Temperatures of hot springs range from barely warm up to the boiling point. Openings where water vapor issues at temperatures higher than boiling, often conspicuous because of plumes of condensed steam rising above them, are called *fumaroles* rather than springs (Plate 3). Hot springs whose channels below the surface are highly irregular or partially blocked may erupt periodically into fountains of boiling water, forming the famous *geysers* of Yellowstone Park, Iceland, and New Zealand.

Most hot springs, as might be expected, are in areas where volcanoes are active or have been active in the fairly recent geologic past. The source of heat is readily imagined as a body of magma within a few kilometers of the surface, perhaps largely crystallized but still hot enough to supply heat to water in its vicinity for thousands or hundreds of thousands of years. The water of hot springs must come in large part from rain, as does groundwater elsewhere, because the flow of hot springs is influenced by changes in rainfall from year to year. One can imagine a complicated sort of circulation, rainfall making its way into the ground over a considerable area, percolating downward until it is warmed by the hot rocks below, then tending to rise because hot water is lighter than cold, perhaps moving up and down many times in response to the temperature differences, perhaps locally vaporized and condensed, and ultimately being channeled into the fissure that feeds the spring.

But is this the only source of the hot water? We recall that molten lava pouring out of a volcano is always accompanied by abundant gas, and that water vapor is nearly always the principal constituent of the gas. Magma underground evidently contains dissolved water, and perhaps hot springs are supplied in part by water rising from the same body of magma that furnishes heat. Should we expect that water coming directly from magma would be a large fraction or a small fraction of the total quantity issuing from a hot-spring area?

Fig. 9-13. The Geysers, northern California. This area is misnamed, because there are no real geysers here. It is an area of numerous fumaroles, which now has been developed for geothermal power. The steam jets in this picture are not natural, but come from holes drilled into the reservoir of steam in the hot rocks beneath the surface. (Thermal Power Company)

This is a question that has worried geologists for many years, and to which there is still no definitive answer. The most promising attack has been a study of the isotopes of hydrogen and oxygen that make up hot-spring water. The ratios of heavy to light isotopes of these elements are different in rainwater from the ratios in water associated with magmas, and the difference is sufficient to identify at least roughly the kinds of water in hot springs. For the hot-spring areas where this method has been tried, isotope ratios indicate that the water is at least mostly derived from rain, the possible addition of magmatic water being no more than a few percent.

Hot water reacts with rock more rapidly than cold water, so that many hot springs are surrounded by rock showing more intense alteration than ordinary weathering—bleaching, iron-oxide staining, softening by conversion of feldspar to clay minerals. Such alteration is especially prominent in the hot-spring areas of volcanic regions, because here the water is generally accompanied by minor amounts of such corrosive gases as hydrogen sulfide and hydrogen chloride, in addition to CO_2. If you have had a chance to visit a large area of active hot springs, you will have vivid memories of the bare, white to yellowish rock, the bubbling springs of clear water, the fumaroles with their clouds of steam, and the pervasive odor of gases containing compounds of sulfur and chlorine.

A few hot springs (or warm springs) are found in areas remote from any known volcanic activity of the present or recent past. White Sulfur Springs in Virginia and Hot Springs in Arkansas are good examples. What heat source can we imagine for such occurrences? One can always postulate a body of hot magma at depth, even without volcanoes at the surface; if the body is deep enough, its existence is hard to prove or disprove by any surface observations. Or one can suppose that frictional heat has been generated by movement along faults. Or perhaps rock structures permit groundwater to penetrate to unusual depths, where it is heated by the general rise in temperature found everywhere at deep levels in the earth. For any given spring a choice among these possibilities may be difficult to make.

Hot springs have long had commercial interest because of the supposed curative power of their sulfur-containing waters, and in some places (Iceland and Japan, for example) they are extensively used for domestic heating. In recent years a more exciting commercial application has been under study: the steam that can be tapped by drilling into hot-spring areas is a possible source of energy for generating electric power. One such plant for producing *geothermal power* has been set up in northern California, and active search is under way for other suitable sites. Not all areas of hot springs can be used for this purpose, because some have too low a temperature or too small a reservoir of groundwater to provide a continuous supply of steam. The evaluation of possible sites depends on the same sort of geologic knowledge of groundwater movement that is useful in locating wells, plus additional information about the nature and probable durability of the heat source. As more familiar sources of energy like coal and petroleum become scarcer and more expensive, the search for hot-spring areas suitable for power development will become more active.

Fig. 10-1. A stream valley bent by movement on the San Andreas fault, as seen from an airplane looking westward across the fault. The land on the west side, in a long succession of earthquakes, has moved northward (to the right in this picture) several hundred feet, producing two nearly right-angle bends in the course of a stream that once flowed straight across the fault line. (R.E. Wallace, U.S. Geological Survey)

10

Architecture of the Crust

Tilted beds, folded beds, beds broken by faults: these have figured prominently in past discussions. Together with bodies of igneous rocks they make up the fabric, the structural elements, of the earth's crust. Just as we can describe a building in terms of its walls and ceilings, its supporting members, its foundations, the angles and curves of its floor plan and roofline, so we may speak of the architectural relations of the rocks beneath our feet. How are different rock units put together? What sort of patterns can we discern in this grand-scale architecture? An inquiry of this sort leads into the branch of geology called *structural geology* or *tectonics.*

We may look at the earth's crust with the eye of an architect, but we go beyond the usual architectural questions to ask how the rock structures were produced. Sedimentary beds that are bent and broken, or lifted high above their place of origin, bespeak the action of forces far greater than any in our immediate experience. Can we get any hint from study of the structures as to the nature of these forces? We think, for example, of the flat-lying strata of the Middle West, with their marine fossils showing that the middle of the continent once lay beneath the sea, and then was lifted bodily more than a thousand feet; or of the sandstones and limestones of the Appalachian ridges, their steep angles showing that these thick beds of solid rock have somehow been crumpled like sheets of paper; or of the intricate patterns of folds and faults in some of the western mountains, so complex that forces acting at different times and in different directions must be assumed. The architecture of rock masses preserves a record of forces that have racked the crust in past ages, and one of our aims in studying tectonics is to interpret this record.

Another aim is the more practical one of seeing what kinds of structures may serve as traps for the metallic ores and combustible fuels on which our industrial civilization depends. Prospecting for petroleum and natural gas is in large measure a search for structures in which permeable strata have been sealed off from access to the earth's surface. Successful mining ventures often depend on the ability of geologists to follow ore-bearing beds or veins through complex rock structures. Even if you never plan to look for oil or prospect for gold yourself, some knowledge of the relation of these resources to rock structure is necessary background to evaluate the increasingly frequent predictions about imminent shortages of energy and industrial metals.

The language of tectonics

The literature of tectonics, even more than most parts of geology, is studded with long words of Greek origin. The words are useful to specialists, but they make the subject seem forbiddingly difficult to a beginner—and even to other geologists. We will bypass most of the longer terms here, but we do need a basic vocabulary of the simpler and more commonly used expressions.

Think of any mountain range you have seen, in which sedimentary rocks are prominent and exposures are reasonably good. You want to describe the structures—the flat-lying beds, the uptilted beds, the beds bent into folds, the fault lines where beds appear to be broken, perhaps a granite dike that cuts the beds, perhaps the direction of foliation in beds that have been metamorphosed. In large part you will be describing the orientations of surfaces, such surfaces as bedding planes, foliation planes, faults, and intrusive contacts. How will you record such orientations, either to keep them in order in your own mind or to compare them with observations made by someone else?

An architect faces much the same question when he sets out to express his ideas for a new building. His solution is to draw detailed plans, showing surfaces by conventional patterns and describing relations of surfaces by means of floor plans and vertical sections through the building as seen from various directions. Correspondingly a geologist records his observations on rock structures by making a *geologic map,* on which surfaces are represented by lines, and one or more *cross-sections* (imagined vertical cuts through the area) to show the surfaces from different angles. Compared with the architect he works under a disadvantage because he cannot actually see his surfaces except where they crop out, so that his maps and sections must involve a good deal more conjecture than the architect's plans for a building (Plate 29).

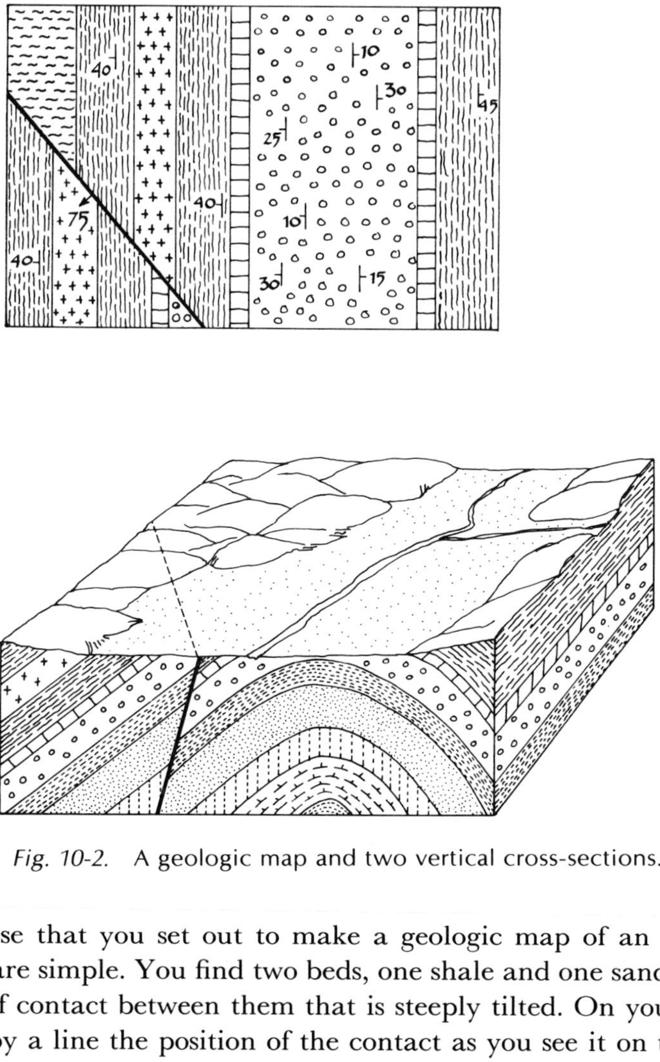

Fig. 10-2. A geologic map and two vertical cross-sections.

Suppose that you set out to make a geologic map of an area where structures are simple. You find two beds, one shale and one sandstone, with a surface of contact between them that is steeply tilted. On your map you can show by a line the position of the contact as you see it on the ground, but how will you indicate its slope? This could be done by drawing a cross-section, but it is convenient to have a way of representing slopes on the map itself.

What do you need in order to specify the "slope" of a surface? Two things come to mind: the steepness of the slope, and its direction. In technical language, to specify the *attitude* of a surface or a bed we need to know its *dip* and its *strike.* To visualize what these words mean, think of a slab of wood held at an angle in a basin of water. The surface of the water against the wood defines a horizontal line on the wood, and the direction of this line (the compass direction, measured from a north-south line) is the strike of the slab. Then the angle between the water surface (a horizontal surface) and the slab, measured in a direction perpendicular to the strike, is

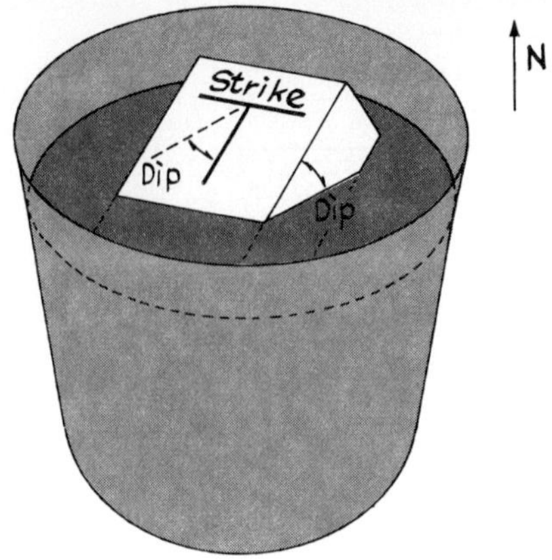

Fig. 10-3. Dip and strike illustrated by a slab of wood in water. The *strike* of the wood surface is the direction of a horizontal line, here defined by the line of intersection with the water surface. The *dip* is the angle between the wood surface and the water surface, measured at right angles to the strike. The combination of strike and dip gives the orientation or *attitude* of the wood surface.

the dip. For example, the strike of the slab's surface might be N30°W, meaning that it was oriented in a direction 30° west of north, and the dip might be 70°W, meaning that the slab was tilted steeply toward the west. These two quantities give all the information needed to specify the orientation of a plane in space. The dip is a precise measure of the steepness of the slope, and the strike is a measure of its direction.

On geologic maps the dip and strike of bedding planes at any point are represented by a symbol like this: 72 ⅄. The position of the long bar is the strike (no number need be given, because it is drawn in the correct compass position), and the short crossbar with attached number shows the direction and amount of dip. If you have looked at geologic maps you have doubtless noticed many of these small symbols scattered over some of the colored areas representing sedimentary rocks. For faults and contacts the strike is given (or can be inferred) from the direction of the line on the map, and the dip is shown by arrows and numbers at intervals along the line.

Dip and strike are concepts that often seem a major hurdle to beginners in geology. Once grasped, they are very simple—just a convenient way of showing the position of things in space. Unless you are going to try your hand at actually making a geologic map, their precise meaning

Fig. 10-4. How the dip of bedding is measured. The geologist is holding a compass equipped with a clinometer (a device for measuring angles). He looks at a bed directly along its strike and holds the edge of his compass parallel to the bedding plane; then a glance at the clinometer gives him the angle of dip. (J.R. Stacy, U.S. Geological Survey)

need not be too much of a worry. The important things to remember for future discussions are simply that *dip* refers to amount of slope, in the direction where the slope is steepest, and that *strike* is a horizontal direction.

Joints

All rocks are cracked, and geologists refer to the cracks as *joints.* In some rocks the joints are so closely spaced that outcrops fall apart into tiny fragments. In others, particularly some varieties of granite and marble, the spacing is so wide that large flawless blocks can be quarried for building stone or sculpture. Joints may cut a rock in random directions, but often many joints are conspicuously parallel, forming a *joint set.* Some rocks are cut by two or more joint sets, so that the debris of weathering or blasting may consist of angular, roughly geometric shapes. The importance of joints in speeding up weathering and in aiding the formation of limestone caves has been emphasized in earlier chapters.

Why should rocks be so universally cracked? Several causes of jointing can be suggested, for example shrinkage by loss of water, expansion during weathering, and strain resulting from temperature changes. Probably the chief cause is simply minor movement in solid rock. Very little movement is required, for rocks in the upper part of the crust are brittle, and any slight disturbance should be capable of producing cracks. Joint sets in folded and faulted rocks often show by their orientation a clear relation to the major structures, hence also a relation to the forces that produced the folds and faults.

If you have traveled through areas where volcanic rocks are abundant you have probably been struck by another conspicuous kind of jointing. Lava flows, particularly basalt flows, commonly have joints perpendicular to the flow surfaces that divide the rock into roughly hexagonal columns—thus the name *columnar jointing.* The columns have diameters from several centimeters to a few tens of centimeters; they may extend completely through a flow, or may die out toward the center. In places their hexagonal shapes are so perfect as to suggest the forms of crystals, and it is small wonder that several centuries ago they were thought by some to be related to the six-sided crystals of quartz.

What sort of origin for these columns would seem reasonable? We recall that a basalt flow was once very hot, and that as the solidified lava cooled it must have shrunk. Perhaps the joints between columns are just shrinkage cracks, developed as the uniform mass shrank toward randomly spaced centers. You can see a similar phenomenon occurring when a mud-puddle dries: shrinkage of the mud (here due to drying rather than cooling) causes cracks to develop in sharp-angled geometric forms, many hexagonal but some with more or fewer sides. The analogy is far from perfect, in that the cracked layer of mud is generally no more than a centimeter or so thick, but one can readily imagine that the slower shrinkage in a lava flow would permit the cracks to grow downward and would favor more regular hexagonal shapes.

Another kind of jointing that often catches the eye of a traveler is the horizontal or gently curved slabbing of exposed granite surfaces. The joints in places can be shown to parallel the original upper surface of the granite intrusive, so that a possible origin is simply expansion of the rock following release of pressure as the original overburden of metamorphic or sedimentary rock was removed by erosion. This speculation is supported by the observation that flat floors of deep quarries, either in granite or in sedimentary rocks, are sometimes observed to arch slowly upward, or even to break suddenly with destructive effects, evidently as a result of release of pressure by the quarrying operation.

Quite possibly shrinkage and release of pressure are responsible for other joints besides these fairly obvious examples, but almost certainly minor movements in solid rock are a more important general cause for the formation of cracks.

Faults

Joints result from the simple breaking of rock, without much displacement along the joint planes. If the two sides of a fracture are displaced, indicating that one side has moved with respect to the other, we call it a *fault* rather than a joint. The distinction between faults and joints may be hazy if the displacement is very slight, but ordinarily the shift in position along a fault is obvious. It may amount to a few centimeters or tens

Fig. 10-5. Columnar jointing in a basalt flow, Devil's Post Pile, near Mammoth Lakes, California. The remarkably regular five- and six-sided columns were produced by shrinkage cracking as the hot basalt cooled. (H.R. Cornwall, U.S. Geological Survey)

Fig. 10-6. Gently curved joints in granite (*exfoliation*), perhaps caused by relief of pressure as erosion stripped the granite of its original cover. (G.K. Gilbert, U.S. Geological Survey)

of centimeters, so that the displacement is readily visible in outcrops, or it may be measured in kilometers or even hundreds of kilometers.

Suppose you are examining an outcrop where beds of sedimentary rock are obviously broken and displaced along a fault surface (Fig. 10-7). What can you infer about the geologic events that have taken place? That movement along the fault has occurred is evident, simply because the broken ends of the beds are separated. With careful inspection you may find other proofs of movement. There may be a band of fine-grained material (*fault gouge*) along the fault surface, or a band of angular rock fragments (*fault breccia*), representing material ground up as the movement took place; the broken ends of beds near the fault may be bent, evidently because the beds were dragged by the movement; if the actual fault surface is exposed, you may find part of it highly polished (*slickensides*), with scratches and grooves produced by the scraping of grains along the tightly squeezed plane of movement. What guesses can you make about the direction of movement? Your first glance would probably suggest that the right side has slipped down along the inclined fault surface. But is this a

Fig. 10-7. A normal fault cutting nearly horizontal beds of sandstone and shale. The fault surface is oblique, and the side above the surface (here the side on the right) has apparently moved down *relative* to the side below the surface. (Actually the left side may have moved, or both sides may have moved; all we see is evidence of *relative* motion.) The amount of displacement on the fault, found by matching beds on the two sides, is four or five feet. The beds are cut by several smaller faults in addition to the main one. (Mary Hill, California Division of Mines and Geology)

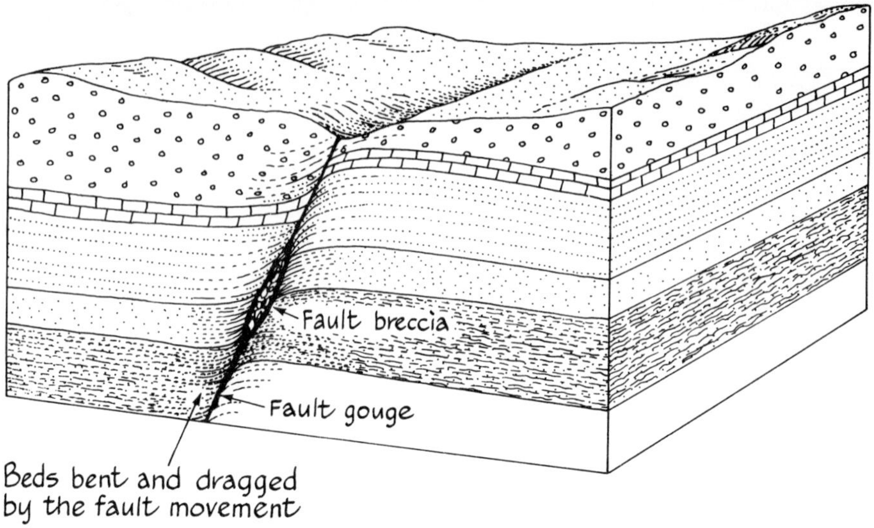

Fig. 10-8. A normal fault, showing fault breccia, fault gouge, and drag of the beds by the fault movement.

necessary conclusion? One might equally well maintain that the left side has moved up, or that both sides have moved. We speak casually about "up" and "down," but if you think about it a minute these words in a situation like this are not well defined. Up and down with respect to what? To adjacent unfaulted rocks? To the present-day earth's surface? To the earth's surface at the time faulting took place? Unless we can fix a frame of reference, our question is meaningless.

Interpreting fault movements always involves this sort of ambiguity. All we see in outcrops, or in reconstructions of larger faults from geologic maps, is evidence of *relative* movement. The only accurate statement we can make is that the rocks on the right side have apparently moved down *with respect to* the rocks on the left. To say that movement was *actually* downward implies that we have other information about the state of the earth at the time faulting took place—information that we generally do not have.

Still looking at the figure, we say that the relative movement was *down* on the right. In slightly more technical language we describe the movement as down the dip of the fault plane, and such a fault is therefore labeled a *dip-slip* fault. (If you are truly skeptical you will see that this doesn't end the matter. In the outcrop it looks as if the relative movement was directly down the dip, but this isn't necessarily right; the motion may have been obliquely downward, so as to involve some slip along the strike of the fault plane as well as along its dip. To settle this additional question would require more information than can be read from the figure alone.)

Now, by way of contrast, look at Fig. 10-10. Again sedimentary beds are cut by an inclined fault surface, but here it looks as if rocks on one side

Fig. 10-9. A cluster of small normal faults (emphasized by ink lines) cutting sandstone beds in southern Utah. (Arvid Johnson)

(right) have been pushed up over rocks on the other (left). We reflect, however, that the same structure could be produced by downward slipping of the left side, or by simultaneous movement of both sides. Regardless of the actual movement, the *apparent relative movement* is again dip-slip. The difference is that now the rocks *above* the fault have apparently moved *up* along the dip of the fault surface, whereas in Fig. 10-7 the rocks above the fault moved apparently downward. We call the fault in Fig. 10-10 a *thrust* fault, that in Fig. 10-7 a *normal* fault. The distinction depends on the fact that the fault surface is oblique. If a fault is vertical it can be neither normal nor thrust, and is called simply a *vertical* fault.

The contrast between normal and thrust faults is important because of what it tells us about the forces responsible for the faulting movements. In forming a normal fault a part of the earth's crust has been stretched and lengthened, hence has been subjected to tensional forces (Fig. 10-11). A thrust fault means that a section of the crust has been squeezed and shortened, hence has undergone compressional forces. The faults tell us nothing about the origin of the forces, but they at least show that some parts

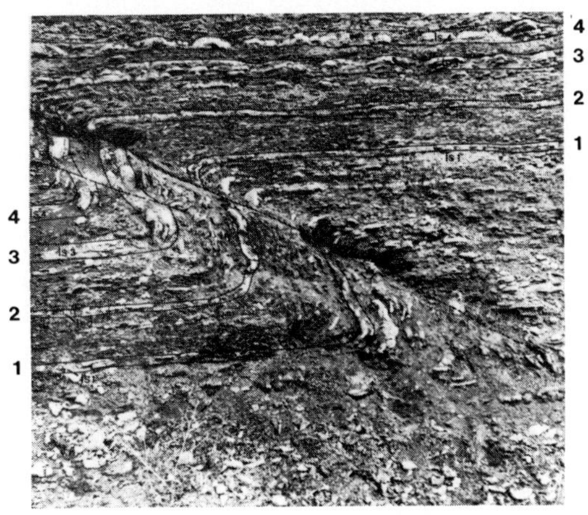

Fig. 10-10. A thrust fault in shale with thin beds of limestone. Four of the limestone beds are identified on the two sides of the fault. On this oblique fault surface, the side above the surface (on the right) has apparently moved up relative to the side below the surface. Note the drag of beds near the fault and the development of a small fault roughly parallel to the main one (left-hand side). (Arvid Johnson)

of the earth's crust have been under horizontal compression and other parts under horizontal tension at some time in the geologic past.

The small faults shown in Figs. 10-7 and 10-10 are the sort that might conceivably be produced during a single earthquake. We have no way of knowing whether they were actually produced in this fashion; it is equally likely that the visible displacement is the result of many small earthquakes occurring at intervals over a long period of time. Large thrust faults and normal faults, with displacements of hundreds of meters or kilometers, were almost certainly formed by repeated small movements over long intervals. Such repeated movements along faults are responsible for the formation of some impressive mountain fronts, as we noted in Chapter 3. The movements have been accompanied by erosion, so that the cliff produced directly by faulting has been greatly modified. The interplay of continuing fault movement and erosion is often strikingly shown by small fault scarps cutting alluvial fans at the base of some of the ranges in the Great Basin.

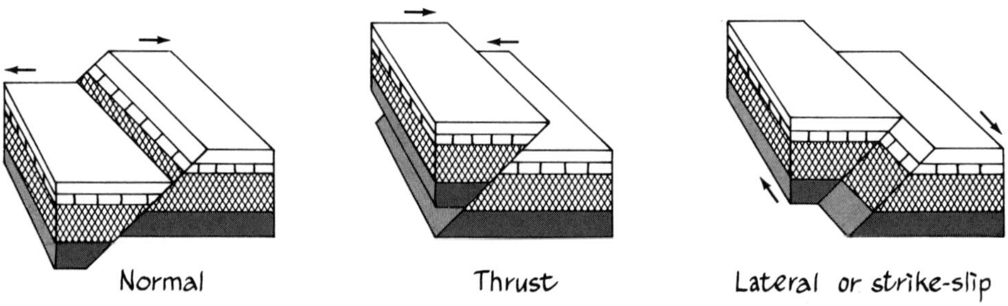

Fig. 10-11. Diagrams of normal, thrust, and strike-slip (lateral) faults. Movement to produce a normal fault involves lengthening of a part of the crust; thrust faulting involves compression; lateral faulting requires neither extension nor compression, but sidewise shearing.

Mountain fronts produced directly by faulting are such easy prey to erosion that they must disappear in geologically short times unless fault movements persist. The fact that such mountain fronts are common in Nevada and Utah and parts of adjacent states tells us that this must be a region of presently active faults, or at least of faults active during the last few million years (Fig. 3-17). Large faults of older vintage, which have been inactive for tens or hundreds of millions of years, may not have any obvious effect on the present topography because erosion has long since obliterated the mountains that they once produced. For example, thrust faults with displacements of many miles cut rocks of the Appalachians in eastern Tennessee and Kentucky, but their existence is not obvious from the present mountain landscapes. Movement on these faults ceased more than 200 million years ago, and they can be detected only by careful geologic mapping of a large area.

Faults may, of course, bring together rocks of differing hardness, and erosion working on the softer rocks may then leave the harder ones as conspicuous ridges or mountains. Thus an ancient fault may lie at the base of a present-day mountain range, but the existence of the range is an accident of erosion rather than a product of fault movement. For example, the spectacular Northern Rockies of Glacier National Park consist of Precambrian sedimentary strata that have been moved eastward along a gently dipping thrust fault over Mesozoic rocks that underlie the high plains. The fault movement ended some 60 million years ago, and the present mountains are in large part simply a result of the difference in resistance to erosion of the Precambrian and Mesozoic rocks.

The most common and obvious faults, either those visible in outcrop or those that can be inferred from the form of mountain ranges, are dip-slip

Fig. 10-12. Fault scarp cutting alluvial fans. The small cliff that can be traced across the two alluvial fans is the result of fault movement so recent that erosion has not yet destroyed the surface break. The displacement was evidently the result of uplift of the mountains with respect to the large valley into which the alluvial fans extend. The conspicuous horizontal lines in the middle of the picture are shorelines of a lake that filled the valley during the Pleistocene, when the climate was more humid. Slate Range, southeastern California. (G.I. Smith, U.S. Geological Survey)

Fig. 10-13. Chief Mountain, Glacier National Park, Montana. The mountain consists of Precambrian rock, lying on top of Mesozoic rocks which crop out in the wooded area in the foreground. The contact is a nearly horizontal plane of thrusting, along which the Precambrian rocks have moved many miles eastward. The mountain exists not because of the thrust faulting, but because the Precambrian rocks are more resistant to erosion than the Mesozoic rocks. (B. Willis, U.S. Geological Survey)

faults. Along some important faults, however, the movement is dominantly in the direction of the strike of the fault plane. The most famous example of a *strike-slip* fault is the San Andreas fault of California, a vertical break in the crust that can be followed for more than 500 miles down the west side of the state and into the Gulf of California. Movement on this fault is horizontal (along the strike of the vertical fault plane), the region west of the fault moving persistently northward with respect to the east side. Evidence for such movement is abundant: in many historic earthquakes, including the San Francisco quake of 1906, the displacement was in this direction; at some places along the fault slow horizontal movement is visibly displacing roads, walls, and fences without much accompanying earthquake activity; and older movement in the same direction is shown by displacement of streams and ridges that the great fault cuts (Fig. 10-1). Offhand one would expect this kind of movement to have little effect on the topography, but actually the San Andreas fault is marked over much of its length by a conspicuous valley. The valley is not a direct result of fault movement, but is carved out by streams cutting into the soft fragmental material (fault breccia and fault gouge) that has accumulated from many millions of years of grinding of rocks along the fault.

Fig. 10-14. A fence broken and offset by horizontal movement on the San Andreas fault at the time of the San Francisco earthquake in 1906. (G.K. Gilbert, U.S. Geological Survey)

To return for a moment to the question of forces, strike-slip movement poses a real puzzle. One region of the earth's crust moves past another region *horizontally:* this cannot result from compression (as do thrust faults) or from tension (as do normal faults), but must indicate a huge sidewise shearing force. Compressional forces and tensional forces do not seem particularly mysterious. With a bit of imagination, one can think of the earth's interior as somehow contracting or expanding to generate forces of this sort in rocks of the crust. But how account for horizontal shear on a large scale? What could be going on in the interior to push segments of the crust past one another continuously for periods of millions or tens of millions of years? This is one of the great conundrums of geology, and it is only very recently that a possible answer has come to light. We will look at this suggested answer in a later chapter.

Folds

If you drive through mountainous country where rocks are well exposed you will occasionally see outcrops in which sedimentary beds are bent into folds—sometimes broad folds with gentle slopes, sometimes tight folds with sides steeply dipping or vertical. Even where folds are not directly visible in outcrop, you can often detect their existence by constructing a geologic map: if you find a series of sedimentary strata all dipping south in one area, and then in another area to the north of the first you recognize the same sequence of beds dipping northward, a reasonable guess is that you are mapping the two sides of a large fold the top of which has been

Fig. 10-15. The valley along the San Andreas fault north of San Francisco, looking northwest. The fault zone here is a mile or more wide, and the valley is a result of erosion of the material broken and pulverized by millions of years of fault movement in different parts of this broad zone. The epicenter of the 1906 earthquake and the area of greatest horizontal displacement (7 meters) are in the middle ground of this picture. (R.E. Wallace, U.S. Geological Survey)

Fig. 10-16. Anticlines and synclines. (H.E. Gregory, U.S. Geological Survey)

eroded. In most places where beds show dips much different from horizontal, it is a safe assumption that some folding has taken place.

The two descriptive terms most commonly used in talking about folds are *anticline* and *syncline,* the first referring to an arch or up-fold and the second to a down-fold. In a region of folded strata, anticlines and synclines generally follow one another in long succession like waves at sea, the anticlines being analogous to the wave crests and synclines to the troughs. Where the beds in an anticline are folded in two dimensions rather than one, we speak of a *closed anticline* or *structural dome,* and correspondingly a syncline folded lengthwise as well as crosswise is a *structural basin.*

Active folds—anticlines and synclines that are demonstrably growing at the present time—are less common (or perhaps less easily detected) than active faults. Most of the folds we see are only the eroded remnants of structures that were formed long ago, bits of planetary architecture that have been stable for thousands or millions of years. Folded strata are common in mountainous regions, but this does not mean that folding was directly responsible for the mountains we see today. In a few places mountains are still being elevated by the slow growth of folds, but much

Fig. 10-17. Folds, like thrust faults, imply horizontal compression of a part of the earth's crust.

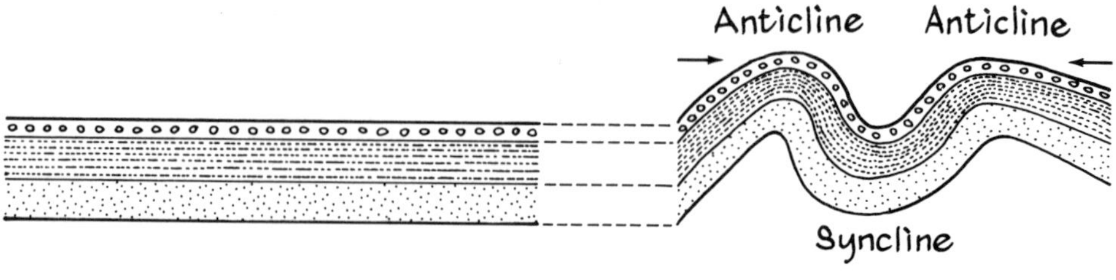

Fig. 10-18. Oblique air photo of a structural dome near Lander, Wyoming. The beds dip outward in all directions away from the center of the dome, and their eroded edges form roughly circular ridges. (John Shelton)

more commonly mountainous topography is developed simply by erosion wearing down folded strata of differing hardness.

Take, for example, the part of the Appalachian Mountains stretching from Pennsylvania to northern Alabama. The outstanding characteristic of these mountains, as we noted in a previous discussion, is the long, straight, parallel ridges, many of them conspicuously steeper on one side than the other. The ridges consist of hard sedimentary rocks—sandstone and limestone—tilted at steep angles, and the intervening valleys are underlain by softer layers of shale. Erosion of tilted beds, as we have noted, accounts nicely for the existence and shape of the ridges. Now we go a step further and identify particular beds by their fossils and their rock types, plotting on a map their positions and their dips and strikes. We find that the beds can be connected in a series of huge folds, the upper parts of which have been removed by erosion (Figs. 10-19, 3-2, and 3-23). Note that anticlines and synclines *as such* have nothing to do with the heights of the mountains: a

Fig. 10-19. Diagram of Appalachian structures. Oblique air photos of these structures are shown in Figs. 3-2 and 3-23.

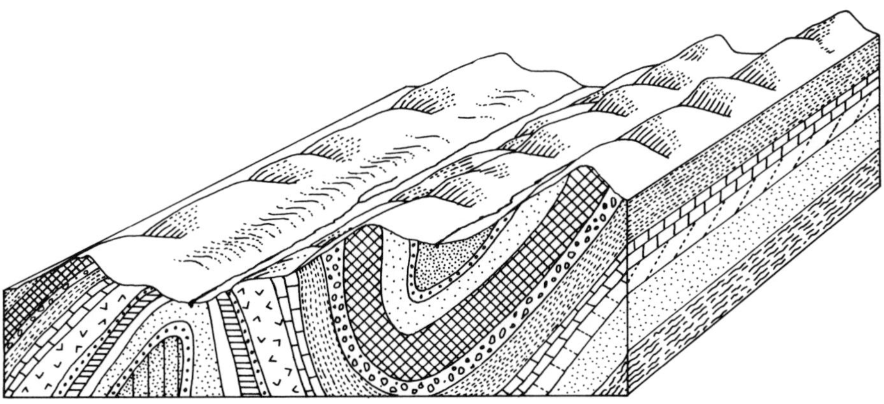

Fig. 10-20. Folding in bedded chert. (Mary Hill, California Division of Mines and Geology)

high ridge is just as likely to be developed near the trough of a syncline as near the crest of an anticline. Heights are determined only by relative hardnesses of different strata. In other words the folding here is very ancient (actually some 250 million years old), erosion has worn the folds down to their stumps, and the shapes of the mountains depend on how fast various rocks succumb to weathering and stream cutting.

The Appalachians are a particularly beautiful example, because here the old folds are fairly simple, regular, and on a large scale. But the same sort of structure—deep erosion of folded sedimentary rocks—is a common pattern in hilly and mountainous regions elsewhere.

What can folds tell us about forces operating in and under the crust? Clearly, any sort of folding requires that beds which were once horizontal now occupy a smaller area than they did originally, in other words that the beds have been compressed horizontally. Thus folding is a response to the same kind of force that produces thrust faults. This leads to a query as to why compressional forces should cause thrust faults in one area and folds in another. On the simplest sort of mechanical principles one might expect that the answer would depend on the brittleness of the rocks subjected to the force, on their depth of burial while the force was acting, and perhaps on their temperature and water content. In particular one could guess that a massive igneous rock like granite would be hard to fold and thus might yield by faulting, and that at the other extreme layers of shale would be easily deformed by folding. As confirmation of such guesses we often find combinations of folds and thrust faults in the field, a series of strata showing brittle beds offset by faulting and more pliable beds folded around the fractured ends. Predictions from mechanical properties of rocks are often remarkably successful, but many puzzles remain. It is hard to see, for example, how a very brittle rock like chert can be bent into the intricately contorted folds that one frequently finds in outcrops.

Parallel anticlines and synclines like those of the Appalachians are easily visualized and readily explained as a result of simple compressive forces—much like the crumpling of a stack of papers placed on a flat surface and pushed together from the two ends. In most areas of folded rocks, however, the pattern of folding is much more complicated. Different orientations of folding suggest the action of compressive forces in various directions, and some of the folding may be in response to vertical or oblique forces rather than horizontal. Sometimes even in a single outcrop one finds thin beds folded and broken in a pattern so complex that it looks like the result of stirring with a gigantic spoon. Distinguishing separate anticlines and synclines in such a mixture is hopeless, and relating structures to forces is possible only on a large regional scale.

As a single example of the possible complexities of structure in a region of strong folding, Fig. 10-21 shows an interpretation of the structure of part of the Alps. This mountain range, one of the most complicated on earth, has been the subject of long and intensive study by European geologists, but many questions about its structure remain unanswered. Note on Fig. 10-21 that the main elements of structure are huge folds and thrust faults; that deformation is so extreme that many of the folds have been pushed into a horizontal position with their sides nearly parallel; that the folding and thrust-faulting are in the same direction, as if the beds were first folded and then as deformation continued the folds were broken and pushed one on top of another. A huge complex fold thrust for a long distance over other deformed rocks is called a *nappe,* and Alpine structure is described as consisting of a series of piled-up nappes. Similar structures are recognized in other mountain ranges, including parts of the eastern Appalachians, but probably nowhere in the extreme form found in the Alps.

Ups and downs of crustal blocks

Faults and folds are the common, obvious, easily grasped elements of the earth's architecture. Yet clearly they are not sufficient for a complete picture of the way the crust is put together. To describe the crust entirely in terms of faults and folds is just as impossible as to describe the structure of a building in terms of the curvature of its walls and the angle between walls and roof.

To appreciate what is missing, think back for a moment about the geology of the Mississippi Valley, as pictured in previous chapters. This broad segment of the crust is underlain by sedimentary rocks in layers that are nearly as flat as when they were formed on the floor of a shallow sea. Faults of small displacement cut the layers in a few places, and careful mapping brings to light broad folds whose sides have very gentle dips, but certainly faults and folds play only a minor role in the structure of this huge area. The outstanding structural fact about the Mississippi Valley is just

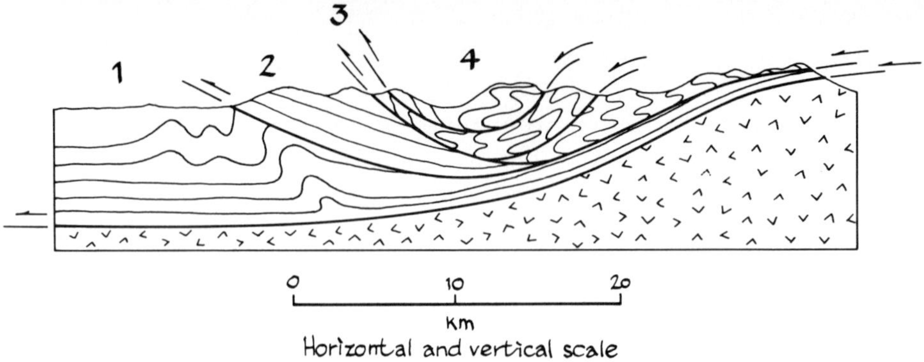

Fig. 10-21. An example of Alpine structures. (After Albert Heim)

that it has been raised, as a whole, a thousand feet or so above the level at which some of its beds were deposited. Such up-and-down movements of large crustal segments, much less obvious than faults and folds, are an important part of the architectural picture.

The broad Mississippi Valley is hard to discuss in detail just because of its size, so we look first at a smaller uplifted area represented by the Black Hills of South Dakota. We make another of our imaginary journeys, starting in the eastern part of the state and driving westward across the seemingly endless prairies until the dark hills take shape in front of us. Near Rapid City the topography and the geology change abruptly. The road goes through a gap in a steep north-south ridge, and we note that the ridge consists of sandstone in beds that dip steeply eastward in contrast to the monotonously flat-lying beds that we have seen in a few outcrops while crossing the plains. As we head into the hills from Rapid City we see more ridges parallel to the first and more east-dipping sedimentary layers. Then the geology changes, and we find ourselves driving through canyons cut in schist and granite. It is these rocks that make up the spectacular spires and sheer precipices in the higher part of the hills (Fig. 8-7).

If we continue driving west across the state line into Wyoming, we will find the same sequence of geology and scenery in reverse. Emerging from the granite-and-schist core of the hills, we traverse a series of ridges with sedimentary layers now dipping west, and then start across a sagebrush plain where most outcrops show layers roughly horizontal. If we go back into the hills and try roads leading north or south we find a similar succession, the sedimentary beds of the foothills everywhere seeming to dip away from the central core. As a whole, then, the Black Hills apparently consist of a dome of sedimentary layers arched over a core area of igneous and metamorphic rock, and eroded in its central part so that the core rocks are exposed.

Now what can we make of a geologic situation like this? One guess that comes to mind would focus attention on the granite: we know that this

Fig. 10-22. Diagram of the Black Hills uplift as if looking obliquely down on a relief model, cross-sections, and index map. Both cross-sections are somewhat simplified and idealized. The first (b) is a section into the hills from Rapid City west, and the second (c) is a smaller-scale cross-section through the entire Black Hills area southwest from Rapid City.

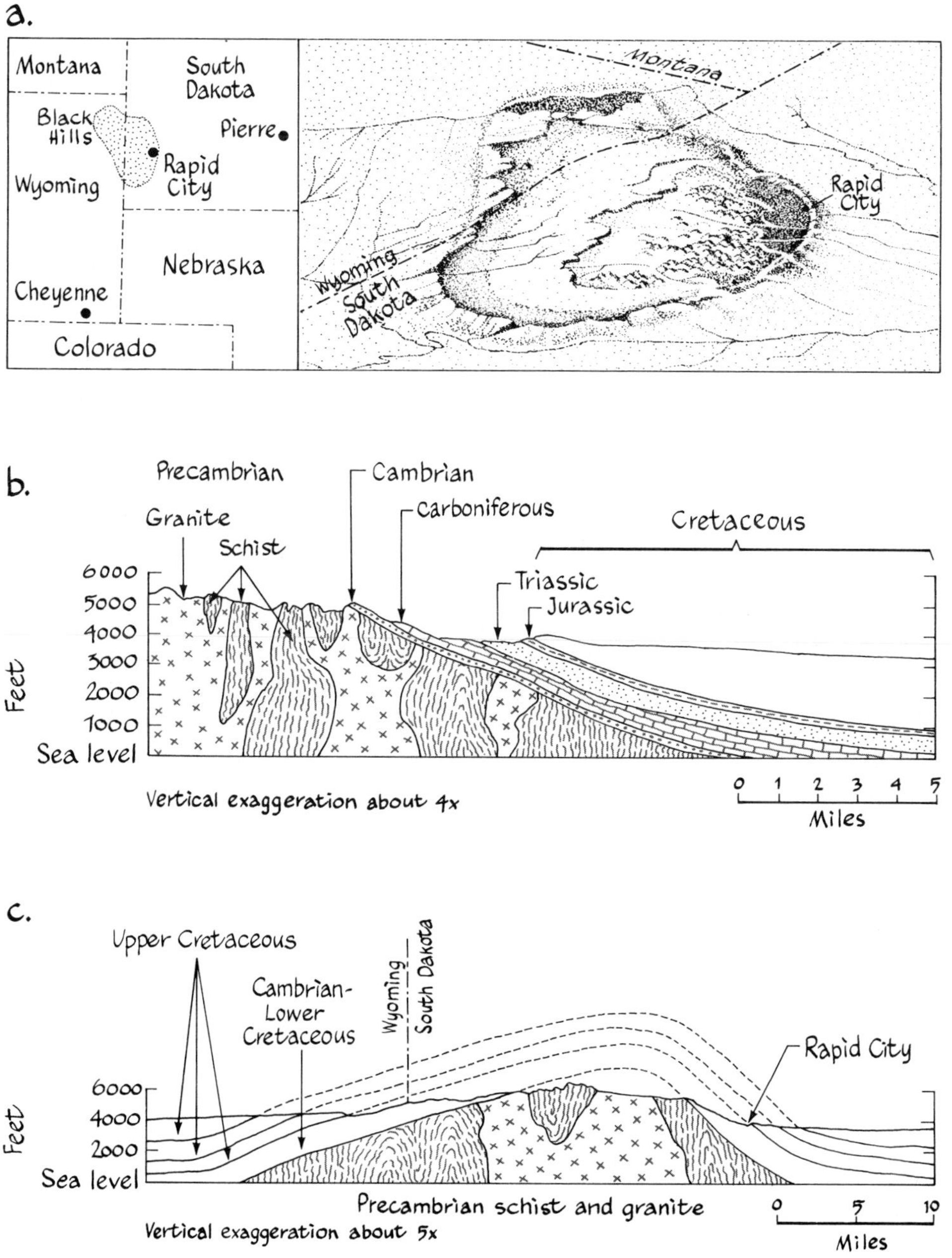

is an intrusive rock, so perhaps the granite has pushed its way up from somewhere below and raised into a dome the rocks that once covered it. Several observations speak against such an hypothesis. At contacts between sedimentary rocks and granite, we find no sign that the granite has intruded the sedimentary layers or metamorphosed them; rather, the contact is an unconformity with the layered rocks lying on an irregular, eroded granite surface. This evidence should be convincing, but if we want corroboration we can send a sample of the granite to a laboratory where its age will be determined from its content of radioactive elements. We would learn that the granite is Precambrian, far older than the sedimentary rocks which can be dated from their fossil content as Paleozoic and Mesozoic. Obviously a granite cannot have intruded rocks that are several hundred million years younger. The granite has invaded the schist of the central core, but it cannot be responsible for arching up the sedimentary cover.

So we try another guess. The overall structure is what we have called earlier a closed anticline or structural dome. Why not, then, regard this as an ordinary anticline, formed by horizontal compression of the Black Hills region? With this guess also there are several difficulties. For one thing, the Black Hills dome is isolated—there are no secondary folds adjacent to it, as one would expect from a regional compressive force. Also there are no minor folds of layers in the dome itself, no crumpling of softer strata such as usually accompanies large-scale folding. Then the shape is not right: this dome is broad and flat-topped, not a rounded arch like most anticlines that result from compression. Evidently our second hypothesis is no better than the first.

Perhaps we have been trying to make things more complicated than they actually are. A cross-section through the hills (Fig. 10-22) looks as if a portion of the solid crust has simply been pushed upward, bending (and locally breaking) the sedimentary rocks that once lay flat across it. Why not just accept this obvious explanation? A reason for hesitation is the fact that the mechanics of movement are unexplained: it is hard to imagine what might be going on in or under the crust that would cause a large segment to rise vertically. But then we reflect that causes for *any* kind of movement in the solid crust remain mysterious. Uplift of crustal blocks is really no more mysterious than other kinds of movement, so our best move is just to accept the evidence at face value and keep in mind that we will be seeking a general explanation for crustal movements later on.

An important feature of crustal architecture, then, is large segments that have changed their elevation with respect to adjacent segments. The Black Hills are a particularly good example, and several other ranges farther west in Wyoming are almost equally clear. But the concept of vertical uplift is by no means limited to small isolated ranges with a fairly simple geologic history. The Appalachian Mountains, as we have seen earlier, are a region of folded strata that have been deeply eroded; for the erosion to take place requires that the region be well above sea level, so we infer that the whole upland area stretching from Alabama into New England and on to Newfoundland has moved vertically upward. Details of the mountain topography, in fact, suggest that the upward movement was not limited to a single episode but has recurred during a long part of geologic history. On a still larger scale the entire central part of the continent has been raised enough to permit active erosion of its ancient marine sedimentary cover.

What happens at the edges of an upward-moving block? No single answer can be given, because vertical movement takes place on many scales and involves many kinds of rock. Around the Black Hills the edges of the block are well defined, and movement has resulted in abrupt bending and local breaking of sedimentary layers. Some of the desert ranges of Nevada have steep normal faults on each side, hence represent blocks that have been raised along clean breaks with their surroundings. On the other hand the upward movement of the Appalachian region or the Mississippi Valley has no well defined edges; one can most reasonably picture a broad warping or gentle arching with no precise limits.

Vertical movement of crustal blocks is probably as common downward as upward. The simple fact that some of the layers of marine sedimentary rocks in the Mississippi Valley are separated by unconformities means that this broad area has not only been elevated at various times to permit erosion, but then has subsided to permit new accumulations of marine sediment. The existence of long estuaries along the Atlantic Coast (Fig. 3-21), as we have noted earlier, tells us that this region has sunk with respect to sea level in fairly recent geologic time. Very large-scale subsidence of the Gulf of Mexico and parts of the adjacent states is demonstrated by the enormous thicknesses of shallow-water sediments of Cenozoic age that have been penetrated by drillholes in the search for oil and gas. On a smaller scale, many of the broad valleys of the Great Basin, with Death Valley as the most famous example (Figs. 3-7 and 3-11), have subsided along normal faults that can be traced along mountain fronts on either side. The boundaries of subsiding crustal segments, like the boundaries of uplifted blocks, may be clearly marked by faults or steeply tilted beds, or they may lack definition because the downward movement is a gentle warp on an enormous scale.

Thus blocks of solid rock that have moved up or down for distances sometimes as much as several kilometers are major parts of crustal architecture. The movement involved, let us emphasize once more, is exceedingly slow on a human time-scale. Probably it is not accompanied by earthquakes, except for movement of small blocks that are bounded by faults. Abundant evidence indicates that vertical movement of crustal segments has gone on all through geologic history and is continuing at present, although it is so slow and so generalized in its effects that it plays a less prominent role in most tectonic discussions than the more obvious motions recorded by faults and folds.

Continental patterns

We have been talking about the kinds of structures that make up the architecture of the earth's crust, and we have speculated about movements and forces that might have created these structures. To find examples we have jumped from place to place, inspecting areas that illustrate different structures particularly well. If we now take a broader point of view and look at the North American continent as a whole, what sort of overall pattern do its structures form?

On a map of the United States, two mountainous areas are prominent: the Appalachians in the east, and the thousand-mile-wide belt of mountainous country stretching westward from Colorado to the Pacific Coast. To the north the Appalachians extend into the Maritime Provinces of Canada, and the western mountains project through British Columbia and the Yukon into Alaska; to the south the western belt becomes the high central plateau and bordering ranges of Mexico. Between the mountains is the great interior lowland, stretching from the Gulf of Mexico northward to Hudson Bay and the Arctic Ocean. This broad geographic picture leaves out many details, for example the Black Hills and the Ozark Mountains which rise out of the central plain, but it gives the general framework for which we need a structural explanation.

The most casual look at rock outcrops in the mountainous areas and the plains will convince you of one essential difference. In the mountains rock layers show conspicuous folds and faults, whereas strata under the plains are in large part flat-lying. Furthermore, most of the folds in the major mountain ranges have axes lined up roughly parallel to the lengths of the ranges, and many of the conspicuous faults—both thrust faults and normal faults—have strikes in the same direction. The folds and the thrust faults suggest that rocks of the mountains have moved under the influence of compressional forces, in other words that major mountain ranges resemble huge wrinkles in the crust produced by forces acting at right angles to their elongation. Other characteristics of mountainous areas are the presence at many places of intrusive igneous rocks, especially granite in large masses, and the frequent occurrence of metamorphic rocks—characteristics we interpret to mean that rocks of mountains have been deeply

Fig. 10-23. Major structural elements of North America.

buried, heated, and invaded by molten material. Most of the plains area between the Appalachians and the Rockies, on the other hand, is underlain by sedimentary rocks which have remained practically undisturbed since they were formed, except for gradual up-and-down movements on a continental scale.

This summary of structure is greatly oversimplified. Tensional forces as well as compressional forces have helped to shape parts of some mountainous areas; we have mentioned in several previous discussions, for example, the numerous large normal faults in the Great Basin between the Rockies and the Sierra Nevada. Although sedimentary rocks underlie most of the Mississippi Valley and the Great Plains of the United States, the flat interior of central Canada has huge expanses of igneous and metamorphic rock. Many such exceptions to the general rules could be pointed out. But despite the oversimplification, the usual contrast between rock structures underlying mountains and plains is a fundamental aspect of crustal architecture. To repeat: major mountain ranges are elongated belts where rock layers have been folded and faulted by sidewise compression, deeply buried, and invaded by magma; extensive plains are areas which for much of geologic time have undergone little deformation but may have changed their elevation with respect to sea level.

One can extend this line of argument by noting that mountains may have different ages, depending on the time at which their rocks were deformed. After the folding and faulting have taken place, a mountain area may be elevated by broad crustal movement and subject to deep erosion, as has happened in the Appalachian region. These mountains are old, in the sense that folding of the sedimentary beds took place in the latter part of the Paleozoic Era; what we see today is simply the stumps of the old folds, raised up repeatedly and worn down by some 300 million years of erosion. Mountains along the Pacific Coast, by contrast, have steeper slopes and ridges of soft sedimentary rock hardly touched by erosion; these are young mountains, still in process of being folded and faulted—as the frequent earthquakes eloquently testify. The igneous and metamorphic rocks of central Canada mentioned above, including gneiss and schist intricately folded and intruded by granite, represent the roots of mountains far older even than the Appalachians, so old that erosion planed them down long ago to a nearly level surface on which later sedimentary strata were deposited. Thus mountains can be recognized by their structures, both mountains that still are elevated above their surroundings and mountains from a distant past that now form the foundation of a level plain.

This discussion, it should be emphasized, is concerned with *major* mountain ranges—like the Appalachians, Rockies, or Sierra Nevada in this country, or the Andes, Alps, and Himalaya abroad. Many less extensive elevations are commonly called "mountains"—a volcano like Mt. Rainier, for example, or an uplifted fault-block like the Ruby Mountains of Nevada. These smaller mountains may be impressive and geologically interesting, but it is the great ranges that are fundamental parts of continental patterns. The large-scale folding and faulting recorded in a major range constitute an important episode in the shaping of crustal architecture, and any attempt to explain how our planet functions must assign a basic role to the making of mountains.

What makes things move?

Faults, folds, uplifted and down-dropped crustal blocks, all are mute evidence of motion in the solid rocks of the crust. The geometry of the structures often helps us to see the direction of relative movement, and it is a fascinating game to work out patterns of movement on both large and small scales. Movement requires the existence of forces, so we try to interpret the patterns we see in terms of forces supposedly operating in or under the crust. We speak glibly of "compressional" forces, "tensional" forces, "shear" forces, "vertical" forces. We know that the forces are unimaginably large, because rocks are strong and the masses moved are sometimes continental in scale. Yet if a skeptic asks the obvious question, "What generates these huge forces?", we find ourselves forced back against yet another of the great geologic unknowns.

We can make some feeble guesses. The earth's interior is hot; perhaps the temperature changes under large regions, causing material to expand or contract, and the crust is stretched or crumpled in trying to accommodate itself to this change. Not a very attractive hypothesis, because the forces so generated hardly seem sufficient for the observed displacements, and because such forces provide no explanation for the observed patterns of mountain ranges. Tidal forces produced by the sun and moon are known to deform solid rock as well as the ocean; perhaps changes in the tidal pull over time could cause net movement of the crust. Again, continents should experience more centrifugal force than ocean basins because they are farther from the earth's center; hence they should be pushed toward the equator, perhaps crumpling the crust as they move. Yet these forces too seem insufficient, and difficult to reconcile with structural patterns. We know that larger forces exist in the crust because we see their effects, but their origin remains an enigma.

The situation is not quite this bleak, because a new hypothesis has come to light recently that seems to offer more promise than these earlier ones. But the new ideas cannot be profitably introduced until we have looked at some further properties of our planet.

Fig. 11-1. A mass spectrometer. This instrument is used to determine abundances of isotopes. Atoms of an element are given an electric charge (ionized) and caused to move through a strong magnetic field; in the field they are deflected from a straight-line path, the amount of deflection depending on their masses; hence isotopes of different masses are separated, and their amounts can be measured. Abundances of isotopes of radioactive elements and their daughter products provide the information needed for calculating geologic ages in years. (P.M. Hurley, Massachusetts Institute of Technology)

11

Dating

Back in Chapter 6 we had our first brush with the problem of finding dates for geologic events. There we concentrated on the method that geologists employ in everyday field work: the search for fossils in sedimentary rocks, and then the careful examination of the fossils to see where they fit in a recognized worldwide sequence that has been worked out during two centuries of geologic study. The fossil sequence represents stages in the development of living creatures over time, so that rocks can be described as "early" or "late" depending on the kind of fossils they contain. Details of organic development are so well known, in fact, that we can divide geologic time into intervals—the major intervals called eras, the shorter intervals called periods, and many finer subdivisions (table, pages 168–169). We date a sedimentary rock, then, by assigning it on the basis of fossils to the "early Paleozoic *Era*," or to the "Triassic *Period*," or to the "middle part of the Miocene *Epoch*."

This is a simple and straightforward method of dating, but it has obvious limitations: it applies only to sedimentary rocks, and it does not give ages in years. Igneous rocks and most metamorphic rocks are devoid of fossils, and even among sedimentary rocks there are many in which fossils are absent or so poorly preserved as to be useless for dating. Besides dating rocks, we would like also to be able to date geologic events—the time or times when glaciers formed in eastern Canada, the time when rocks of the Appalachian Mountains were folded, the time when strike-slip movement began along the San Andreas fault. And of course it would be much more satisfying if we could go beyond the relative ages provided by fossils to an estimate of the actual number of years that have elapsed since a rock was formed or a geologic event occurred.

So we have two problems to investigate: (1) how can we use fossils to date non-fossiliferous rocks and tectonic events, and (2) how can we estimate ages of rocks and events on an absolute scale of years?

Dating non-fossiliferous rocks and geologic happenings

The answer to the first question is simple. It is embodied in the word "interpolation," which means that we look for fossils in sedimentary rocks that are definitely older and definitely younger than the rock or event in question, and then say that the age we want is somewhere between these two known ages. In Fig. 11-2, for example, the granite dike must be younger than the lower sedimentary beds because it intrudes them, and it must be older than the beds above because they lie on its eroded surface. If the youngest of the lower beds can be dated by its fossils as belonging to the Devonian Period, and the oldest of the upper beds contains fossils of the Triassic Period, the intrusion of the granite must have happened at some time between the Devonian and the Triassic. This is not a very precise age, because the time between Devonian and Triassic represents at least half of the Paleozoic Era, but it is the best information that fossils can give us. For another example, Fig. 11-3 shows a fault that displaces rocks containing early Jurassic fossils, and that does not displace the overlying rocks which contain late Jurassic fossils. Movement on the fault must have occurred, then, sometime during the Jurassic Period. Here the age assignment can be much more precise, because the date is bracketed between two parts of the fossil sequence that are close together in time.

For a more complicated age problem, we go back to a question from a previous chapter: how old is a mountain range? The question is difficult because the expression "age of a mountain range" is not well defined. We recall that present-day mountains are largely the handiwork of stream erosion (some of them modified in their higher parts by the carving of glaciers), and that for erosion to take place a mountain area must be

Fig. 11-2. A granite dike is younger than the beds it intrudes, and older than the beds that lie unconformably above it.

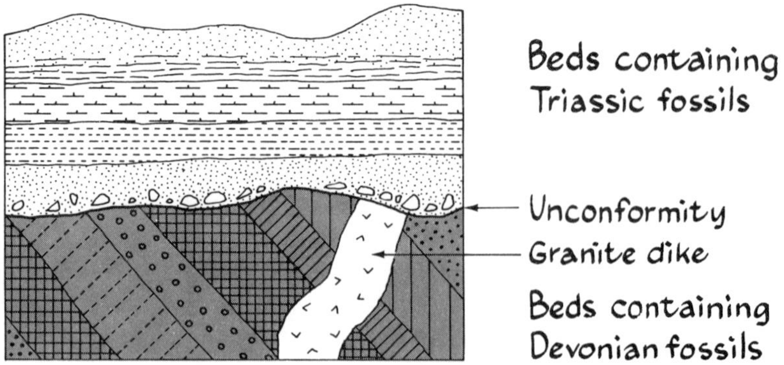

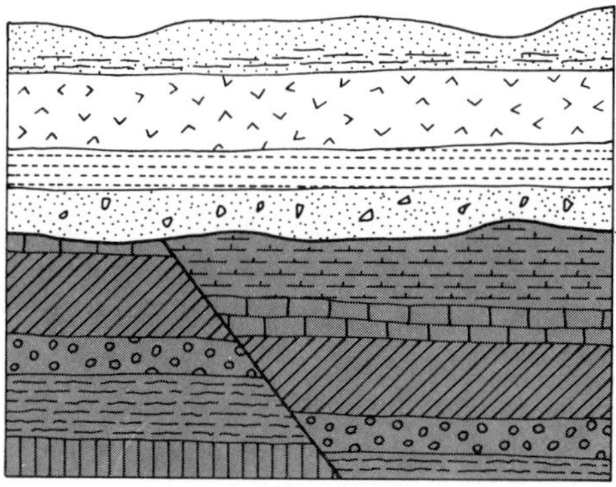

Fig. 11-3. A fault is younger than the beds it displaces, and older than beds above it which are not displaced.

elevated above its surroundings. One possible "date" for a mountain range, then, is the time when the elevation occurred. This date can be roughly fixed by interpolation, if we look for (1) the youngest fossil-containing beds in the mountain structure and (2) the oldest fossils in sediments near the range that contain pebbles and boulders derived from rocks in the mountains. The uplifting of a mountain mass is commonly slow, however, so that the "time of formation" so estimated may turn out to be a very long period during which uplift and erosion went on simultaneously.

Another meaning for the time of origin of a mountain range goes back to its inner structure. The earth's large mountain chains, we have seen, have a complex architecture consisting of folded and faulted sedimentary rocks invaded by igneous material. An important date in the history of a range, therefore, is the time at which the deformation of strata and intrusion of granite took place. Again, the date can be bracketed by means of fossils: the tectonic movement must postdate the youngest deformed fossiliferous bed, and predate the oldest bed that escaped the deformation. This date may also be somewhat indefinite, because the tectonic movement may have continued for a long period. But in many ranges the distinction between the last of the deformation and the almost undisturbed overlying beds is marked by a well defined and well dated unconformity.

It is this second meaning that a geologist usually has in mind when he talks about "young" or "old" mountains. This often seems confusing to a layman, who is quite naturally more interested in the age of the mountain he sees in front of him than in the date of an obscure tectonic event sometime in the distant past. The original deformation doubtless produced a mountain range, but if the date of the event is more than a few million years ago the primordial range probably bore little resemblance to the present one. The range we see today is the result of long erosion acting on an area of the earth's surface that has been raised intermittently to various heights during the time after its rocks were deformed.

As an example, we look once more at the Appalachians. Here the thick folded sedimentary strata contain fossils that place them in the Paleozoic Era. The amount of deformation is less in the uppermost Paleozoic beds than in the lower ones, suggesting that folding started well before the end of the era but continued to its end. In parts of the eastern states adjacent to the mountains, early Mesozoic sandstones and shales appear in beds that are somewhat tilted but not folded. We conclude that the major Appalachian deformation occurred between late Paleozoic and early Mesozoic, and we call the mountains "old" because this time of deformation is so far back in geologic history. After it was deformed by folding, the entire Appalachian region has been subject to recurrent elevation, and erosion has worn down the mountains to low hills between the times of uplift. The long ridges of the present-day range are the result of deep erosion following the latest rise.

Or take the Black Hills of South Dakota (Fig. 10-22). Here the deformation is more recent, because the youngest of the arched-up beds contain late Mesozoic fossils, and we therefore call the Hills a "younger" range than the Appalachians. But the present landscapes of the Black Hills have been carved by deep erosion after later uplift of the entire region, and probably have little resemblance to the original hills that took shape when erosion started work on the newly deformed rocks at the beginning of the Cenozoic Era.

Fig. 11-4. A cross-section through the Appalachian Mountains of Pennsylvania, idealized and greatly simplified. Strata from Cambrian through Permian are folded, and Triassic beds lie unconformably on the eroded folds. The "age" of the mountain range is commonly taken as the age of the folding, which must be late Paleozoic or earliest Mesozoic. The present-day mountains are the result of much later uplift and erosion of the stumps of the old folds.

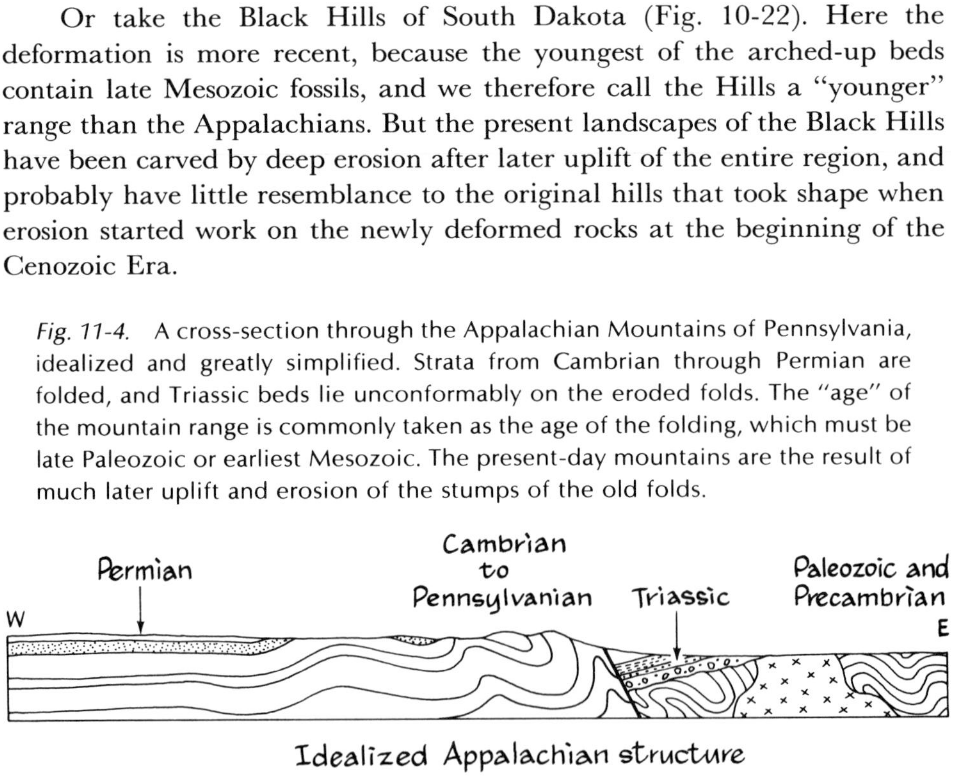

Thus folds and faults and igneous intrusions can be roughly dated by interpolation between ages given by fossiliferous strata. The procedure is basically simple, hardly more than applying everyday horse-sense. It is more complicated when used to date an entire mountain range, but the complication arises chiefly from haziness in defining what the "date of a mountain range" means. At best, such dates have the same disadvantage as other dates derived from fossils: they are only *relative,* telling us what rock or what event comes after another, but giving no idea about absolute ages in years.

So we turn now to the second of the two questions raised at the beginning of the chapter, the problem of assigning numbers to geologic dates.

The decay of elements

Absolute dating, as we have noted several times previously, depends on radioactivity. This is a phenomenon with an aura of strangeness about it, because in ordinary life radioactivity is not a part of day-to-day experience, and because we know vaguely that it is somehow related to the awesome release of energy in atomic bombs and nuclear reactors. We start our discussion of dating with a few background details about radioactivity that should help to make it less mysterious.

If you wear a watch with a luminous dial, you can get a vivid demonstration of radioactivity very simply by taking the watch into a dark closet, waiting until your eyes are thoroughly accustomed to the darkness, and looking at the dial with your handlens. Instead of a uniform glow you now see myriads of tiny flashes, like the sparks from a rocket, each lasting only a moment. The paint on the dial contains a compound of a radioactive element together with a material that fluoresces (glows) when struck by certain kinds of radiation. The element is producing radiation, evidently in tiny pulses, and each of the soft splashes you see is the response of the fluorescent material to a single pulse.

Rather than a single radioactive element, the luminous paint generally contains a mixture of several. Two that are commonly present are the heavy metals uranium and thorium. In the pure state, or in the form of pure compounds, these would not be satisfactory for luminous paint because their radioactivity is too feeble, but they are generally accompanied by small amounts of more active elements.

Uranium and thorium are the two most abundant strongly radioactive elements in nature, and many other naturally occurring radioactive elements are found in ores of these two. The search for radioactive material, to be used for luminous paint or for any other purpose, is thus essentially a search for uranium or thorium. The occurrences and methods of handling uranium ores are better known than for thorium. If you set out to look for uranium minerals, the ones you would most likely find are the canary-yellow mineral *carnotite* (a compound of uranium, vanadium, and potassium), common in certain sandstones in Utah and Wyoming, and the black mineral *pitchblende* (uranium oxide), which is found in these same sandstones and also in some quartz veins and coarse-grained granitic dikes. Several of the rare uranium minerals are prized collectors' items because of their bright colors.

Uranium itself, if you take the trouble to purify it, is a metal that resembles steel in appearance but is considerably heavier. No ordinary observation of a piece of uranium would make you suspect that there was anything strange about it, but the fact that it emits radiation would become evident if you left it for a few days on a piece of photographic film (it darkens the film—i.e., it takes a picture of itself) or if you brought it near a Geiger counter (an instrument for detecting the presence and amount of radiation). If you become interested in the character of the radiation, you could leave the uranium in a sealed box and sometime later analyze the surrounding gas; with a sufficiently sensitive analytical method you could detect the presence of the light gas helium, which was not present at the start. Thus a part of the radiation from uranium consists of atoms of helium. (There are other kinds of radiation also, but it is the helium that concerns us most directly here.)

Now this conclusion, if you think about it, is very odd. The original metal contains nothing but atoms of uranium, yet from these atoms are produced atoms of another element. Ordinarily we think of elements as fixed, immutable, impossible to convert into others; yet here before our eyes uranium atoms are decaying. The decay is extremely slow, only a very tiny fraction of the atoms present changing in any given time. But the decay is inexorable: day after day, year after year, uranium atoms disappear and helium atoms are formed. We might think back to the luminous paint on the watch dial, and guess that each light flash represents the decay of a single radioactive atom, the helium atom released (actually a helium *ion* He^{++}) serving as a projectile to activate the fluorescent material momentarily. As we watch the flashes in the paint we are witnessing the transmutation of an element, the spontaneous conversion of one kind of atom into others. We do not see the individual atoms, of course, but we do

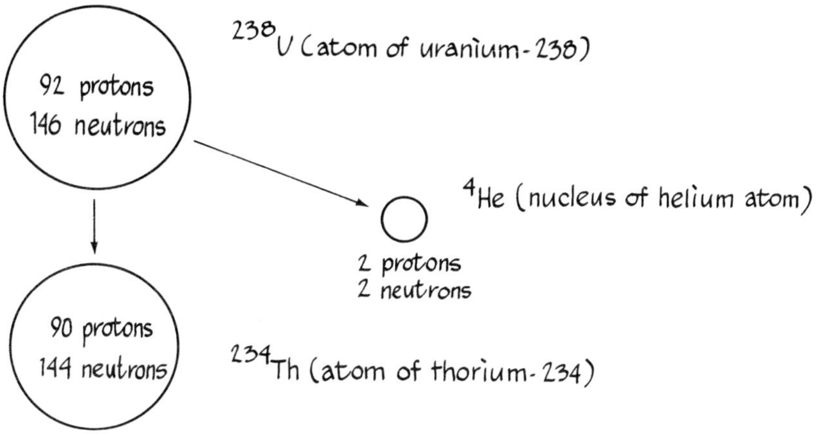

Fig. 11-5. The radioactive decay of a uranium atom, producing helium and an isotope of thorium. The thorium atom decays further to form atoms of other radioactive elements, ultimately becoming an atom of nonradioactive lead.

see the disturbance that the decay of each atom produces. In this very simple observation we have a striking demonstration of two important facts, that atoms exist and that some kinds of atoms are inherently unstable.

Further study of the radioactive decay of uranium turns up two characteristics that have far-reaching significance in geology. For one thing, the decay produces energy. This is evident just in the glow of your watch dial, because light is one form of energy. Most of the energy, however, appears in the form of heat, and the light energy also goes into this form as the flashes die out. Now uranium minerals themselves are rare, but uranium atoms occur in minute amounts in many of the minerals of ordinary rocks. This means that a source of heat is widely disseminated in rocks of the crust—a very feeble source of heat, to be sure, but possibly important when we think of the long reaches of geologic time. Rocks are excellent insulators, and any heat produced by radioactivity under the earth's surface must simply accumulate. Over periods of millions of years this storing up of tiny increments of heat energy may have profound consequences for the behavior of rocks in the crust.

Heat from radioactivity will call for attention later, but with reference to dating a second characteristic of uranium decay is more pertinent: the production of helium goes on at a perfectly constant rate. More helium forms per second from a big chunk of uranium than from a small one, of course, but the amount generated per gram is always the same. Another way of saying this is that the fraction of the total number of atoms which decay in a given time remains constant. Do what you will with the uranium—cool it to the temperature of liquid air, heat it in an electric arc, put it in chemical compounds—still the uranium atoms go on decaying at the same fixed rate. You cannot stop it, you cannot speed it up. The decay rate is simply a property of the element, and nothing you can do will change it.

Consider what this means. If uranium is put in a gas-tight box, the amount of helium generated in a given time depends only on the amount of the metal present. If someone gives you such a box and asks at what date the uranium was put in, you need only measure the weight of uranium and the amount of helium in the gas and you can tell him the answer directly—regardless of what the history of the box may be. To translate this into geologic terms, if you find a rock that contains uranium you need only analyze the rock for its uranium content and its helium content, and you can tell immediately how many years ago or how many millennia ago the uranium was trapped in the rock. The rock may have been deeply buried and heated since its formation, it may have been scraped by glacial ice, it

Fig. 11-6. Diagram to show the rate of decay of any radioactive element. If N atoms are originally present, after time T only 1/2 N will be left, and after another time T only 1/4 N (the heavy solid curve). For a very active element the time T will be short (for some elements only a fraction of a second), and for a less active element it may be very long (for uranium-238, for example, T is 4.5 billion years). Thus T is a convenient measure of the degree of activity; it is called the *half-life* of the radioactive isotope. As the radioactive element decays, the number of atoms of its daughter element increases (the light dotted line).

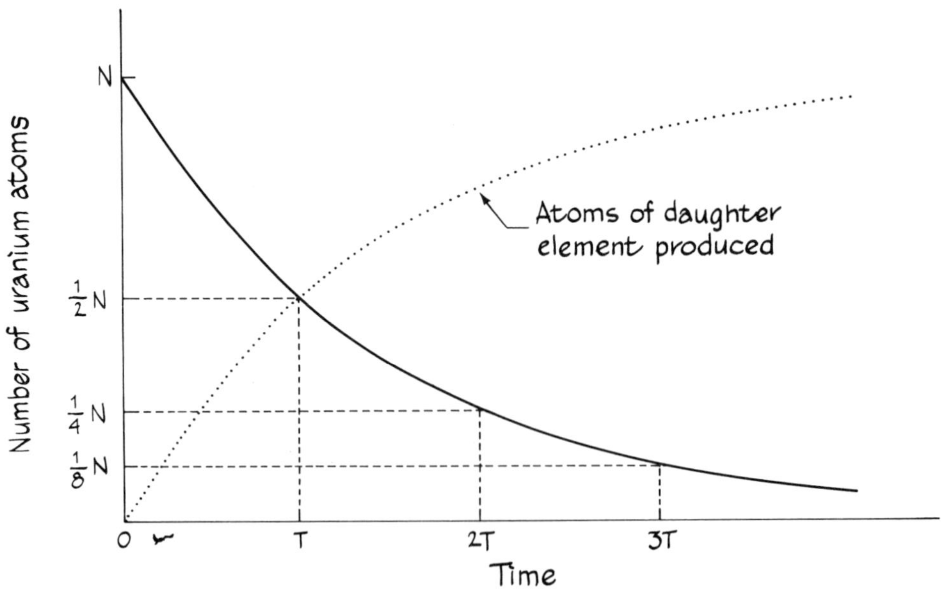

may have been folded and faulted—all this matters not a bit. Regardless of geologic accidents the uranium will have gone on blithely decaying and forming helium, and the ratio of the two elements gives a measure of the age of the rock. This is the secret of using radioactivity for the absolute dating of past events.

Dating difficulties

How nice it would be if the method were in fact so beautifully simple! One drawback to the procedure just outlined has probably occurred to you: helium is a gas, and how can we be sure that the helium produced from the uranium in a rock has not leaked out? In other words, can we really consider a uranium mineral to be analogous to a gas-tight box? There is no good way to answer this question except to try and see, which means to analyze various minerals for the two elements. The results are erratic, as we might expect if part of the helium has escaped. Some minerals retain surprisingly large amounts, but the uncertainty is so great that this method cannot be depended on for more than a rough estimate of minimum age for a uranium-bearing mineral.

We go back to consider the process of decay in more detail. What is left over when a uranium atom decays and a helium ion escapes? The answer is a third kind of atom, an atom with atomic weight four units less than that of uranium (since the atomic weight of helium is four). This atom, it turns out, is also radioactive, decaying much more rapidly than uranium. A whole series of radioactive atoms results, each decaying to the next in line at a fixed rate but all faster than the parent uranium. The end product is an atom of lead, which is not radioactive. In effect, then, uranium decays ultimately into two substances, lead and helium. Any mineral of uranium ought to contain some lead as well as helium, and the amount of lead should be larger in old samples than in young ones. In addition to lead, a uranium mineral must contain also the intermediate elements of the decay series, but the rates of decay are so great that their proportions are always minor and remain approximately constant since these elements are formed as fast as they decay. Thus lead and helium are the only decay products that increase steadily in amount, and the ratio of lead to uranium should serve as well as the ratio of helium to uranium in measuring ages. It should, in fact, serve much better, because lead is a solid and hence would not escape from the uranium-bearing mineral.

Can we then heave a sigh of relief and consider the time-measuring problem solved? Not quite. It seems to be a rule in geology that solutions to problems are never quite complete. One finds a promising way to make a measurement, and then in trying to improve its accuracy one meets unsuspected difficulties. When these are overcome, and further refinement is attempted, more troubles come to light. A simple measurement of the lead-uranium ratio in some minerals containing abundant uranium can give reliable figures for geologic age, but for other minerals, especially those with low contents of uranium, results may be erratic.

One difficulty is the fact that lead is a fairly common element, so that a mineral containing uranium may also have included some lead at the time of its formation. Analysis would give the sum of original lead plus the lead from radioactive decay, and the calculated age of the mineral would thus turn out to be greater than its true value. Is there any way around this obstacle? The answer is yes, but it depends on a still closer look at the process of decay.

To say that uranium atoms decay to form lead is true, but it glosses over an important complication. There are two kinds of uranium atoms which decay, and two kinds of lead produced from them. The two kinds of uranium, called *isotopes* of the element, differ in the weights of their atoms. One kind of atom is roughly 238 times as heavy as the hydrogen atom, and is symbolized ^{238}U; the other, ^{235}U, has a relative mass of 235. An atom of ^{238}U breaks down through a series of steps to form an isotope of lead called ^{206}Pb (note that this means a total loss of 8 He atoms, each with a weight of 4), and ^{235}U decays to form ^{207}Pb (with a loss of 7 He atoms). Lead has two other isotopes, one with a mass (on the atomic weight scale) of 208, formed by the radioactive decay of thorium, and finally ^{204}Pb, which is not the product of radioactive decay. Naturally occurring lead is a mixture of the four isotopes. In most samples of lead the proportions of the isotopes are nearly constant, but if much of the lead in a sample has been produced by radioactive decay the amounts of ^{206}Pb, ^{207}Pb, and ^{208}Pb will be abnormally large in relation to the non-radiogenic isotope ^{204}Pb. Thus measurement of the isotopic ratios as well as the total amount of lead makes it possible to estimate the fraction of the lead in a uranium mineral that has been produced by radioactive decay, hence to get a reliable figure for the number of years since the uranium was imprisoned in the mineral.

Analysis for isotopes has the further advantage that several different figures can be obtained for the age of a single sample, thus providing an internal check on the accuracy of the analysis and on the geologic assumptions. One can measure separately the ratios $^{238}U/^{206}Pb$ and $^{235}U/^{207}Pb$; if the mineral contains thorium, as many uranium minerals do, the ratio $^{232}Th/^{208}Pb$ can also be obtained. If all three give the same age, the result can be accepted with more confidence than an age calculated from a single ratio.

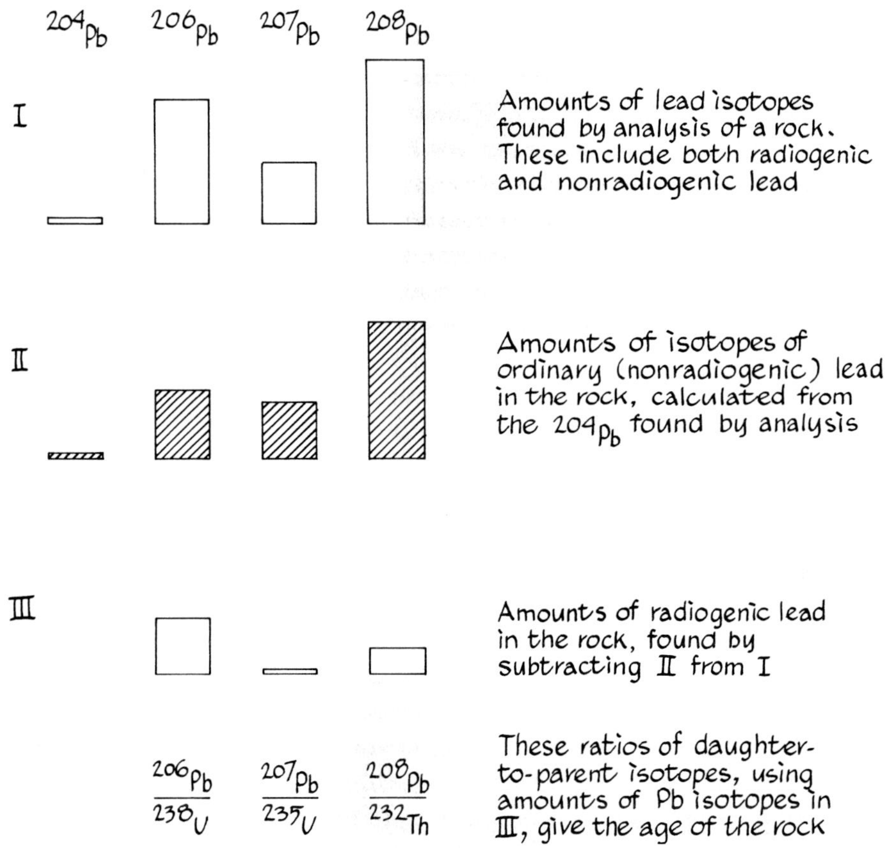

Fig. 11-7. The lead in a uranium mineral is the sum of lead originally incorporated in the mineral and the lead produced by radioactive decay. The amount of original lead can be estimated from the concentration of ^{204}Pb, and this amount is then subtracted from the total to give the amount due to radioactivity.

Is there any reason why three such ratios might *not* give the same age? Think for a moment of the hidden assumptions that lie behind this method of age measurement. We find a uranium mineral, and we ask how old it is. By this question we mean the date at which the mineral was deposited from solution or crystallized from a melt. We think of the uranium, and perhaps thorium also, as being freed of all contamination as they settled down into the chemical compound that makes up the mineral. If they are accompanied by a little ordinary lead it will make no difference, because we can detect this lead by isotopic analysis, but can we be sure that they are not accompanied also by some of the radiogenic lead from their own previous decay?

Then we assume that the mineral has remained inert and unaltered for geologic ages, from the time of its formation to the day when we bring it into the laboratory. Is this reasonable? Remember that most rocks are at least somewhat permeable to solutions, so we cannot be altogether sure that the mineral has remained free from chemical attack. Reaction with solutions would not affect isotope proportions for the individual elements, because all the isotopes of any one element have practically identical chemical behavior; but the chemical properties of lead, uranium, and thorium are different enough so that one of the three might have been leached more than the others. It is possible also that circulating solutions might have added more of one of these metals to the amount present in the original mineral. Thus our assumptions about the pristine purity and later stability of uranium minerals are shaky enough so that we cannot be surprised if age determinations are sometimes anomalous.

Despite these theoretical possibilities, and despite occasional anomalous measurements, geologic dates determined from lead-uranium and lead-thorium ratios are sufficiently consistent, both among themselves and when checked against other methods, to give us great confidence in their accuracy.

Other ways of getting dates

To find the age of a rock or mineral from its content of uranium and lead requires that enough uranium be present so the amount can be determined very precisely. Unfortunately uranium is a rare element, and

Fig. 11-8. The radioactive decay of potassium. The isotope ^{40}K has two ways of changing to another element: the nuclei of its atoms can lose electrons to form atoms of ^{40}Ca, or pick up electrons to form ^{40}Ar. Only the latter is useful for dating purposes.

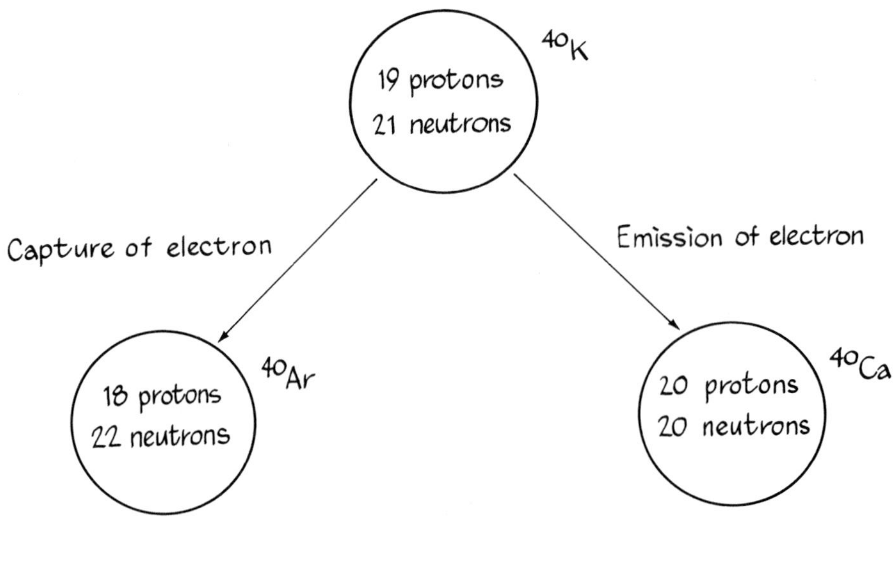

minerals containing appreciable amounts are scarce. A radioactive element of more common occurrence would make this method of age determination much more geologically useful.

The number of naturally radioactive elements is small, but fortunately one of them is the abundant and widely distributed element potassium. This metal has three isotopes (^{39}K, ^{40}K, ^{41}K), of which only ^{40}K is radioactive. Its activity is feeble compared with that of uranium, but sufficient to make age measurement possible. ^{40}K decays in two ways: its atoms can give up electrons from their nuclei, producing the calcium isotope ^{40}Ca, or they can gain electrons to form atoms of the argon isotope ^{40}Ar. (Losing or gaining electrons, in contrast to loss of helium ions, has practically no effect on the weight of an atom.) Once a potassium mineral has formed, both argon and calcium should steadily increase as a result of these processes. The age of the mineral, then, should be given by measurement of either the ratio $^{40}K/^{40}Ca$ or $^{40}K/^{40}Ar$. The difficulty with the first of these ratios is that calcium is an abundant element and ^{40}Ca is its most common isotope; almost any potassium mineral contains a little ^{40}Ca as a contaminant, and there is no way to distinguish this original calcium from that produced by radioactive decay. Argon, on the other hand, would not normally be present when a potassium mineral is formed, so that the $^{40}K/^{40}Ar$ ratio is by far the more promising for dating purposes.

Argon has the disadvantage, however, that it is a gas, and the same question arises as for helium. How can we be sure that the argon generated in a potassium mineral will not in part escape, thus making our age determinations erratic and always too low? The question can be answered only by experiment, and experiment shows that minerals differ greatly in their ability to retain argon. Potassium feldspars lose the gas so readily that they are of little use for age measurements. Micas are better and are frequently used, but slight loss of argon makes many estimates of mica ages somewhat low. The best minerals for potassium-argon determinations, surprisingly, are the ferromagnesian minerals hornblende and pyroxene, minerals whose potassium content is so low that the element is commonly not indicated in their chemical formulas. Analysis for very small amounts of potassium and argon requires special techniques, but methods have become standardized and are widely used. Of all the ways to find radiometric ages, the potassium-argon method has proved the most generally informative, for the obvious reason that it can be applied to so many varieties of ordinary rocks.

One other technique needs brief mention, the determination of age from the decay of rubidium. This rare metal, a close relative of potassium, has a feebly radioactive isotope ^{87}Rb which changes by loss of an electron to an isotope of another rare metal, strontium. The feebleness of the radioactivity restricts application of this method to older rocks, but it has proved particularly useful for age determinations in the Precambrian.

Carbon dates

The methods we have discussed up to now give an impressive array of dates reaching far back in geologic time—back for millions and billions of years. But suppose your interest lies in more recent things—the retreat of the last glacier that covered New England, for example, or the migrations of Neolithic man in Europe. Can these same methods be employed for dating events, say, of the last 100,000 years?

As we try to use these methods for more and more recent material, the results become increasingly uncertain because the products of radioactive decay have not had time to accumulate. Analytical methods can be made amazingly sensitive, but the sensitivity necessarily has limits. Until a radioactive element has built up an appreciable amount of its decay products, even the most sensitive instruments cannot find them. The potassium-argon method, for example, gives fairly reliable results on rocks as young as about 100,000 years, but in more recent material the amount of argon that has accumulated is too small for accurate measurement.

Fortunately a different sort of radiometric technique has come to light that permits determination of recent dates. This method depends on the element carbon, which is not normally considered radioactive.

The story begins with nitrogen, the most abundant gas in air. Cosmic rays from outer space penetrate the atmosphere, and fast-moving particles in the rays occasionally strike the nuclei of nitrogen atoms. By such collisions the nitrogen atoms are converted to carbon—not ordinary carbon, which consists of two non-radioactive isotopes with masses of 12 and 13, but a heavier isotope ^{14}C which is radioactive and ultimately decays back into nitrogen. The new-formed ^{14}C atoms are quickly oxidized by the surrounding oxygen to carbon dioxide, and some of the carbon dioxide eventually finds its way into plants by photosynthesis. Thus radioactive carbon becomes a part of organic material, and the amount in living substance remains constant because it is determined by a balance between the taking up of carbon dioxide in photosynthesis and its loss by respiration. When a plant dies, however, no more ^{14}C is added; the amount present gradually decays to nitrogen, and the diminishing radioactivity can thus be used as a measure of the time since death occurred.

The decay of ^{14}C is so rapid that after about 40,000 years the radioactivity becomes too feeble to measure. Within this span of time,

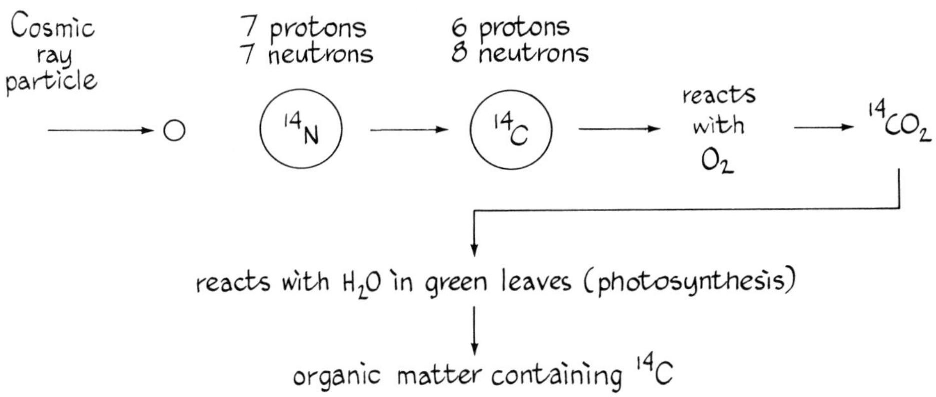

Fig. 11-9. The formation of radiocarbon (^{14}C) from nitrogen.

however, dating with carbon-14 has proved widely useful. The remains of an old peat bog buried by glacial outwash, wooden implements and fibers used by ancient Indian cultures and preserved in the arid climate of the Southwest, charcoal from a cave-dweller's fire in southern France—all these and many more have been assigned accurate dates from the radioactivity of their contained carbon. The material to be analyzed need not even be organic: the time of formation of calcite crystals can be determined from the amount of ^{14}C in their structure. The only requirements for use of this method are (1) that a sample contain carbon which at a specific time in the past was cut off from interaction with atmospheric CO_2 and (2) that the sample be no more than 40,000 years old.

These simple requirements are met by a great variety of materials, so it is small wonder that carbon-14 dating has become indispensable to archeologists and to geologists interested in events of the late Cenozoic Era.

How good are radiometric dates?

Radioactivity thus provides a means of going beyond the simple arranging of rocks and structures in sequence according to their contained fossils and their geometric relations. By analyzing rocks and minerals for their lead and uranium, or their potassium and argon, or their rubidium and strontium, or in geologically young materials for their radioactive carbon, we can attach numbers to geologic events that express the lengths of time, in years, that have elapsed since the events occurred. All the numbers quoted in previous chapters giving ages in millions or billions of years were obtained by such analyses. Before we accept these ages wholeheartedly, a little healthy skepticism is in order. The method looks plausible enough, but we must remember that the numbers, after all, are obtained from a complicated chain of reasoning based on results of laboratory tests on materials that have had a long and often strenuous history. Just how dependable can we expect such numbers to be? Are there restrictions on their use? Are they accurate to the nearest thousand years, or million years, or ten million years? Are they more reliable for some geologic situations than for others?

We note first that radiometric methods cannot be applied to all geologic materials. Suppose, for example, that you are studying an area in which pure quartz sandstone is a prominent rock, and you would like to know the age of the sandstone layer. A radiometric determination would be impossible, simply because the amount of radioactive material in such a sandstone would be too small for reliable measurement. The best you could do would be to estimate the age from the relation of the sandstone to adjacent rocks which contain enough of a radioactive element to make a measurement possible. Perhaps the sandstone rests unconformably on a diorite cut by dikes containing pitchblende, and is overlain by an andesite tuff containing hornblende crystals. If a lead-uranium determination on the pitchblende gives an age of 400 million years, and a potassium-argon determination on the hornblende gives 350 million years, the age of the sandstone must lie somewhere in the 50-million year interval. Because the unconformity must represent a long period of erosion you might guess that the age would be closer to that of the tuff than to that of the diorite. A

Fig. 11-10. Dating a layer of quartz sandstone. The sandstone itself contains so little radioactive material that it cannot be dated directly. If it rests unconformably on rocks dated at 400 million years, and is overlain by tuff dated at 350 million years, the age of the sandstone must lie between these two figures.

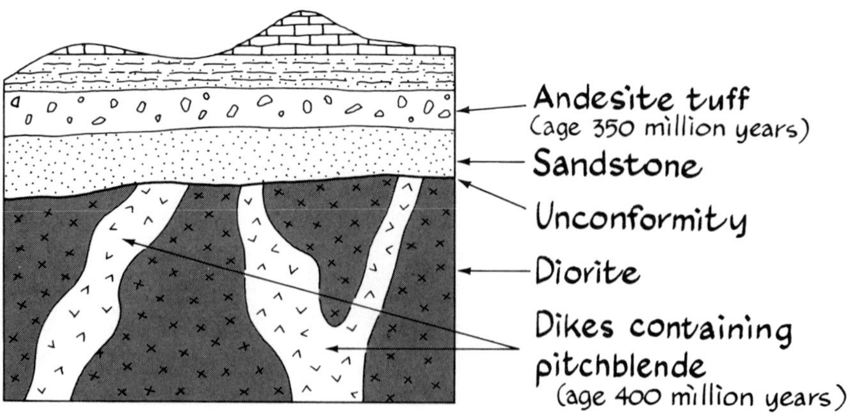

plausible estimate might be 370 million years, with the proviso that it could be anywhere between 350 and 390.

Dates of a great many rocks must be found by this sort of interpolation between known ages. Tectonic events are dated similarly. We have no way of directly determining the age of a fault, for example, but can say only that the movement occurred between the formation of the youngest datable rocks cut by the fault and the formation of the oldest rocks overlying it. Geologic ages that are not found directly by analysis for a radioactive element and its decay product are necessarily uncertain because we can do no more than bracket them between maximum and minimum figures.

What are the reasons for uncertainty in a direct determination? Some have already been noted in preceding paragraphs. A satisfactory radiometric date (except for ^{14}C dates) requires that the radioactive parent element be imprisoned in a mineral with no admixture of its decay products, or that we be able to distinguish decay products originally present from those formed in the mineral. A further requirement is that the mineral be unaffected during its history by chemical changes which might add or remove some of the radioactive parent or one of its daughter products. For determinations by the potassium-argon method we must assume further that the mineral structure is capable of retaining molecules of a gas. These stringent requirements are by no means always fulfilled, a fact that accounts for much of the uncertainty in radiometric ages. A good way to test how well the requirements are met is to use two or more element pairs (e.g. K-Ar and U-Pb) for finding the age of a single rock; if the date obtained by one method agrees closely with that obtained by the other, much greater confidence can be attached to the result. Not many rocks, of course, are well suited to the application of two different methods, but the excellent checks that have been obtained on a few rocks of widely different ages provide one of the best kinds of evidence that radiometric dates can be reliable.

A further source of uncertainty lies in the analytical techniques employed. No analysis can ever be completely accurate, just because of the limitations of machines and their human operators. Analytical uncertainties grow larger as the amount of material sought becomes smaller, and in radiometric determinations the amounts are often very tiny. No precise general figure can be given for analytical uncertainty, because it varies with the technique used, with the kind of rock or mineral to be analyzed, with the amount of radioactive material present, and of course with the skill of the operator. The uncertainty in the best determinations is commonly of the order of a percent or so: thus a date of 10 million years means somewhere between 9.9 and 10.1, a date of a billion years somewhere between 990 million and 1010 million. Needless to say, an uncertainty of 20 million years in the age of a billion-year-old rock is not of great importance for geologic arguments.

In addition to geologic uncertainties and analytical uncertainties, radiometric ages are subject to possible questioning on the basis of physical constants and physical processes relating to radioactive decay. To calculate an age from the ratio of potassium to argon in biotite, for example, requires that we know very exactly the rate at which atoms of ^{40}K decay. The rate for this process, and the rates for all the other processes used in dating, have been the subject of careful study and are known with considerable accuracy. But like most physical constants they are determined solely by experiment, and as experimental methods are refined the value of a constant may need adjustment. When a decay constant is given a new value, all geologic ages calculated from it must change also, and a very slight adjustment in a constant may mean considerable alteration in the calculated ages.

There is also the disturbing possibility that physical constants may change with time. Normally we regard constants like those for radioactive decay as immutable, and we have little reason to question this assumption. But geologic times are very long, and our easy assumption that the constants remain fixed over such immense periods has no wholly convincing proof. For the carbon-14 method, another necessary physical assumption is a guess that the cosmic radiation striking the earth has not changed in amount or character over the last 40,000 years. Any change in the radiation would alter the amount of ^{14}C formed in the atmosphere, and hence would affect the calculated ages. We do, in fact, have reason to suspect slight changes of this sort, and ^{14}C ages are subject to a corresponding slight uncertainty.

Thus geologic ages determined from radioactive decay must be looked on with at least a grain of skepticism. The order of magnitude of geologic time given by such measurements is hardly subject to doubt. Ages determined in many different ways give generally concordant results, and the ages are consistent with times required for such geologic processes as sedimentation and organic evolution and also with times estimated from astronomic phenomena. Sequences of numbers for the geologic events during an era or period can likewise be checked in many ways. But the precise number of years calculated for a particular geologic event is always to some extent uncertain, as is demonstrated by the frequency with which the numbers are revised in geologic literature.

Radiometric dates, useful and impressive though they are, have one final handicap: their determination requires elaborate apparatus and great experimental skill. No amateur can pick up a rock and by any reasonable procedure find its age in years. He, and most geologists also, must send the rock to a specialist. Many facts in geology can be worked out, or can be made reasonable to an amateur, with little equipment and with straightforward arguments. But the important matter of absolute dating must be left in the hands of those with special training.

A look at the dates

Radiometric dating, a branch of earth science called *geochronology* (literally "earth-time study"), occupies the attention of an increasing number of geologists. Many university and government laboratories are equipped for age determinations, and several commercial laboratories have been set up for the express purpose of dating specimens on order. What sort of numbers come out of this activity, and why are they considered so important?

One reason for the interest is the purely philosophical satisfaction of knowing at just what time in the past certain events took place. When we visit an ancient building or look at ornaments made by an ancient tribe our first question is usually, "When was it made?" In much the same spirit we ask of a rock or a mountain, "When was it formed?", and we feel more secure if the answer can be given in years. It is somehow more satisfying to know that the first trilobites appeared about 600 million years ago, and that dinosaurs flourished between 100 and 200 million years ago, than to be told merely that primitive trilobites are found in early Paleozoic rocks and dinosaur remains in rocks of the Mesozoic Era.

But beyond the simple satisfaction of knowing numbers, we can make use of absolute dates in correlating rocks and geologic events. Fossils give us the first clue to arranging events of the past in sequence, but in places the fossil record has awkward gaps or is subject to different interpretations; radiometric ages are widely used to check conclusions based on fossils and to supplement the record where fossils are scarce or poorly preserved. The absolute dates find their greatest usefulness, however, in working out the geologic history and structural relations of large rock masses where fossils can offer no useful guide. Two examples are bodies of igneous rock and areas where the only rocks exposed are Precambrian.

To illustrate the application of radiometric techniques to igneous rocks, we look at some recent work on the Sierra Nevada batholith of eastern California. This huge mass of granitic rock intrudes sedimentary layers containing scarce fossils of the early Mesozoic Era and is overlain by layers with fossils of the late Mesozoic. Careful mapping shows that the

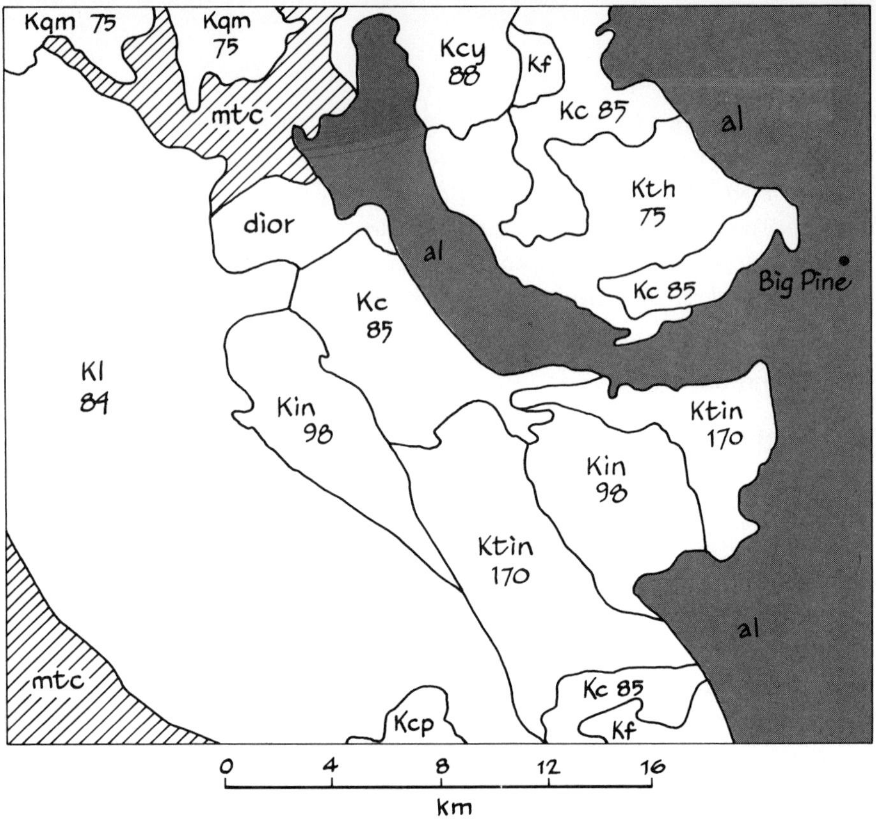

Fig. 11-11. Simplified geologic map of intrusives in part of the Sierra Nevada, California. The separate intrusives or *plutons,* designated by symbols like Kep and Ktin, consist of rocks ranging in composition from diorite to granite. Such intrusives make up the huge granitic mass called the Sierra Nevada batholith. Ages determined from ratios of radioactive elements to daughter elements are given in millions of years. Note that the ages even in this one small area span a period of about 100 million years, more than half the length of the Mesozoic Era. (After Paul Bateman, U.S. Geological Survey)

batholith is not a single body of granite but consists of many small masses that differ slightly among themselves in grain size and mineral composition. Contacts between the small masses show that they were intruded at different times, generally those on the west side of the batholith being older and those on the east side younger but with many irregularities. This much we get from ordinary geologic study: a general Mesozoic age for the whole batholith, and a rough sequence of intrusion of its constituent parts. Further details would be useful, not only from a theoretical standpoint but also because important mineral deposits, especially of tungsten, are associated with some of the intrusions and not with others. We would like to know, for example, at what times during the Mesozoic the different bodies were intruded, whether they were all formed during a single short interval or were widely spaced, whether differences in composition can be correlated with differences in age. For answers to such questions radiometric ages are ideally suited.

Many age determinations on Sierra Nevada rocks have been made in recent years, most of them potassium-argon ages measured for both biotite and hornblende in the same specimen to insure greater reliability. The ages have a wide spread, the majority in the latter half of the Mesozoic but some going back to its beginning. Evidently formation of the Sierra Nevada batholith was not a single episode, but a long series of separate intrusive events going on over some 150 million years. Beyond this the conclusions from the radiometric dates are still a subject of controversy. Some geologists think the dates cluster in five major episodes separated by 20–30 million-year intervals of no igneous activity, while others interpret the dates as an almost continuous sequence through the entire Mesozoic. Presumably this question of detail will be answered by additional age determinations.

For geologists interested in the ancient rocks formed before the Paleozoic Era, radiometric dates have proved a special boon. Precambrian rocks contain no fossils usable for correlation; in a few places remains of primitive life have been recognized (algae, bacteria, worms), but no distinctive shells or bones by which sedimentary strata can be traced over long distances. Precambrian geologists of past decades have been forced to rely on such criteria as geometric relations, rock compositions, and degree of metamorphism in their attempts to reconstruct earth history before the Paleozoic, and it is small wonder that the attempts have been plagued with uncertainty and bitter controversies. All this has been changed, suddenly and dramatically, by the accumulation of radiometric dates for Precambrian rocks. No part of geology has shown more rapid progress over the past ten or twenty years than the deciphering of Precambrian history on all the continents.

In North America, for example, the great area of ancient rocks underlying central and eastern Canada can now be divided with confidence into belts and provinces of various ages reaching back to more than 3 billion years—an important enterprise economically because it guides the search for new mineral deposits similar to those already known in many of the belts. The Precambrian rocks of Canada extend down into the northern fringe of the midwestern United States, and are known from drill cores to underlie the Paleozoic rocks of the Mississippi Valley. One of the oldest dates so far obtained, 3.7 billion years, comes from a Precambrian area in Minnesota. Farther west exposures of Precambrian rocks are spotty—at various places in the Rocky Mountains, at the bottom of the Grand

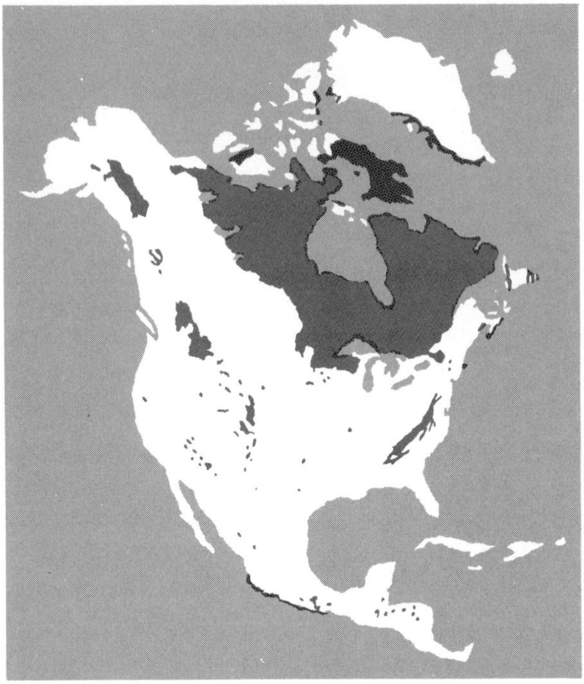

Fig. 11-12. Exposures of Precambrian rocks in North America. The large Precambrian area in eastern Canada, with fringes extending down into the northern U.S., is often called the *Canadian Shield* in reference to its shape. (University of Michigan Press)

Canyon, in many of the desert ranges of the southwestern states and northern Mexico. These western occurrences give dates in the later Precambrian, from 600 million back to about 2 billion years. An important goal for future work is to find an explanation for the apparent absence of more ancient rocks in the western half of the continent.

So we can summarize the results of radiometric dating by noting that Precambrian rocks go back at least 3.7 billion years, that the Paleozoic Era began nearly 600 million years ago, that the earliest Mesozoic rocks are about 225 million years old, and that the Cenozoic Era dates from about 70 million years ago. It is easy to quote these figures, as geologists are fond of doing, but it is hard to give them real meaning. When our own lifespan is less than a hundred years, and all recorded history amounts to less than 6000 years, what possible significance can we attach to time intervals measured in millions and billions of years?

A gross contraction of scale may help. Suppose that a long-lived science-fiction monster on Mars had trained a camera with a telescopic lens on the earth back at the start of the Paleozoic, and had taken pictures at regular intervals down to the present. Suppose that he puts these pictures together into a movie that will run for just one hour, and invites us to watch this greatly accelerated survey of the development of our planet. The events of 600 million years will be compressed into 60 minutes; each minute will represent 10 million years, and each second about 170,000. For the first 35 minutes we watch the events of the Paleozoic Era unfold: widespreading shallow seas over the continents much of the time, teeming at first with trilobites and brachiopods and later on with fishes of unfamiliar kinds; land areas at first bare but becoming clothed with vegetation, at times a very dense growth of strange-looking plants in huge swamps, with sluggish amphibians and later small reptiles crawling through the brush. Then for nearly 20 minutes we watch the more lively scenes of the Mesozoic—the growth of huge cephalopods in the sea, the development of giant lizards on land, and toward the end of the era the appearance of birds and flowering plants. Now only 7 minutes of the hour remain, and in this brief time we watch mammals grow to large size, at first unfamiliar forms but later assuming the recognizable shapes of horses, elephants, cats, and dogs. In the last half-minute, if we watch carefully, we can begin to spot creatures somewhat resembling ourselves, but the resemblance will become close only in the very last second. And the whole story of the civilization of which we are so proud will be lost completely, because it will occupy much less than the last tenth of a second of the movie.

We reflect that this hour-long show gives us only the last part of the geologic record, and we ask our accommodating monster if he started his movie before the Paleozoic. He brings out other reels, and we sit down for a much longer session. If the record is complete from the earth's beginning, 4.6 billion years ago, the movie will run for 8 hours. Much of the first 7 we will not find very entertaining. The lands will be bare, the seas will have only slimes and perhaps small soft-bodied creatures, and we will tire of a long succession of volcanic eruptions and the up-and-down movements of land masses. Only during the last hour will we see again what we saw in the previous showing, the record of life that makes the movie seem pertinent.

The last half-second of an 8-hour show—this is the story of our race from a planetary perspective. All the battles of our history books, the migrations of tribes, the rise and fall of empires, the music and poetry and sculpture and painting, all these belong to a still shorter time. Surely the earth was a long time in preparing for the human race, and humanity so far has walked its surface for only a brief moment.

Fig. 12-1. Damage to freeway overpasses during the San Fernando earthquake of 1969. (R.E. Wallace, U.S. Geological Survey)

12

Makers of Catastrophe: Volcanoes and Earthquakes

This planet on which we live is by and large a friendly place. It provides man with sites for his homes and his cities, with soil to grow his food, with resources of rock and metal and energy to build and maintain his industrial civilization. It can even absorb much abuse from man's careless treatment of his environment. Soil lost when erosion is accelerated by over-grazing or strip-mining is gradually restored, the bomb craters of war are slowly filled and hidden by vegetation, streams and lakes purify themselves from large amounts of sewage and industrial pollutants. Only when man's thoughtlessness permits abuses on a very large scale, or concentrated in particular areas, can the planet be damaged beyond its ability to recover in a reasonable time. If man can control the increase of his numbers and can learn to live in harmony with his environment, instead of overtaxing its restorative powers and exhausting its bounty, the planet will remain a friendly place for a long future.

But occasionally the earth can seem anything but friendly. Talk to the victims of a flood, a great storm, or a landslide, and nature will appear of a sudden malevolently destructive. The natural catastrophes that beset mankind are ample reminders that the earth is not as placid as it usually seems, that huge forces lie in wait to wreak havoc from time to time on man and all his works. Catastrophes are of many kinds—meteorological, biological, geological. Of all the strictly geological calamities, the ones most devastating and most feared are volcanic eruptions and earthquakes.

Eruptions and earthquakes have little in common, but both obviously result from the build-up of some sort of strained condition within the earth.

In a violent eruption the strain can be linked to increasing gas pressure, in an earthquake to accumulation of stress in solid rock. If we can look behind the surface phenomena of catastrophes to see how these predisposing unstable conditions are brought about, we will be making a start toward possible prediction and control of their destructive effects. At the same time we should gain some new insights into the mysterious geologic processes that operate far beneath our planet's surface.

Explosive volcanoes

The kind of volcanic eruption we considered in Chapter 2, by far the commonest kind the world over, is not usually very destructive. Molten lava pours from a crater, or out of fissures on a mountainside, and flows out over adjacent flat country; it may ruin good agricultural land, it may inundate villages in its path, but the area covered is generally not large and the motion is slow enough so there is no loss of life. Accompanying the flow is a great cloud of steam and dust and cinders billowing up from the crater, in pulses marked by booming explosions, impressive to watch but not very dangerous except to foolhardy climbers who venture too close. To the local inhabitants such eruptions are a nuisance, but their limited capacity for damage is attested by the fact that large populations have persisted in living near volcanoes of this kind since the beginning of history.

Fig. 12-2. Part of the city of St. Pierre after it was overwhelmed by a nuée ardente in 1902. (Underwood and Underwood)

Entirely different is the sort of calamity that befell the city of St. Pierre on the Caribbean island of Martinique on a May morning in 1902. The volcano above the city, Mount Pelée, had been belching dust at intervals for a few weeks, and the island had been shaken by a succession of earthquakes, but the disturbances were not sufficiently different from past episodes of activity to cause great alarm. Then without warning came a huge explosion from the mountain. A turbulent cloud of dust, hot gas, and incandescent rock fragments rolled down a valley and over the city, and within a few seconds all but one of the 25,000 inhabitants were killed.

Mount Pelée has become the classic example of a dangerously explosive volcano. Detailed study of its 1902 eruption has given us a clear idea of what happened on that fateful morning in St. Pierre. The principal rock of the mountain is a silica-rich lava called *dacite,* with a composition between that of rhyolite and andesite. In the molten state this lava is exceedingly viscous, scarcely able to flow at all, so that during an eruption a plug or dome of dacite simply rises in the throat of the volcano and stays there rather than flowing downhill. The lava plug acts as a trap for gases coming up from below; like the cork of a champagne bottle it blocks the exit while pressure mounts to the point where an explosion must take place. When it comes, the explosion has a different order of magnitude from the explosions of a more ordinary volcano. Large masses of glowing, viscous lava are blown into tiny shards of glass, carried along with the suddenly expanding hot gases, and the whole mixture becomes a churning, wildly turbulent mass that can flow downslope much like a liquid. The explosion takes place, of course, at the point where the plug in the volcano's throat is weakest, and when this is at its side the cloud may be projected laterally rather than upward. Whether the motion is simple gravity flow or aided by sideward projection from the explosion, the cloud moves so fast that no living thing can escape. The French geologists who described the eruption in Martinique called the destructive cloud a *nuée ardente,* which has become the usual designation for this phenomenon in all languages. The English translation is simply "glowing cloud."

Volcanoes showing the same kind of eruptive activity as Mount Pelée are numerous, but most of them are fortunately not as close to large communities. Lassen Peak in northern California, for example, during an eruption in 1915 produced a small nuée ardente that devastated a few square miles of forest, but no one was close enough even to observe the event from a distance.

Turbulent masses of ash and hot gas take many forms, ranging from the spectacular nuées ardentes to quieter flows that move much like liquid lava and follow depressions in the topography. A probable example of the latter type is the deposit of ash and pumice fragments that filled a valley

Fig. 12-3. Photomicrograph of welded tuff. The phenocrysts (white) are mostly feldspar; the matrix consists of glass fragments partly fused together. Note the elongated fragments of pumice, deformed and partly collapsed by the weight of overlying material while they were still hot. About 3 times enlarged. (P.W. Lipman, U.S. Geological Survey)

during the eruption of Mount Katmai in southwestern Alaska in 1912. The region was so difficult of access that the valley was not seen until many months later, when it was widely publicized as "The Valley of Ten Thousand Smokes" because of the clouds of steam and gas rising from cracks in the surface of the deposit. The origin of the deposit has occasioned much argument, but recent investigation has shown convincingly that it formed from a tongue of pyroclastic material propelled and buoyed up by hot, turbulent gases, the material coming not from Katmai itself but from a smaller volcano that was active at the same time. Turbulent ash-plus-gas mixtures, whether of the violent, far-spreading Pelean type or quieter ones like that at Katmai, go by the general name of *ash flows* or *pyroclastic flows.*

Volcanic mountains that produce such flows commonly lack the symmetrical conical shape that the word "volcano" usually calls to mind, but consist of a rounded or irregularly shaped dome flanked by steeply sloping piles of debris from many explosive eruptions. If lava flows are present they are short, thick, steepsided, and covered with jagged, glassy blocks. Typically the rock of both domes and flows has a composition rich in silica, such as rhyolite, dacite, obsidian, or pumice. Violently explosive eruptions are not limited to volcanoes with felsic lavas, however. They are not uncommon at andesitic volcanoes, and the famous nuée ardente that buried the Roman city of Pompeii in 79 A.D. came from the volcano Vesuvius, whose lavas are mostly basaltic.

What sort of material would you expect to see in a deposit dropped by a nuée ardente or a quieter pyroclastic flow? Abundant fine ash, certainly, probably accompanied by chunks of pumice and fragments of solid lava. The pumice and rock fragments should diminish in size away from the source, because the violence of the turbulent motions must decrease as the flow spreads and cools. The debris from a small flow would be a loose aggregate of ash and fragments, because the material would cool rapidly as the gas dissipates. But a deposit laid down rapidly by a large ash flow would consist of fragments that are still very hot and exuding gas; as successive layers are buried, the still-viscous fragments become flattened, partly grown together, and in places altered by hot gases ascending from layers beneath. This process results in a hard rock called *welded tuff*—like an ordinary tuff in its essential materials, but with fragments conspicuously deformed and so firmly welded together that the rock may closely resemble solid lava. Historic ash flows have not been large enough to produce much welded tuff, but huge volumes of this rock in many parts of the world, for example in western Nevada, are evidence of eruptions on a grander scale in the not very distant geologic past.

Destructive volcanic activity is not limited to gas explosions from volcanoes with mouths choked by felsic lava. The long-dormant volcano Bandai-san in central Japan, for example, was blown apart in 1888 by an explosion so loud that it was heard in Tokyo, 150 kilometers away, and so violent that half the mountain was spread over the surrounding countryside—yet after the eruption no new lava, ash, or pumice could be found. Evidently a mass of warm but solid rock remained beneath the apparently quiet mountain, and groundwater over a period of years was converted into steam that eventually built up enough pressure to explode. Like the eruption of Pelée, this one was triggered by the accumulation of gas under enormous pressure—but the source of gas was just heated groundwater, whereas the source at Pelée was gas that had been dissolved in magma at great depth.

Questions about volcanoes

If we forget the human misery that violent eruptions can sometimes cause, and look at them with the cool eye of a scientist, what do they tell us about processes going on under the surface? The answer, unhappily, is "disappointingly little." There is no sharp distinction, of course, between volcanoes that occasionally erupt destructively and the better behaved volcanoes we have looked at in earlier chapters. In all kinds of volcanoes we see evidence for the ascent of molten rock or hot gases or both from somewhere down below, and the different sorts of eruptive activity can be nicely correlated with different compositions of magma, different temperatures, and different amounts of dissolved or added gases. But the more violent manifestations give us little more information than quiet eruptions about the really interesting questions—where does the magma come from, how does it get hot, why does it contain so much dissolved gas, why do most volcanoes erupt intermittently instead of all the time, why do volcanoes eventually grow old and die?

Volcanoes do tell us that a potent source of energy exists within the earth. Coupled with other information about the earth's rigidity, they tell us that the energy must somehow be localized in a few areas, and not many tens of kilometers under the surface. They bring us samples of material from depth, and they help us make guesses about temperatures and pressures at different levels. They give evidence that gaseous materials in enormous quantity must somehow be imprisoned within the earth. Because volcanic gases sometimes deposit coatings containing common ore metals on the sides of fissures, volcanic activity provides a possible insight into processes of ore formation. Such information is useful, assuredly, but it doesn't lead far toward solution of the more difficult questions about the earth's interior.

A more practical kind of question relates to the possibilities of predicting and controlling volcanic activity. If people persist in living near volcanoes, as they evidently will, can they be warned about impending eruptions? Our understanding of volcanic behavior is still so limited that warnings based on theoretical knowledge are impossible, but progress is being made in the development of instruments to detect signs of lava movement under a volcanic structure that may presage an eruption. Slight swelling of a mountain, indicated by delicate measurements of the tilt of its slopes, may mean that molten material is rising into fissures. An increase in the temperature of hot springs, or in the emission of gas from fumaroles, has long been used as a possible warning. Localized earthquakes often precede an eruption, and very minor ground tremors detected by seismographs may give more accurate information about lava movement underground. Empirical data of this sort have been moderately successful in predicting changes in activity of Hawaiian and Japanese volcanoes, and much further refinement seems possible.

Control of volcanic eruptions is probably only a dream for the distant future. The energies involved are so great, and the difficulties in bringing any adequate human energy source to bear directly on an eruption would be so fantastic, that development of warning systems seems a more practical approach to minimizing damage from volcanoes. Even the simplest sort of control—attempts to divert lava flows by bombing or constructing barricades—has so far proved discouragingly ineffectual.

What happens in an earthquake?

In their destructiveness to man and his works, and also in their usefulness as probes of the earth's interior, earthquakes deserve a far higher rating than volcanoes. If simple loss of life is used as a measure of destructiveness, the difference is overwhelming: during the 500 years in which anything like accurate statistics have been kept, the deaths from earthquakes number in the millions, as opposed to tens of thousands killed by volcanic eruptions. In the United States alone there is no certain record of any deaths except suicides as a direct result of volcanic activity, but several hundred have lost their lives in earthquakes. For the future the chances of death from volcanoes in this country are very small, but an earthquake comparable to the San Francisco quake of 1906 or the New Madrid, Missouri, quake of 1811 could cause prodigious loss of life in one of our big cities—despite all the recent advances in earthquake-resistant design for buildings. On the other side of the ledger, the interpretation of records of earthquake vibrations that have traveled through the earth are by far the most revealing source of information we have about conditions in the deep interior. So we approach the study of earthquakes with a divided mind: in awe at their destructive capabilities, and in expectation that they will fill many gaps in our knowledge of the planet's overall structure and functioning.

We begin with the record of an actual earthquake. If you had been in Anchorage, Alaska, in the late afternoon of Good Friday in 1964, you would have returned to your home or your hotel in the gathering dusk of a cool, calm day, with a few snowflakes floating down from the low gray overcast. Perhaps on your way home, perhaps after you had made yourself comfortable before your television set, the world would have suddenly gone mad. The ground beneath your feet, or the floor of your room, would have been wrenched with a mighty jolt, and for four or five endless minutes everything about you would have been in violent motion. If you had been on a downtown street, you might have had to dodge pieces of walls and cornices from the swaying buildings. If you had been in a home on the bluff on the west side of town, you would have felt the house tilt and perhaps break apart as the ground beneath it became part of a huge slide.

Next morning you would have learned from the radio that you had

Fig. 12-4. Damage from a landslide triggered by the Alaska earthquake of 1964. (M. Bonilla, U.S. Geological Survey)

Fig. 12-5. Effects of a tsunami after the Alaska earthquake (M. Bonilla, U.S. Geological Survey)

lived through the strongest earthquake recorded in North America during historic time. That day, and for many following days, you would have felt the ground shake again at intervals, but never as strongly as on Good Friday. And gradually from news reports you would have pieced together the magnitude of the disaster: the port facilities of Seward and Valdez torn loose and submerged by underwater sliding, other coastal towns devastated by huge waves following the earthquake, and over a great area the level of the land relative to the sea changed permanently by several feet.

The great Alaska earthquake was unique in some respects, particularly in the long duration of the shaking, but in many ways its effects were typical of earthquakes anywhere. A large quake is one of the most terrifying of human experiences. We are so accustomed to the solid earth remaining fixed and immovable that when everything about us starts shaking we feel unnerved and helpless. Some observers have described a wave motion of the ground, and huge cracks opening and closing, but the more usual sensation is simply jolting and rapid shaking. Buildings are set in motion, and tall ones may sway crazily; broken windows and cracks in walls are common, and poorly constructed buildings may collapse. Landslides are often triggered by the shaking, especially if soil and alluvium are saturated with water. An earthquake near the sea coast may set in motion large waves or *tsunamis* (popularly but wrongly called "tidal waves"), generated by the sudden disturbance of the sea floor. Tsunamis can travel long distances, and may wreak immense destruction where they strike the shore. In all these aspects a great earthquake is obviously a menace to human life.

The casualty list for the Alaska quake was surprisingly small—115 killed, most of these by the tsunamis following the shock—but the small number was a fortunate accident of the time of day, the season, the occurrence on a holiday, and the sparse population of south central Alaska. If an earthquake of similar severity should strike a major urban center during the busy hours of the day, the possibilities are not pleasant to contemplate.

What starts an earthquake?

The intense vibration of the ground during a strong earthquake means that something must have jarred a portion of the solid earth. What sort of natural process could cause so strong a jarring that areas of hundreds of square miles are set in motion? We think first, perhaps, of explosions: it is common experience that blasting in a quarry or mine makes the ground shake in a manner reminiscent of earthquake tremors. One possible kind of natural explosive process is the violent escape of gas that characterizes some volcanic eruptions, and we recall that earth tremors often accompany and sometimes precede volcanic activity. Can we, then, account for all or most earthquakes by linking them to movement of magma and gases underground?

This kind of hypothesis was a common explanation for earthquakes in ancient times, and it seemed credible because the Mediterranean region is noted both for frequency of earthquakes and for a number of active volcanoes. Examined more critically, however, the hypothesis has serious drawbacks. Granted that volcanic activity does in fact produce earthquakes, sometimes severe enough to be damaging, still these quakes are felt only in small areas near volcanic centers and are clearly linked to eruptive activity. Major earthquakes, on the contrary, often occur in regions far from any sign of recent eruptions and affect huge areas of the earth's surface. To ascribe an earthquake like the one that shattered Charleston, South Carolina, in 1886, or the New Madrid earthquake that shook

Fig. 12-6. The "elastic rebound" hypothesis for the mechanism of earthquakes. Slow movement of rocks on either side of a fault builds up strain, represented by progressive bending of the line AOC; when the strength of the rock is exceeded the strain is suddenly relieved and an earthquake results. A line like AOC snaps back to a straight position, but its ends are now offset along the fault.

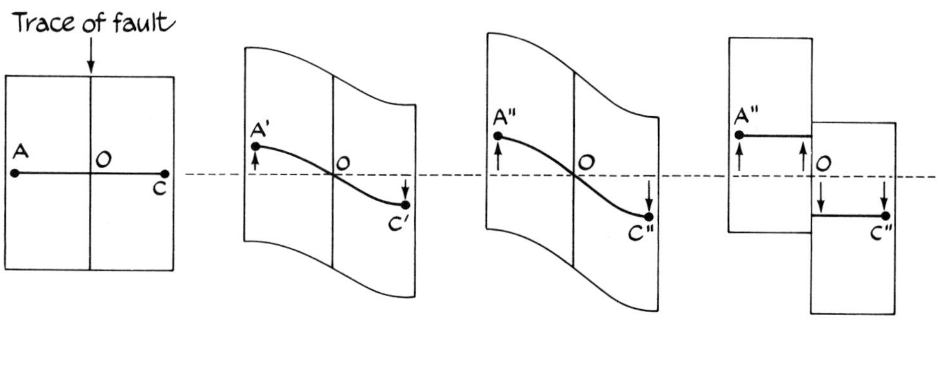

Missouri and adjacent states in 1811—both in areas where no volcanic rocks are known younger than early Paleozoic—to movement of magma and gas underground would be completely unreasonable.

So we look for another cause of shaking. We know that solid rocks in the crust are subject to slow movements of various kinds—the folding, faulting, uplift, and subsidence that we talked about in Chapter 10 are ample evidence. Suppose that continued slow movement of the bedrock in a given area gradually builds up strain, and that ultimately the strength of the rock is exceeded so it suddenly fractures. If the fracturing is on a large scale the adjacent rocks will be strongly jarred, and vibrations are set up that may travel long distances—just as vibrations travel through a long plank if one end of it is bent until it breaks. This is the generally accepted explanation for earthquakes. Note that it is far from a complete explanation, for it depends on the assumption of extensive movement in solid rock beneath the surface. Why such movement should be taking place is a deeper question that we have not yet looked into.

Direct evidence that some earthquakes, at least, are the result of sudden fracture is the appearance of conspicuous cracks at the ground surface. We have noted earlier the long fault scarps observed after earthquakes along the bases of some mountain ranges in Nevada—ragged cliffs a few meters high, indicating that during an earthquake a range front is elevated with respect to the adjacent valley. The famous San Francisco earthquake of 1906 gave evidence of a different kind of movement: over a distance of more than 100 miles the ground surface was broken by cracks along which one side was displaced horizontally with respect to the other (strike-slip movement as opposed to the dominantly dip-slip movement on the faults in Nevada). Not all earthquakes, however, provide such readily observable evidence of fracturing. Most small quakes produce no cracking

Fig. 12-7. A fault scarp produced by the Alaska earthquake of 1964. Despite its impressive size in this picture, the scarp could be followed for only a few miles and was actually only a minor result of the great earthquake. Another fresh fault scarp is shown in Fig. 3-16. (U.S. Geological Survey)

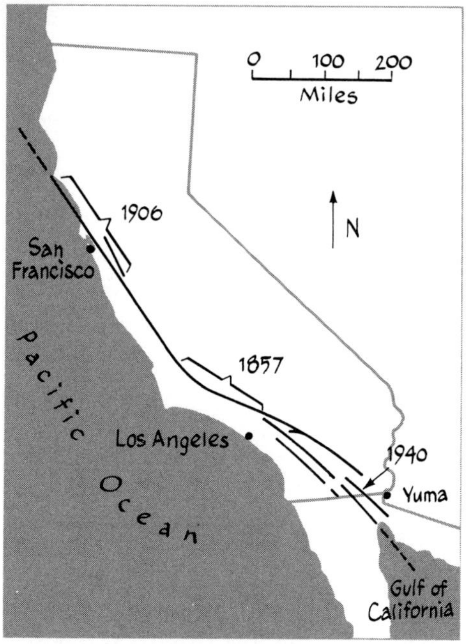

Fig. 12-8. The San Andreas fault can be followed diagonally across California for a distance of more than 500 miles. The dates show the parts of the fault that moved during the earthquakes of 1857, 1906, and 1940.

of the ground at all, and even many large ones give rise only to minor irregular cracks at the surface that cannot be directly related to a line of fracture at depth. Alaska's great Good Friday earthquake, for example, generated much local cracking associated with landslides, and two small fault scarps traceable for a few miles, but no visible fracturing at all commensurate with the amount of overall disturbance (although it is possible that extensive fracturing occurred on the sea floor, and hence was not observed). Where no surface manifestation of fracturing appears, it is reasonable to suppose that the breaking of rock was deep enough or on so small a scale that the overlying soil and alluvium could adjust itself to the displacement without visible deformation.

The association of earthquakes with fracturing is also indicated by the frequent concentration of quakes along particular lines, lines which by geologic mapping can be shown to represent boundaries between different kinds of rocks. The San Francisco quake, for example, is only one of many historic disturbances along the line of the San Andreas fault, a major structural belt in the California Coast Ranges that separates rocks of very different kinds. A given earthquake along this fault affects only part of it, but a long succession of movements on different segments over some tens of millions of years has produced an overall displacement amounting to hundreds of miles. Many other faults are known to be active, in the sense that frequent earthquakes can be attributed to movements along them.

Other possible kinds of sudden movement in the crust can cause earthquakes, but it is well established that by far the greater number—and certainly all the major earthquakes—have their origin in the breaking of rocks, or more specifically in the accumulation of strain and sudden release by displacement of rocks along faults.

Measuring earthquakes

To get very far with a scientific study of earthquakes requires that we make our descriptions more quantitative. How strong is one earthquake compared with another? How large an area does an earthquake affect? How long does the shaking last? Precisely where does an earthquake start? How fast do vibrations travel away from the place where rock is fractured? Such questions can be answered only by assigning numbers to various properties of an earthquake.

The most obvious kind of number would be a designation of the extent of damage that an earthquake causes. We could label an earthquake "1" if the shaking is barely perceptible, "2" if loose objects are knocked off tables and shelves, "3" if chimneys topple and windows are cracked, and so on. A scale of this sort, with carefully specified gradations from one number to the next, has been widely used. Technically, the numbers are said to represent the *intensities* of different earthquakes. Such numbers are useful, but have some serious disadvantages. They are meaningless for earthquakes in wilderness areas where buildings do not exist, they depend not only on the intensity of shaking but on the kinds of buildings affected, and they tell nothing about the size of the area damaged. Far more helpful from a scientific standpoint would be a set of numbers comparing the total amounts of energy released in different earthquakes.

To obtain numbers for energy comparisons, students of earthquakes (seismologists) use instruments called *seismographs*. Such instruments may look complicated, but basically they are very simple. A seismograph consists of a heavy weight delicately suspended, so that when its support starts

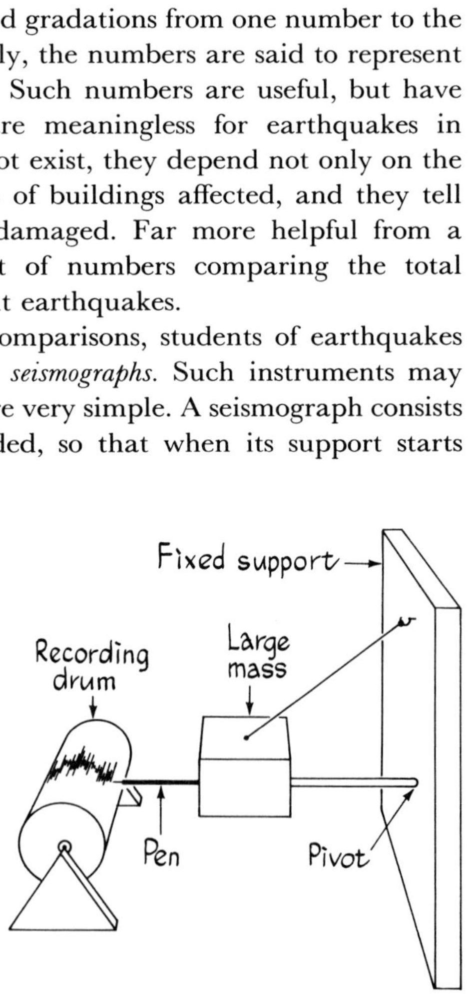

Fig. 12-9. The principle of a seismograph. The large mass is pivoted so that it does not move immediately with its support. During an earthquake the support and the revolving drum move with the earth, but the mass lags behind. The relative motion is recorded as a wavy line on the drum. This kind of instrument would give a record of horizontal motion; an instrument for recording vertical motion would have a mass suspended so as to move freely up and down relative to its support.

vibrating the weight does not immediately respond. If a pen is attached to the weight and arranged so that it touches a slowly revolving drum, the vibration of the support (which is transmitted directly to the drum) is recorded as a wavy line. During periods when no earthquakes are occurring, the line is straight except for minor jiggles produced by nearby traffic or sometimes by storms; when an earthquake starts, the line abruptly becomes a series of waves, the height (or *amplitude*) of the waves depending on the strength of the shaking. Many modern seismographs use a light-beam in place of a pen, and make their records photographically, but the principle remains the same.

The amplitude of waves recorded on a *seismogram* is a measure of the energy of the ground vibrations at the point where the seismograph is located. This energy must depend not only on the actual energy released by the earthquake, but also on the distance between the seismograph and the place where the earthquake occurred. Since vibrations from an earthquake travel away from their source in all directions through the crust, like ripples spreading out from a disturbance on the surface of a pond, the energy falls off rapidly with distance. If seismograms recorded at various distances are available, the amplitude of waves that would have been obtained by a seismograph at some fixed distance, generally taken as 100 km, can be found by interpolation. In this way numbers are obtained for the relative energies released by different earthquakes.

A number on this scale is called the *magnitude* of an earthquake (or the *Richter magnitude*, after the seismologist who suggested the scale). These are the numbers most widely quoted in newspaper reports of earthquakes. An earthquake of magnitude 3 is barely perceptible near the center of disturbance, one of magnitude 5 is generally felt but causes only minor damage, and quakes with magnitudes of 6 or higher can be responsible for serious damage. The strongest earthquakes in modern times have had magnitudes of about 9; the San Francisco quake of 1906 was about 8.25 on the Richter scale, and the Alaska quake just a little higher. (Numbers on the Richter scale seem deceptive, because the energy increase from one number to the next is geometric rather than linear. Thus magnitude 7 represents an energy release about 30 times that of magnitude 6 and about 1000 times that of magnitude 5.)

What else can be read from a seismogram, besides an estimate of relative energy? It turns out that these wiggly lines are remarkably informative, especially about the kinds of rocks through which the earthquake vibrations have moved. Since the vibrations from major quakes can travel through the entire earth, we have here a powerful tool for

Fig. 12-10. A seismogram held by Dr. Charles Richter, the seismologist who devised the Richter scale of earthquake magnitudes. (David Iwerks)

investigating the deep interior. In the next chapter we will look at the information earthquakes have brought to light about the inner structure of our planet, but for the moment our concern is with other earthquake properties.

There is the question, for example, of locating the point where fracturing started. (This information would be needed, of course, in making an estimate of earthquake magnitude as described above.) The question can be approached by a closer look at the details of a seismogram. Typically the wavy-line record of an earthquake can be divided into three major parts: the record begins abruptly with fairly simple waves that gradually diminish in amplitude; then suddenly waves of greater amplitude are superimposed; and still later come the strongest waves of all, making the latter part of the record very complex. In the early days of seismology the first waves were called "primary" and symbolized *P*, the next waves were "secondary" with the symbol *S*, and the third were "long" waves with the symbol *L*. Further study of the transmission of vibrations through solids has shown that the wave motion consists of two parts, waves that are longitudinal like sound waves in air and waves that are transverse like those which travel along a stretched rope. The P and S waves produced by earthquakes can be identified with these two kinds: the faster-moving P waves are longitudinal or "pressure" waves in which vibration is along the direction of wave travel, and the slower S waves are transverse or "shear" waves with vibrational movement perpendicular to the direction of travel. The very strong L waves that come still later represent complex vibrations that have traveled along the surface of the earth, analogous to ordinary waves on the surface of water.

Because S waves are slower than P waves, the length on a seismogram between the starting points of the two kinds of waves is greater for a distant earthquake than for one close at hand. The speeds of both kinds of waves are well known, from study of earthquakes and also from laboratory experiments on vibrations in rocks, so the gap between starting points on a seismogram is a direct measure of the distance to the earthquake center. In other words, when one has obtained a seismogram for an earthquake, he

Fig. 12-11. A seismogram showing first arrivals of P waves, S waves, and L waves.

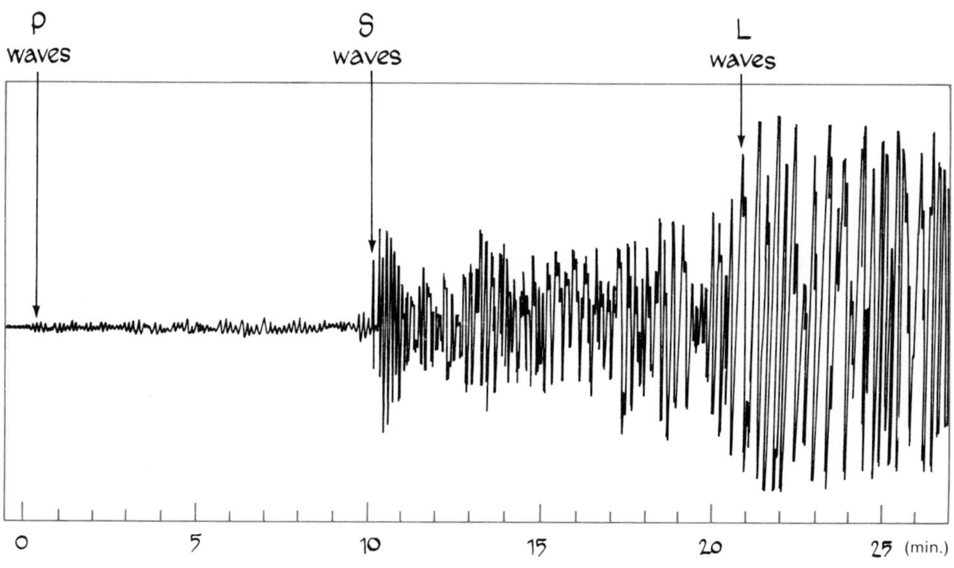

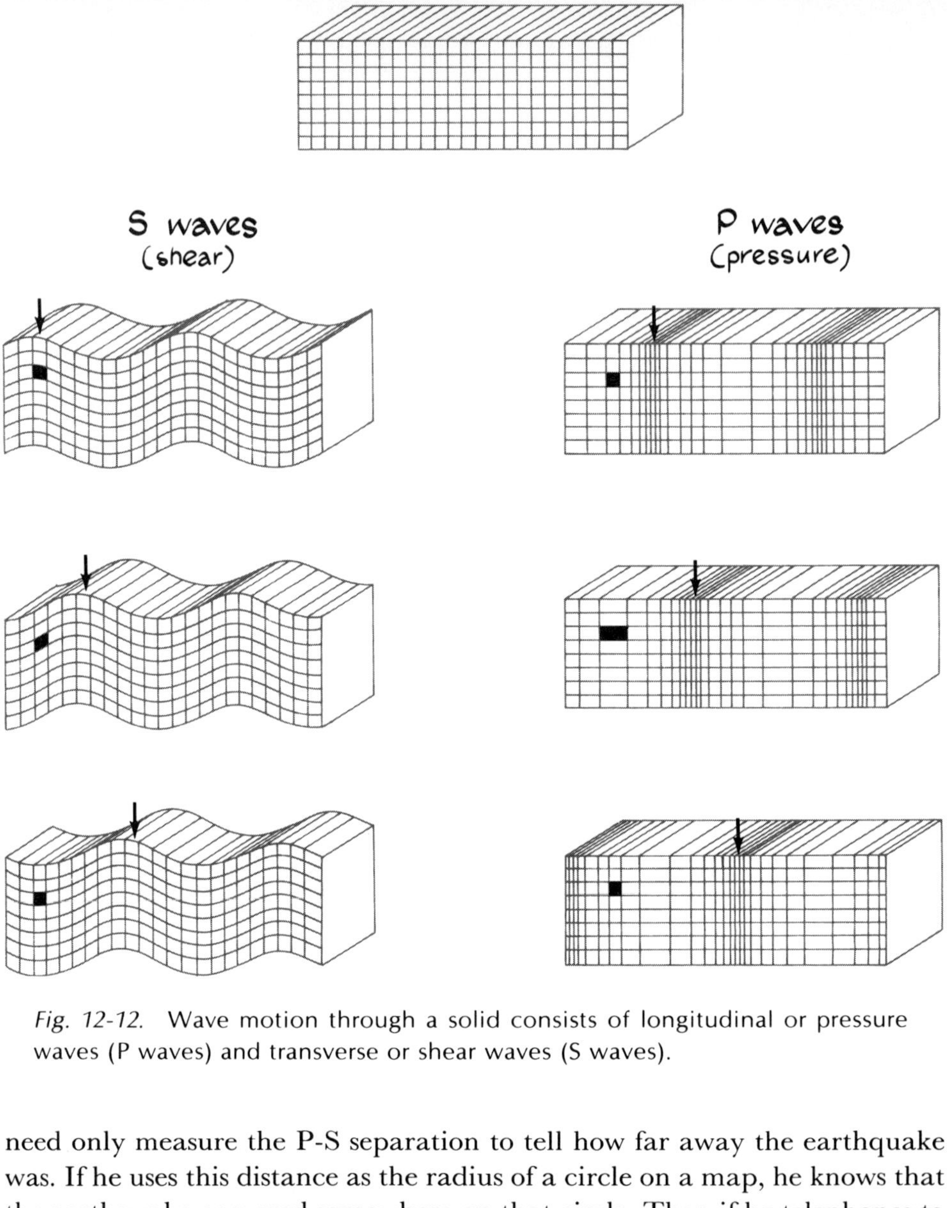

Fig. 12-12. Wave motion through a solid consists of longitudinal or pressure waves (P waves) and transverse or shear waves (S waves).

need only measure the P-S separation to tell how far away the earthquake was. If he uses this distance as the radius of a circle on a map, he knows that the earthquake occurred somewhere on that circle. Then if he telephones to friends with seismographs at two other locations, he can transfer their circles to his map; and the point where the three circles intersect is the center of the earthquake disturbance. This is the technique that enables newspapers to report the occurrence of an earthquake in a sparsely inhabited part of the earth long before actual news of the event reaches the outside world.

The point we have been calling the "earthquake center" or "center of disturbance" is described in more technical language as the *epicenter*. This is simply a point on the earth's surface, specified by its latitude and longitude.

Fig. 12-13. Locating an earthquake. The distance to the quake from each of the three stations can be found from the separation of P and S arrival times on the seismograms, and circles are drawn with these distances as radii. The earthquake must have occurred at the point where the three circles intersect.

The fracture that produces an earthquake, however, is not at the surface but somewhere beneath, so we need also a number to indicate depth. This can be calculated by comparison of records from seismographs near the earthquake. The point so located, called the *focus* of the earthquake, is the actual place where the disturbance started. Once fracturing has begun, the break may travel for long distances along the fault at depth; as the break migrates, new vibrations are set up which complicate the seismograph records. For some earthquakes it has been possible to unravel the complications and so to estimate the speeds with which fracturing is actually propagated along faults.

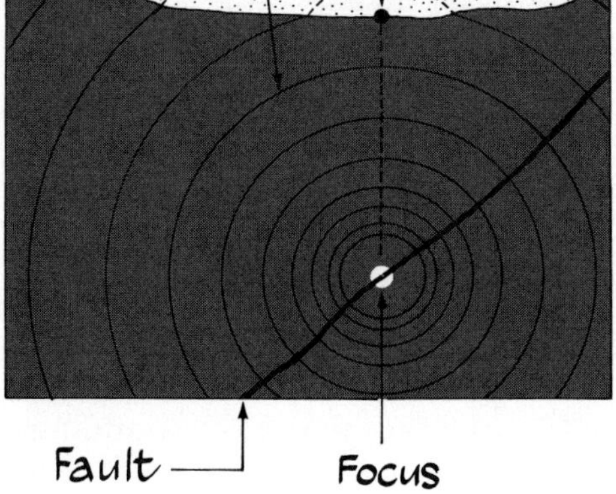

Fig. 12-14. Focus and epicenter.

For future reference some of the numbers obtained from study of seismograms will be of interest. Speeds of P waves in the upper part of the crust are generally 4 to 5 km per second, and in rocks near the base of the crust 6–7 km/sec; speeds of S waves are 1 or 2 km/sec less than these numbers. Frequencies of vibrations of the waves from a moderately strong earthquake are of the order of 1 to 10 per second. Depths to earthquake foci are most commonly in the range 10 to 50 km, but some quakes have been recorded at shallower levels and a few *deep-focus* earthquakes at depths as great as 700 km. The actual energy released during the Alaska earthquake is estimated at roughly 3×10^{24} ergs, equivalent to the energy obtainable from several thousand atomic bombs of the sort that destroyed Hiroshima.

Can earthquakes be predicted?

Earthquakes are rare events, generally occurring without warning and capable of enormous destruction. What can be done to minimize the damage and loss of life?

The most obvious answer is to improve the engineering design of buildings and other structures. People are seldom injured by an earthquake in open country; the great danger is being caught in or near a poorly constructed building. To minimize the effects of an earthquake, a building should be constructed so as to move as a whole with the vibrations, which means that it should be fairly flexible rather than rigid, that its walls and floors should be firmly tied together, and that it should be strengthened by special bracing. Structures built of brick or stone are more hazardous than those of wood or steel, or those of concrete reinforced by steel. The foundations of a building are important: structures anchored in bedrock are generally less damaged than those on alluvium or artificial fill, because earthquake vibrations are amplified in loose material much as vibrations are amplified in a bowl of jelly when the edge of the bowl is struck. Decorative cornices and overhangs on the exterior of buildings are particularly dangerous for people in the street below, and should be avoided or firmly anchored. Elementary rules of this sort have been greatly refined by engineering theory and experimentation, so that structures can now be designed capable of surviving even major earthquakes with only minor damage.

This theoretical possibility is comforting, but it doesn't completely solve the problem of minimizing damage and loss of life. For one thing, the kinds of vibrations set up by earthquakes differ from one event to another, and buildings designed to resist one kind of quake may be severely damaged by a second kind—as is attested by the fact that a few buildings supposedly of the most modern construction have performed poorly in recent earthquakes. Engineering knowledge, in other words, is still inadequate for all contingencies. Then there is the difficulty that earthquake-resistant design adds greatly to the expense of construction. In planning a building, one must often weigh the infrequency of earthquake occurrence against the extra cost of making the building completely invulnerable. And even if a structure is well designed, the natural tendency of builders to cut corners and of building inspectors to overlook details can frustrate the designer's good intentions.

Improvement in engineering practice and more rigid enforcement of building codes are at present the most promising means for reducing the damage and loss of life caused by earthquakes. But theoretically, if we had enough knowledge, we should be able to devise still better means. If we could predict earthquakes, for example, they could be prepared for in advance; or even better, if we knew exactly how earthquakes occurred, we might find ways to prevent them or to reduce their intensity. Much current research is directed toward these two possibilities—toward the prediction and control of earthquakes.

What are the chances of predicting earthquakes? At present the only sure statements we can make are broad generalizations to the effect that earthquakes have been more frequent, and hence are more probable, in some areas than in others, and that an earthquake in a given area becomes more probable as the time since the last major quake lengthens. We can say also that from a scientific standpoint the popular superstitions about earthquake prediction—that earthquakes are preceded by a particular kind of weather ("earthquake weather"), or that dogs bark unnaturally a few hours before a quake—have no basis in anything we know.

One possibility that has been suggested for slightly improving predictions depends on the fact that rocks in the crust are subject to changing forces as the position of the earth changes with respect to the sun and moon. If earthquakes result from the slow build-up and sudden release of strain, perhaps the release can be triggered by these forces. The forces themselves are far too small to account for earthquakes, but conceivably when they rise to a maximum in a part of the crust where rocks are already strained almost to the breaking point, the added force is enough to set things in motion. Checking this hypothesis is a matter of making detailed correlations of past earthquakes with phases of the moon and positions of the earth in its orbit. Such statistical studies show enough correlation to indicate that planetary forces may indeed have some influence on earthquakes, but the correspondence is far from perfect. At best such a prediction would be still very general, and could not forewarn us when the next earthquake of a particular magnitude will occur in a given region.

Where recurring earthquakes can be attributed to movement on a particular fault, as they can along the San Andreas fault in California, other ways of attempting predictions suggest themselves. The San Andreas, we recall, is a strike-slip fault on which all historic earthquakes have resulted in displacement of the land west of the fault to the north with respect to land on the east. We infer that rocks on the west are continuously moving northward, building up strain along the fault surface until a break occurs. It should be possible, then, by careful surveying of lines across the fault to keep track of the continuous movement and perhaps to learn when the movement has gone so far that a sudden readjustment is imminent. The inference has proved to be correct, and movement between the two sides of the fault has been monitored for several decades. It turns out that different segments of the fault show different behavior. About a hundred miles south of San Francisco movement is taking place along the fault surface itself, at a rate of 1 to 2 cm per year, so that sidewalks and walls built many years ago

Fig. 12-15. The peculiar jog in this sidewalk is the result of creep on a branch of the San Andreas fault over a period of a few decades. The fault runs approximately perpendicular to the street and sidewalk, and its motion is horizontal (strike-slip). The movement in this town (Hollister), about 100 miles south of San Francisco, is a continuous slow displacement without major earthquakes. (U.S. Geological Survey)

are visibly offset; near San Francisco, on the other hand, the fault itself is not moving but points at a distance are being slowly displaced. One can guess that near the city the fault is somehow locked in position, so that strain is accumulating, while to the south the steady movement prevents any significant strain from building up. By this line of reasoning San Francisco seems in danger of another major earthquake, but with present knowledge we cannot predict whether the quake should be expected tomorrow or a century hence.

The San Andreas fault is an ideal laboratory for the study of fault movement as related to earthquakes. Besides the movement itself, instruments placed along the fault have shown recent changes in electric and magnetic properties of the rocks and changes in water levels in wells and drillholes. There seems an excellent chance that continued study of such measurements will eventually provide a basis for at least rough forecasts of the timing and magnitude of future earthquakes.

Finding a way to control earthquakes is a more distant hope than developing means for prediction. A possible lead, however, is suggested by observations of recent years that injecting liquids under pressure through drillholes into rocks at depths of several thousand feet often causes a succession of minor earthquakes. To explain this phenomenon one can think of the liquid working its way along joints and into fault surfaces; if strain has been built up along the faults, the liquid may provide enough lubrication to permit slippage and so to relieve the strain. It would follow that injection of liquid at depth along a major fault like the San Andreas might have the same lubricating effect. If the injection were repeated at intervals of a few years, the accumulating strain might be relieved by a series of minor quakes and hence could never build up enough to produce a major shock. In other words, the fault would be artificially lubricated so that all parts could move fairly steadily, as does the segment of the San Andreas south of San Francisco. This is a promising and ingenious hypothesis, but finding a place to test it is difficult. Few faults are far enough from inhabited areas so that local residents would not object strenuously to the prospect of seismologists tampering with one of nature's most violent agents of destruction.

Some measure of earthquake prediction seems possible in the not too distant future, and control is not a completely outlandish dream. But for the moment humanity's best hope for coping with earthquakes lies with the geologist's ability to increase popular understanding of earthquake dangers and with the engineer's ability to design more earthquake-resistant structures.

Back to a planetary perspective

We have been looking at volcanoes and earthquakes in considerable detail, just as we have looked at many other geologic phenomena. Generally, after a detailed examination, we have expanded our viewpoint to include the whole earth. It is not enough to talk about erosion in a given area, or the folded strata exposed in a single roadcut, or the origin of a particular gabbro dike. We are concerned also with the effects of erosion, and folding, and intrusion of igneous rocks over the whole body of the planet, and over long periods of time. So now with volcanoes and earthquakes, we abandon the viewpoint of a puny mortal and imagine ourselves once more contemplating the sun's third planet from a point out in space. What do volcanoes and earthquakes have to do with the form of the earth as we see it, with the history of the earth, with the strange interior forces that have changed and re-changed the earth during so many millions of years?

Look first at the simple facts of geographic distribution. Where are the active volcanoes, and the volcanoes that have only recently become extinct? In the United States we think of the great peaks of the Cascade Range, stretching from Washington into northern California, then of the line of volcanoes in southern Alaska reaching out along the Aleutian chain toward Asia, and then of the isolated cluster of basaltic shield volcanoes in Hawaii. Looking farther afield we see the volcanoes of Central America and the long line of active peaks at the crest of the Andes; and on the other side of the Pacific we recall the many volcanoes that dot the island chains stretching southward from Kamchatka through Japan, the Philippines, Indonesia, and on to New Zealand. So many volcanoes are clustered close to the Pacific margin that we often speak of this belt as a "ring of fire." Another major belt can be discerned in the Mediterranean region, stretching from near the Caspian Sea westward beyond Gibraltar to the Azores, and with a little imagination we can project it on across the Atlantic to include the volcanoes of the West Indies. Then we add the famous volcanoes of Iceland and a few other widely spaced volcanic centers in a line down the middle of the Atlantic, and another chain of volcanoes from the Red Sea south through eastern Africa. Nearly all the world's present volcanic activity can thus be assigned to five or six major belts.

Earthquakes show a geographic pattern surprisingly similar to the belts of volcanoes. We think of regions plagued with frequent destructive quakes—the Pacific coast of the Americas, Japan, the Philippines, the Mediterranean—and the correlation with the "ring of fire" and the Mediterranean belt is striking. The correlation breaks down when we look at details, however. Southern California is famous for earthquakes, but it lies in a conspicuous gap, completely devoid of active volcanoes, in the Pacific ring of fire. The devastating earthquakes that have afflicted northern India and western China seem unrelated to current volcanic activity. There is evidently some general, overall association of frequent strong earthquakes with areas of recent eruptive activity, but the relationship is far from simple.

The peculiarities of distribution are brought out more clearly if active volcanoes and epicenters of recent earthquakes are plotted on world maps (front end paper). One of these maps shows the belts of volcanoes we have noted, and the other shows the clustering of many earthquakes, but by no means all, along these same belts. Another striking feature of the maps is the grouping of epicenters along lines far out in the ocean basins, and the spotting of a few volcanic islands along these same lines. Surely these facts of distribution have something to tell us, if we can only put them together properly, about the ultimate causes of earthquakes and volcanic activity. If we call to mind the further fact that many of the places where volcanoes and earthquakes seem to be concentrated are also places where mountains are young—in the sense that the mountains consist of Cenozoic rocks that are strongly folded and faulted—we are evidently getting close to an explanation for the movements and forces within the earth that are responsible for crustal architecture and hence for the foundations of present landscapes.

Volcanoes and earthquakes by themselves give us no more than hints about this explanation. Volcanoes tell us that temperatures a few kilometers below the surface are high enough locally for rocks to be molten, that molten rock contains abundant dissolved gas, and that the liquid-plus-gas mixture can move long distances underground. Earthquakes provide evidence that solid rocks in the crust can move and fracture, and that in some places the movement is continuous and has been going on for a long time. This kind of information is valuable but tantalizing: it tells us a little about conditions under the surface, but it leaves us without an explanation as to *why* liquid rock forms in certain places and *why* solid rocks are in motion. Only by combining these hints from volcanoes and earthquakes, further hints from their distribution and their relation to young mountains, and some other bits of information from measurements of magnetic properties of rocks, heat flow, and the ability of rocks to transmit earthquake waves, will we have enough information to try, in a later chapter, to frame a reasonable theory.

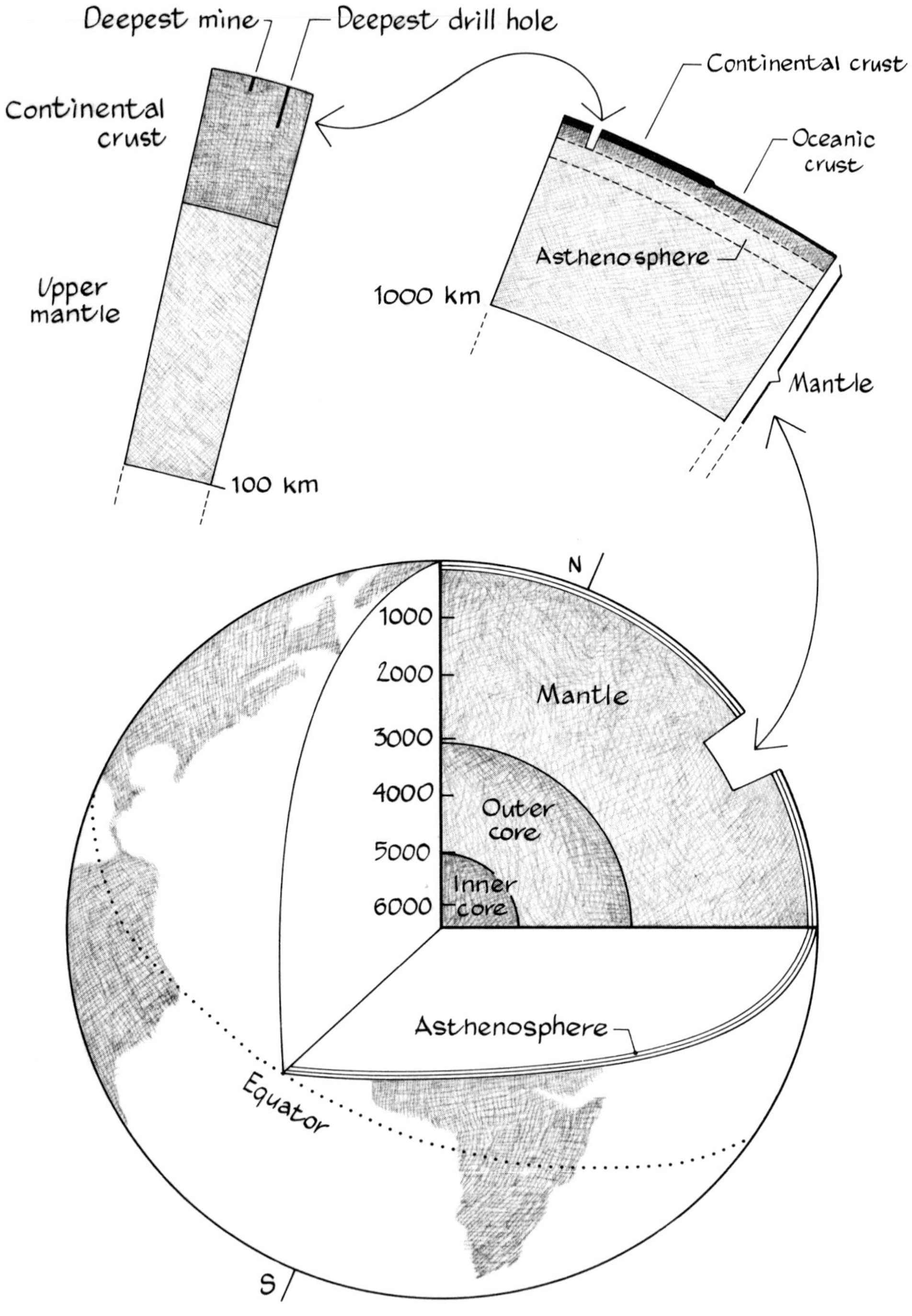

Fig. 13-1. Major divisions of the earth's interior. On the larger drawing the entire crust is included in the thickness of the outer line. To show the crust an enormous magnification is necessary, as illustrated by the two smaller drawings.

13

Deep Inside

The riddles posed by volcanoes and earthquakes, plus others suggested by the folding of sedimentary rocks and the intrusion of granite batholiths, involve processes that take place deep under the earth's surface. Such processes have always been shrouded in mystery, and to a large extent still are. How much do we actually know, and how much can we guess, about what goes on deep inside?

The best modern drilling equipment can poke holes about six miles, or ten kilometers, into the earth's tough hide. In the deepest mines men work nearly two miles underground. These distances are tiny in comparison with the earth's 4000-mile radius, but they are currently the limits of our *direct* observation of the interior. Mechanical improvements may permit somewhat deeper penetration in the future, but a large increase in these figures seems unlikely. The strength of ordinary rocks, as we saw in Chapter 7, sets a limit on the depth at which drillholes or mine tunnels would remain open, even if equipment were invented to make very deep drilling possible. All our information about the deeper interior rests on *indirect* evidence, and in all probability this is the only kind of evidence we will ever have.

Some of the available indirect evidence has been hinted at in previous discussions. We set out now to review these hints and to marshall whatever additional information we can find, so as to get as clear a picture as possible of the depths of our planet.

How heavy is the earth?

One important bit of evidence about the interior comes from a simple comparison of the earth's mass with its volume. How would you find these two quantities? The simple answer is to look them up in tables, but if you choose to measure them yourself the procedure is straightforward. To find the volume you would need the radius, and the radius comes from measurements of the curvature of the earth's surface by elementary surveying techniques. Finding the mass is a bit more difficult: you would have to measure the force of gravitational attraction between two heavy objects, which requires considerable experimental skill because the force is very small; when you have mastered the technique, you need only compare this force with the much larger force (weight) that the earth exerts on one of the objects to get an estimate of the earth's mass. Such measurements would give you a figure of about 1×10^{27} cm^3 for the volume and 6×10^{27} grams for the mass. Comparing the two numbers tells you at once that on an average every cubic centimeter of the earth's material must weigh about 6 grams. And this is extraordinary, because ordinary rocks do not weigh nearly so much; densities of the commoner rocks range between 2.7 and 3.1 grams per cubic centimeter. We are forced to conclude that materials of the deep interior are different from ordinary rocks and must be a great deal heavier.

What can this heavy stuff be? One of its important properties we have mentioned back in Chapter 7: it shows the overall behavior of a rigid solid. This conclusion is backed by several kinds of evidence, of which the two simplest are the response of the earth to tide-producing forces and the behavior of seismic waves. The tidal pull of the sun and moon acts not only on the oceans but on the body of the planet as well; if the interior were largely liquid, rather than solid, it would respond like the water of the oceans and points on the surface would show a marked up-and-down movement twice a day. Seismic waves, which we will examine more closely in the next section, travel long distances through the earth; this is true for *both* longitudinal (P) waves and transverse (S) waves, but transverse waves can move only through solids. These two pieces of evidence show convincingly that most of the interior must be solid. Bodies of liquid may exist in some places—and *must* exist, to account for volcanic activity—but the great mass of the planet is a solid material with a density considerably greater than that of ordinary rocks.

Seeing into the earth with seismic waves

We will probably never be able to look directly at the material beneath the earth's surface layer, which means that we cannot use in our investigations the wave motion of light to which our eyes are sensitive. Another kind of wave motion, however, the motion of seismic waves produced by earthquakes, does travel freely through the interior, and it is this wave motion that has provided our most detailed evidence about the earth's deeper levels. We detect this kind of waves not with our eyes, not with a microscope, but with the seismographs described in the last chapter.

Thus earthquakes are important in geology not only as agents of destruction but as generators of valuable scientific information, and seismographs are indispensable not only to determine locations and magnitudes of earthquakes but as tools for looking deeply into the earth.

The shock of an earthquake jars the rocks for miles around, and a large earthquake may shake the entire earth. The jarring consists of the vibrations we have described before—the P (longitudinal) waves, in which particles vibrate back and forth in the direction of wave travel; the S (transverse) waves, in which particles vibrate at right angles to the direction of travel; and the complex waves that travel along the earth's surface. We think of the P and S waves as traveling away from the earthquake focus in all directions through rocks, and the surface waves as moving like ripples on a pond away from the point of disturbance at the epicenter. The waves travel at different rates, so that a distant seismograph records first the arrival of P waves, then the arrival of S waves, and still later the strong vibrations of the surface waves (Fig. 12-11).

The first P and S waves to reach a distant seismograph must have traveled by the shortest possible path through the earth's interior. One would expect the path to be a straight line, or a chord of the earth's

Fig. 13-2. Chord distances through the earth from the site of an earthquake to seismograph stations. If the earth's material were homogeneous, P and S waves would follow these paths.

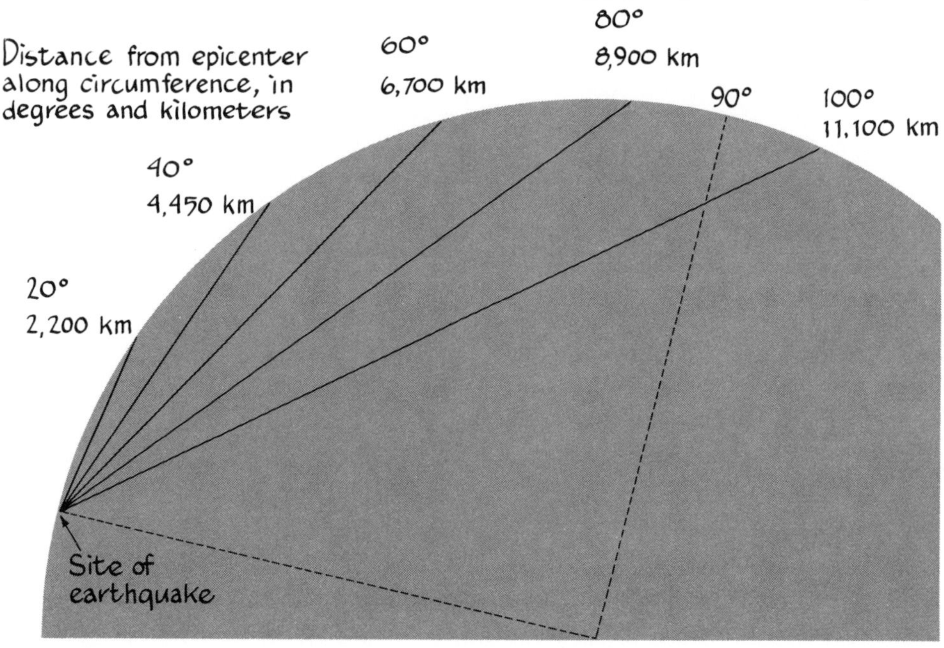

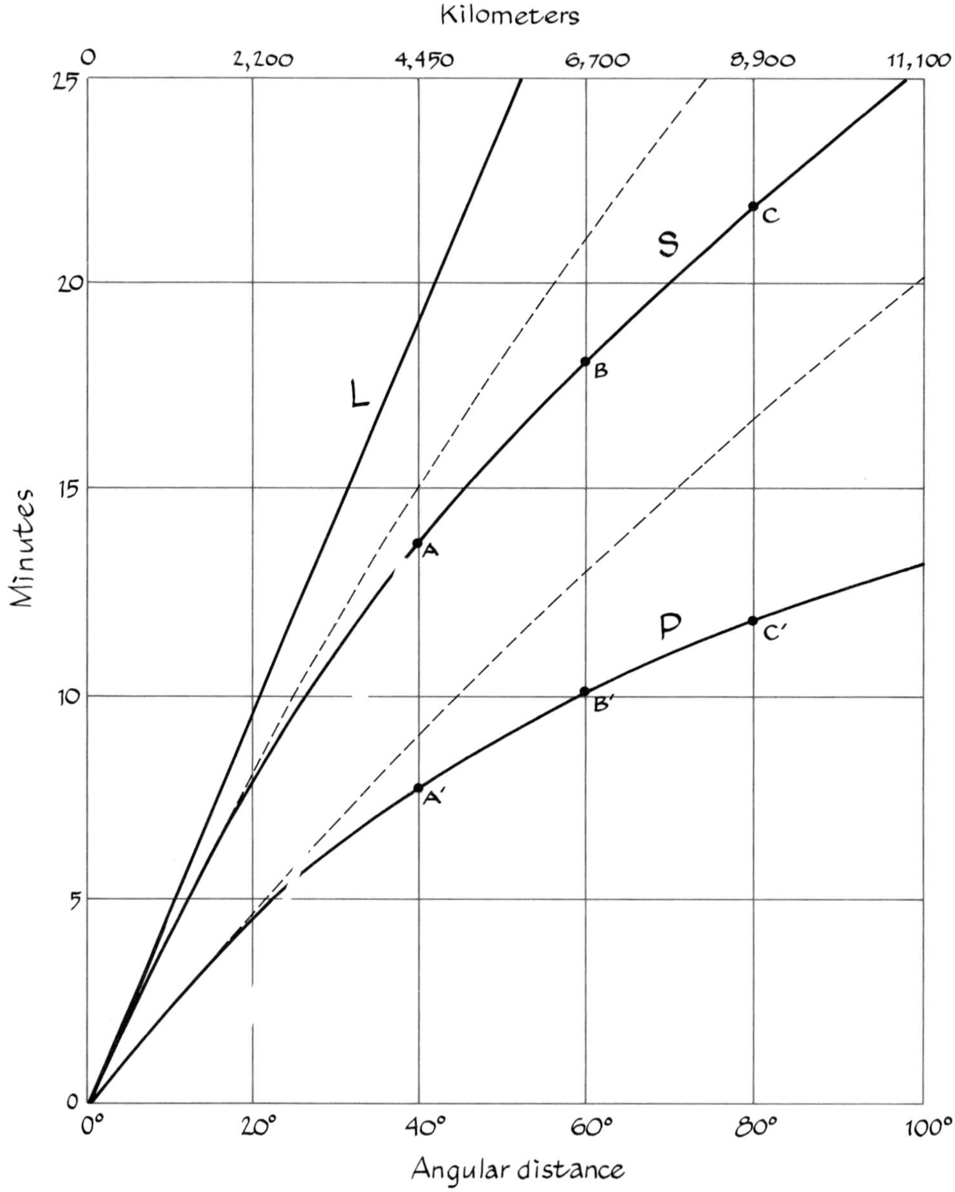

Fig. 13-3. A travel-time diagram. The heavy-line curves are drawn through points representing times of first arrival of P, S, and L waves at seismograph stations located at different distances from an earthquake. If the P and S waves had followed straight-line paths through the earth (Fig. 13-2), arrival times would be represented by the light dashed curves. Because the P and S waves arrive earlier than these predicted times, they must go faster as they penetrate deeper and deeper levels of the interior. Thus the actual paths are not straight lines but curves (Fig. 13-4).

circumference, connecting the focus (or approximately the epicenter) with the seismograph (Fig. 13-2). If this is true, the length of path to seismographs at greater and greater distances should increase just as the lengths of chords would increase if plotted on a circle; and the times of arrival of the two waves (P and S) should show the same increase with distance. In practice, if arrival times are plotted against distances (a *travel-time* curve, Fig. 13-3), the relation turns out to be less simple: the P and S waves always arrive *sooner* than they should from the simple geometric prediction. This is possible provided that the waves travel faster as they penetrate more and more deeply into the earth. Because the waves are moving through non-homogeneous material they must be refracted, just as light waves are refracted by layers of varying density in air or water. Hence the actual paths of the first waves to arrive are not straight but curved, as shown in Fig. 13-4.

The arrival times permit not only a qualitative statement about the curvature of the path but also a quantitative estimate of speeds of the seismic waves at different levels in the interior. The increase of speed with depth is fairly steady, but at some levels the change becomes abrupt, as if the waves are encountering materials of markedly different properties and perhaps different compositions. At such levels of abrupt change the seismic waves are in part refracted and in part reflected, just as any form of wave

Fig. 13-4. P and S waves move more rapidly the deeper they go, so paths of shortest time between an earthquake and seismograph stations are curved lines like these rather than straight-line chords. The amount of curvature is calculated from travel-time diagrams (Fig. 13-3).

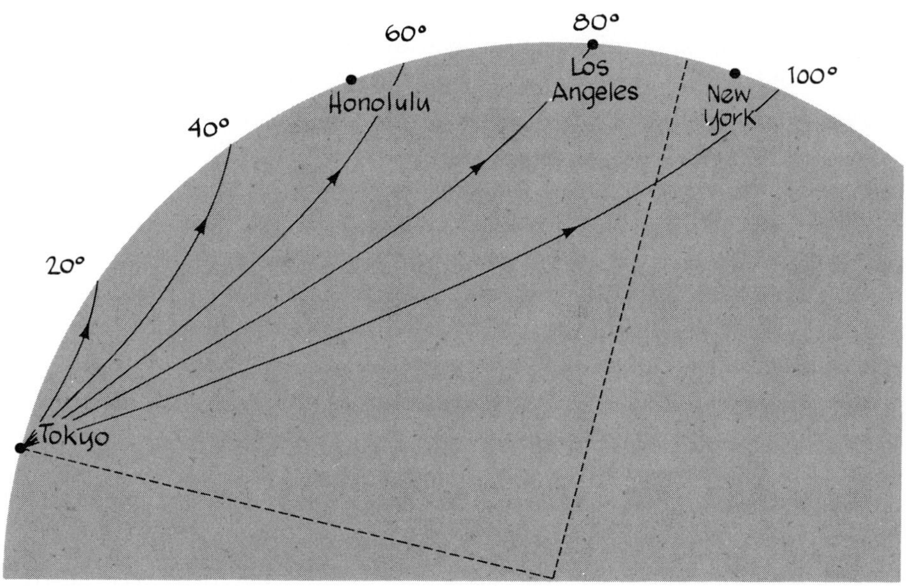

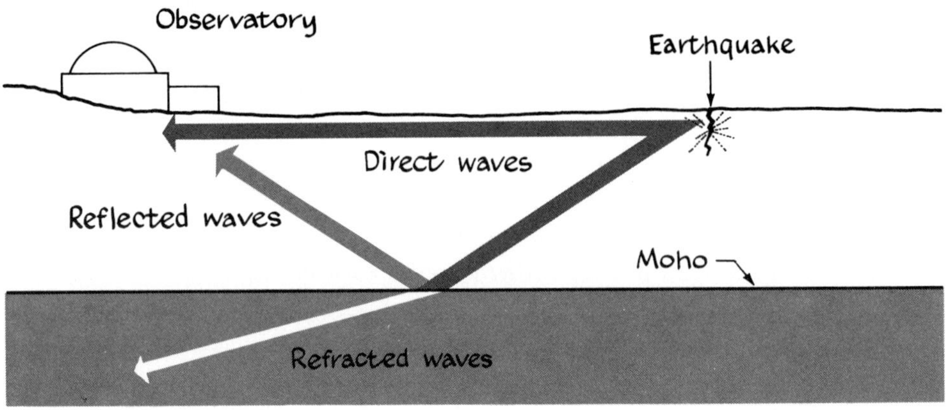

Fig. 13-5. Waves of any kind traveling through a solid are partly reflected and partly refracted when they strike a surface separating materials of different composition.

motion would be at an interface, and the resulting secondary waves produce great complications in the seismograph records. The complications make the seismograms hard to interpret, but on the other hand can provide much detailed information if the interpretation is carried out.

One striking fact that emerges from a study of complex seismograms is the existence nearly everywhere, at a depth ranging from a few kilometers to a few tens of kilometers, of a surface below which P and S waves have conspicuously greater speeds than they do in the rocks immediately above. A surface of this sort, where speeds change sharply, is called a *discontinuity*, and this particular one within a few tens of kilometers of the surface is the *Mohorovičić discontinuity* (often called simply the *Moho*), named for the Yugoslav seismologist who discovered it. Above the Moho speeds of seismic waves are typical of those in ordinary rocks; below it the speeds have higher values, which can be duplicated in rocks found at the surface only by a few rare rocks, chiefly *ultramafic rocks* consisting almost wholly of ferromagnesian minerals. The change in speed therefore probably results from an abrupt change in composition of the rocks—ordinary rocks above, heavier ultramafic rocks below.

The term "crust of the earth" has appeared frequently in past discussions, without a very precise definition. It has signified the near-surface part of the earth's solid material, but we have not tried to specify its lower boundary. The discovery of the Moho provided such a boundary, and it is common in modern geology to *define* the crust as including all rocks above the Moho. The definition gives the crust a variable thickness, for the Moho lies at an average depth of 35 kilometers under the continents but only about 5 kilometers under the floor of the oceans. When plans were formulated some years ago for drilling a hole completely through the crust

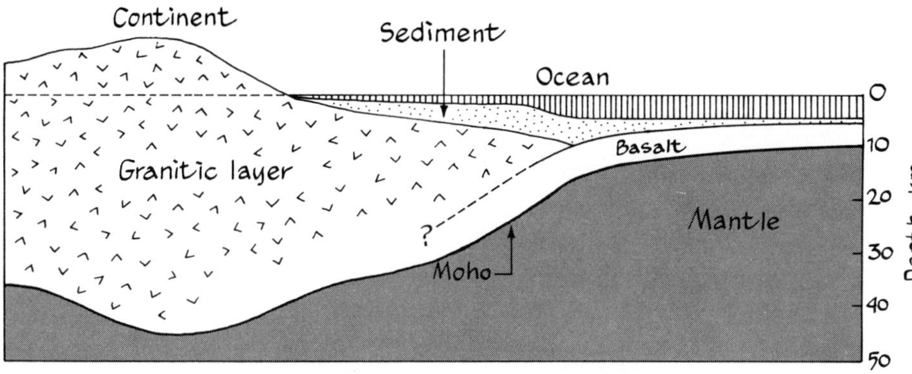

Fig. 13-6. The crust of the earth. If the crust is defined as the material above the Moho, the oceanic crust is about 5 km thick and the continental crust averages about 35 km. The level of the Moho beneath the continents varies considerably from place to place, reaching depths of 60 or 70 km under some mountain ranges.

and into the material beneath (the "Mohole"), the possible sites were all in the ocean. Despite the difficulty of drilling from a floating platform, the crust is so much thinner under the sea that the engineering problems seemed more manageable than they would have been at a site on land. From the standpoint of scientific and engineering progress, it seems unfortunate that financial and procedural difficulties prevented the Mohole experiment from being carried out.

Besides the Moho, the only other major discontinuity revealed by seismic waves is one at a depth not quite half-way to the earth's center. This surface must represent an even more profound change in material, for S waves die out completely below it and P waves have their speeds sharply reduced (Fig. 13-7). The disappearance of S waves suggests that the material below the discontinuity is liquid, for transverse waves can travel only in a rigid solid. Hence our statement in the last section that the earth behaves as a rigid solid needs modification: the material below a point half-way to the center is probably liquid, but so much of the earth is solid that the planet as a whole acts like a rigid body.

The two major discontinuities divide the interior into three concentric regions: the part near the center, called the *core*, probably in large part liquid; the solid *mantle* between the discontinuities, making up nearly half the total thickness and by far the largest part of the total volume; and the thin solid *crust* between the Moho and the surface. The crust is so thin in comparison to the others that on a diagram like Fig. 13-7 or Fig. 13-1 its thickness must be exaggerated for it to show at all. Other discontinuities can be reçognized in all three parts of the interior, but the changes in seismic velocities that they represent are not as sharply defined as across the major breaks. Of the secondary discontinuities perhaps the most important

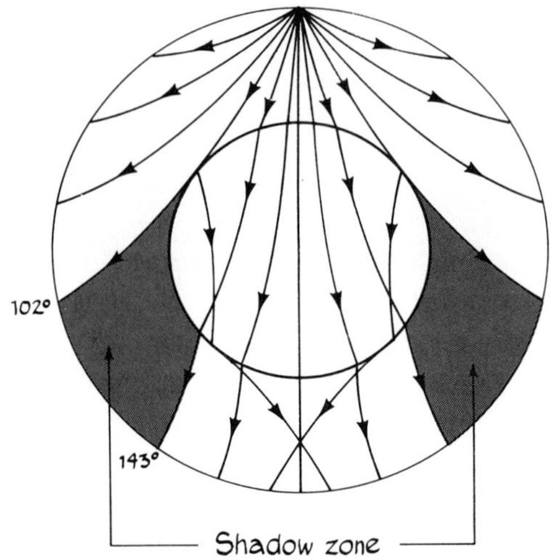

Fig. 13-7. The earth's core. Evidence for the existence of a liquid core comes from the behavior of S and P waves that travel long distances from a major earthquake. S waves are not recorded at all at points more than 102° from the epicenter (except for feeble waves from multiple reflections). P waves are not recorded at distances between 102° and 143°, and those that appear beyond 143° arrive later than expected. These observations can be explained if the material below a depth of 2900 km is liquid, for a liquid would damp out transverse (S) waves and greatly reduce the speed of longitudinal (P) waves. Between 102° and 143° P waves are absent because they are refracted into the liquid, much as light waves moving from air into glass are refracted into the glass, where their motion is slower. The belt between 102° and 143°, where both kinds of waves are absent, is appropriately called the *shadow zone.* (After B. Gutenberg, *Internal Constitution of the Earth,* 1950, courtesy of Dover Publications)

is one in the core, difficult to detect and still somewhat questionable as to its meaning, but probably separating a solid *inner core* from the liquid *outer core.*

What is the earth made of?

From the earth's mass and volume we learn that the interior is heavier than ordinary rocks, and from seismic waves we find that the heavy material is arranged in concentric shells. The earthquake waves tell us further that most of the heavy substance is solid, but that a thick shell between the inner core and the bottom of the mantle is liquid. Still more details can be gleaned from the seismic waves, for their speeds permit calculation of the elastic properties of the materials through which they travel. Yet all this information leaves us with a very imperfect idea of what the heavy interior substance actually is. Many well known materials would fit the requirements of density and elastic properties, so we face the problem of guessing which would be the most likely.

We might try first an appeal to geology. A geologist can see beneath the surface no better than anyone else, but he has ways of telling whether a rock now exposed at the surface has been deeply buried at some time in the past. By careful geologic mapping and constructing cross-sections he can see what rocks have been piled on top of others and what ones have been raised by faulting or folding and cut into by deep erosion. In some places there is good geologic evidence that rocks now exposed were once near the base of the crust, or that they have come even from below the crust in the upper mantle. The commonest of the rocks that can be identified as probable samples of mantle material is the heavy ultramafic rock *peridotite*, a rock with the coarse-grained structure of an intrusive igneous rock and consisting entirely of ferromagnesian minerals (olivine and pyroxene). If we guess that the overall composition of the crust in continental areas resembles granite in its upper part and basalt in its lower part, and that the Moho separates basalt above from peridotite below, we find that speeds of earthquake waves in shallow levels of the earth can be satisfactorily explained.

But this kind of geologic argument takes us only a short distance into the earth. What about materials of the deep mantle, which are never exposed at the surface? What about the mysterious heavy core, with its fluid outer part and solid center? Here we have no recourse but a flight of astronomical fancy. We recall that the earth is one of a family of planets circling the sun. We do not know how this family was formed, but it seems reasonable to suppose that the earth originated at the same time as the other planets and contains the same general sort of material. Nothing is known for sure about the composition of the solid materials of the other planets (except for the samples plucked from the surface of the moon), but we do find on the earth pieces of material that may represent a sort of average planetary substance. These are the *meteorites*, chunks of rock that have fallen from the heavens—their fall marked by the brief bright trails on the night sky that we call *meteors* or *shooting stars*. Meteorites may represent fragments of a shattered planet, or perhaps remnants of the original material from which the earth was made, but in either event they give us a probable sample of planetary substance.

Acting on this hunch, we look at the composition of meteorites. It turns out that there are two principal kinds and many minor varieties. The

Fig. 13-8. A meteorite, with one surface cut and polished to show the inner structure. The black outer "skin" is material that melted from friction during the flight through the atmosphere. This is the commonest kind of meteorite (*chondrite*), with a composition like that of ultramafic rocks and a structure characterized by small spherical masses of crystals called *chondrules.* (Carleton Moore)

principal ones are the iron meteorites, consisting largely of metallic iron alloyed with a few percent of nickel, and the stony meteorites which have compositions resembling that of peridotite. If we try to strike an average meteorite composition (which is difficult to do, because meteorites are so variable and the more unusual ones are favored in collections), we find something less than half metallic iron and more than half ferromagnesian silicates—roughly the proportion we would calculate for the composition of the earth if we imagine that the core is largely iron and the mantle largely peridotite. The correspondence is so striking that this assumption about the general composition of the interior has not been seriously challenged for more than a century.

In modern times much additional evidence has been found as support for the assumption. The densities and elastic properties of iron and peridotite under the high-pressure, high-temperature conditions of the earth's interior can be estimated by extrapolation of laboratory measurements, and turn out to agree nicely with figures for the earth's density and the observed speeds of seismic waves. Temperatures in the interior can be guessed only roughly, but the change in melting point of iron with pressure will permit the outer part of a metal core to be liquid at reasonable values of the temperature. Much argument remains about details of composition, particularly about minor changes from one level to another of the interior, but a steadily increasing mass of evidence corroborates the general picture of a core consisting of iron alloyed with a little nickel, a mantle containing chiefly iron-magnesium silicates, and a crust made up of familiar rock types.

The astronomical argument leads directly to speculations about the make-up of other bodies in the solar system. If the planets and satellites were formed at the same time, all should consist, like the third planet, chiefly of iron-magnesium silicates and metallic iron. Different proportions can account for observed differences in densities of the smaller planets. The large cold outer planets, with very low densities, must contain also large amounts of lighter material such as ice, solid carbon dioxide, and solid ammonia. But it will be an enormous surprise if further exploration of the planets turns up anything as major constituents of their heavy rocky interiors other than iron and the silicate minerals of peridotite.

How hot is the interior?

We have noted that interior temperatures can only be guessed, but it's worth looking briefly at the kinds of guesses that are possible and reasonable.

Measurements in mines and drillholes show that temperatures increase steadily with depth below the surface. We recall from Chapter 7 that the average rate of increase (the *geothermal gradient*) is about 30°C per kilometer, but in some places is as low as 10° and in others as much as 50°, and even higher in volcanic areas. If the gradient continues downward at 30°/km, the Moho under the continents (depth averaging 35 km) should have a temperature of about 1000°C, hot enough for melting to start in many ordinary rocks. At depths of only a few tens of kilometers into the mantle all rock material should be molten. Since we have good evidence that it is not, our assumption of a constant gradient must be faulty. The 30°/km figure is valid only very close to the surface, and at lower levels the rate of increase is much smaller.

It seems likely, nevertheless, that much of the mantle is not far below its melting point. The melting point is probably approached most closely at depths between about 60 and 200 kilometers; in this interval seismic waves travel at speeds slightly less than those either above or below, indicating that the material is less rigid. There may well be pockets and stringers of molten material, and these may be the source of some of the lava that appears in volcanic eruptions. Because this layer in the upper mantle is weaker than layers above and below, it is labeled the *asthenosphere* (from the Greek word for feeble). Under the asthenosphere the temperature continues

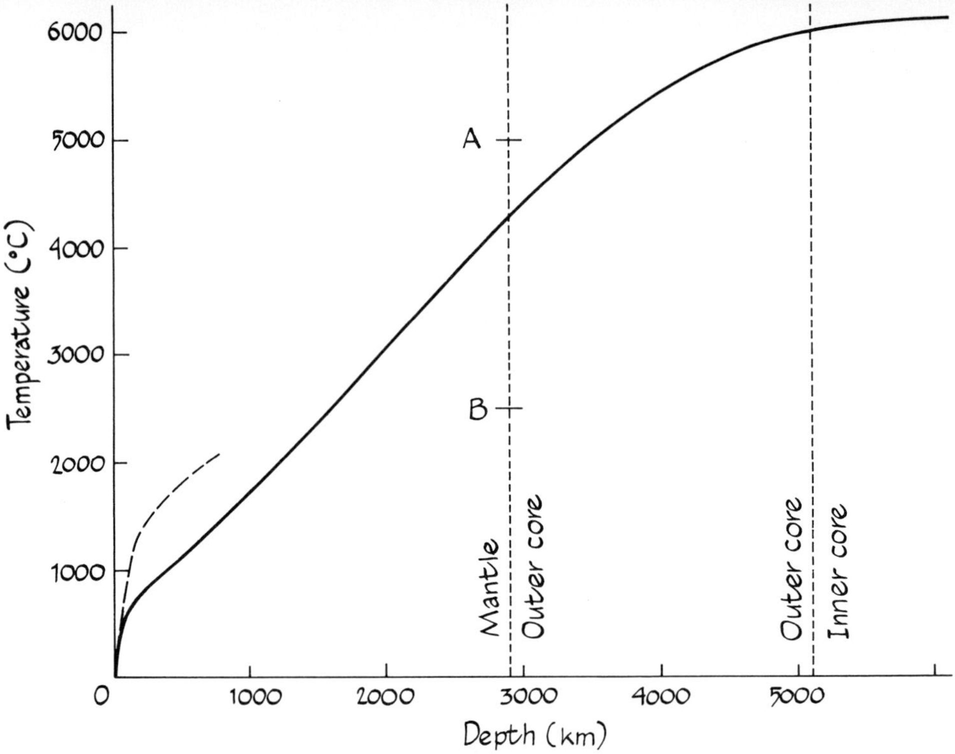

Fig. 13-9. Graph showing a possible distribution of temperatures in the deep interior. The part of this curve at deeper levels may be in error by as much as 1500°. A graph of temperatures at shallow levels from an earlier chapter (Fig. 7-2) is reproduced here as a light dashed line, and the discrepancy between the two curves is an indication of the uncertainties involved in temperature estimates. (After Takeuchi, Uyeda, and Kanamori, *Debate about the Earth,* 1970, Freeman, Cooper, & Co.)

to rise slowly with depth, but the melting point of the rock also rises because pressure is steadily increasing—and evidently the rise in melting point keeps ahead of the rise in temperature, since the material remains solid down to the boundary of the core. Estimates of temperature at this boundary are in the range 2500–5000°C, and for the still deeper interior near the earth's center one or two thousand degrees higher.

Why should the earth's interior be so hot? Two possible reasons come to mind: (1) perhaps the earth was very hot when it was formed, and the interior hasn't yet cooled down, and (2) rocks contain trace amounts of radioactive elements, which produce heat as they decay. To suppose that the original heat of the earth is still preserved seems at first sight an outlandish hypothesis, for one would think that four or five billion years should be ample time for a planet to cool down completely. A little study of heat flow through rocks, however, shows that the hypothesis is far from outlandish. Rocks are such good insulators that if the earth loses heat entirely by conduction to the surface, none of the original heat from depths below about 400 km would yet have escaped. Radioactivity also at first glance seems an inadequate source of heat, because the quantities of radioactive material are so small; but again the insulating properties of rock are so effective that minute amounts of heat would accumulate and over hundreds of millions of years could raise the temperature substantially. The radioactive content of ordinary crustal rocks is well known, and calculation from these figures shows that temperatures in the crust can be accounted for solely by the energy released in radioactive decay. Whether radioactivity is entirely responsible for high temperatures in the mantle and core is less certain, because the distribution there of radioactive elements is not accurately known.

Because the interior is hot, the earth must necessarily be losing some heat through its surface. Laboratory measurements of the thermal conductivity of rocks combined with field measurements of the temperature gradient in drillholes permit an estimate of the amount of heat that is flowing to the surface and is being radiated into space. Figures obtained for different locations are surprisingly uniform, the average loss over the entire earth being about a millionth of a calorie per second from each square centimeter of surface. This is not very much heat; at this rate about three months would be required to melt a layer of ice 1 millimeter thick. It is far less, more than a thousand times less, than the heat received by the earth's surface from sunlight. But over the enormous area of the earth's surface and over geologic times the amount of heat lost by the earth is prodigious.

The uniformity of heat flow from different parts of the earth's surface, if you stop to think about it a moment, is truly astonishing. Measurements have been made in a great variety of geologic situations: in mountains, plateaus, and plains, in all manner of igneous rocks and sedimentary rocks, in rocks of different ages from the ancient schists and gneisses of eastern Canada to the young sediments of the Gulf Coast, in rocks and sediments under the ocean as well as on the continents. There are differences, of course, but most values are within a factor of two or three of the average. Near active volcanoes, as might be expected, the figures are much higher. Many of the minor differences can be explained by differences in concentrations of radioactive elements in rocks near the surface; thus heat flow may be higher from a granite batholith than from an area underlain by thick limestone beds, because granite contains more of the heat-generating elements uranium, thorium, and potassium. But this explanation does not account for the similarity in heat flow between continents and ocean basins. Under the oceans one would expect the heat flow to be small, because the oceanic crust consists of a thin layer of basalt whose radioactive content is far less than that of the thick granitic crust under the continents. The relatively high heat flow from the material beneath the oceans, despite its low concentration of heat-producing elements, is one of the strange facts that must be explained in any theory about the behavior of the earth's interior.

Curiosity prompts another question about temperatures. If heat is being steadily produced in the earth by radioactivity, and if heat is being steadily lost from the surface, which of these is dominant? Is the earth as a whole heating up or cooling off? Before the discovery of radioactivity the answer seemed obvious: the earth was necessarily cooling off, because there was no internal source to make up for the heat radiated out from the surface. Radioactivity supplies such a source, and the answer is no longer at all obvious. The best current guess is that heat production and loss are approximately in balance, so the earth is getting neither hotter nor colder, and the distribution of temperatures in the interior has been constant for much of geologic time. But this conclusion may well change as more information becomes available.

The earth as a magnet

When mariners in the 15th century discovered that magnetized needles of iron could be an aid to navigation when they were out of sight of land, they set in motion a long and fruitful study of another important attribute of the earth's interior.

The earth behaves like a magnet, in the sense that any magnetized object that is free to move tends to swing into a preferred position with respect to geographic directions. One might try to explain such an observation at first by supposing that the magnet was being attracted by a mass of highly magnetic rock immediately below the surface, but such an hypothesis is easily proved wrong by the further observation that the tendency to move is not limited to any one locality but is universal over the earth's surface and conforms to a simple overall geographic pattern. By exploring the pattern with a magnetized needle suspended so that it can move in three dimensions, we find that the needle behaves as if one end is attracted to a magnetic pole deep under the Arctic coast of Canada, and the other end to a pole under the coast of Antarctica south of New Zealand. Clearly it is the earth itself that acts like a huge magnet, and the field of the magnet is the sort that would be produced by a bar magnet near the earth's center oriented at a slight angle to the axis of rotation.

Strictly speaking this is too simple a picture. There are many deviations from the pattern of a bar-magnet field, some explainable by masses of highly magnetic rock near the surface and others evidently having

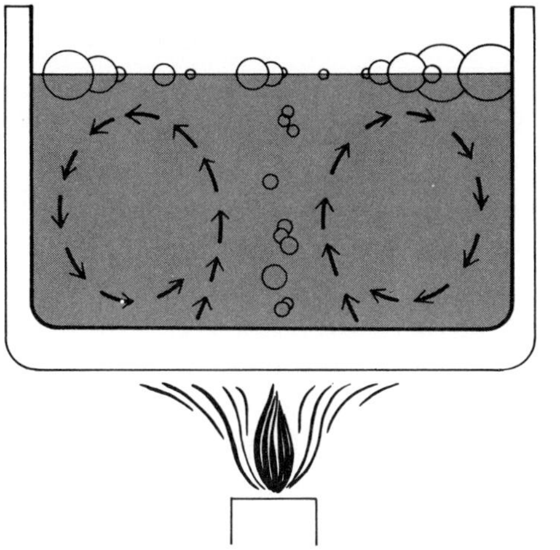

Fig. 13-10. Convection currents in water.

their origin at greater depth. But it is the general field that needs our attention here. In one sense the magnetic field adds to the mystery of the interior, but on the other hand it may serve as an additional tool for exploring far beneath the surface.

How is it possible for the earth to be a magnet? A first guess is easy: the core very probably consists chiefly of iron; iron is the best known of substances that can be strongly magnetized; therefore the core simply functions as a huge mass of magnetized iron. Like so many pretty hypotheses, this one can be easily demolished. When a piece of magnetized iron is heated, its magnetic properties disappear abruptly at a temperature of about 770°C, called the *Curie temperature.* No amount of pressure or impurities can raise the temperature of disappearance much above this. The core is far hotter than 770°, so any iron it contains cannot be magnetic in the sense that a bar magnet is magnetic. Other possible magnetic substances have similar temperature limits to their magnetic properties, all far below the temperatures in the core. We must seek some other way to account for the earth's effect on compass needles.

Electric currents, it is well known, generate magnetic fields. Can we imagine some reason for currents to be flowing in the earth's interior? We recall that the outer part of the core is molten iron. It is hot enough for some of the iron atoms to be ionized, and movement of ionized atoms is one form of electric current. Why should there be movement in the core? If the solid inner core is at a higher temperature than the core-mantle boundary, liquid iron will tend to move from the hot boundary to the cooler boundary because the density of iron decreases on heating—much as water heated in a pan moves from the hot bottom of the pan to the cold upper surface. As the liquid is cooled against the mantle boundary it grows heavier and tends to sink, so that a circulatory motion, called *convection,* or a *convection current,* is set up in the molten part of the core. The rotation of the earth, which of course carries the core with it, would modify such a convectional circulation profoundly, causing a large part of the motion to be horizontal or nearly so. Details are difficult to work out, but a reasonable circulation pattern can be calculated which would account for the major features of the earth's magnetic field.

In effect, this hypothesis pictures the earth's outer core as acting like a sort of huge dynamo. Just as the dynamo in a power plant produces an electric current from the heat energy of burning coal or oil, so the earth's internal heat causes motion of electric charges. The ultimate energy source is the temperature difference between inner core and mantle; any motion of charged particles caused by the temperature difference would generate changing magnetic fields, which would cause additional motion of charged particles; the motions are modified by planetary rotation; and the net effect is the generation of strong electric currents which are manifested at the surface by the magnetic field. The hypothesis is highly speculative, but it is the best answer at present to the question why the earth should act like a magnet.

The dynamo hypothesis not only accounts for the general pattern of the earth's field, but also provides insight into some more detailed aspects. A circulation of molten iron like that pictured should not be entirely stable, but would fluctuate slowly in intensity and position. The fluctuations would of course affect the magnetic field, and serve as a convenient explanation for observed changes that have long seemed mysterious. In the three centuries during which accurate records of the earth's magnetism have been kept, both the directions and intensities of the field at various places have gradually changed. Over most of the United States, for example, between 1850 and the present time compass directions have changed by a few tenths of a degree and the intensity of the field has dropped by about six percent. According to the dynamo hypothesis, however, changes in the positions of the magnetic poles must be limited to about 20° from the geographic poles. Despite the continual fluctuations of the field, when averaged over millions of years the positions of the earth's magnetic axis and rotational axis should be the same.

Magnetized rocks

If the earth's field is continually changing, the traditional geologic question springs to mind: Can we follow these changes into the past? Have we any way of guessing the direction and strength of the field in a given area at various times in geologic history? Up until about 1940 the answer would have been a flat "No!", but in recent decades the question has not only been answered positively but has proved to be one of the best of all guides to unraveling the inner operations of our planet.

Magnetism, in ordinary experience, is limited to a very few substances: iron, the black iron-oxide mineral magnetite ("lodestone"), and some man-made alloys. Many other materials are faintly magnetic. With sufficiently sensitive instruments, in fact, practically all substances can be shown to respond in one way or another to a strong magnetic field. The magnetism of many ordinary rocks is easily measurable, particularly rocks that contain iron-oxide minerals. Concentrations of iron-oxide minerals can make a rock so strongly magnetic that a sensitive instrument ("magnetometer") will detect its presence even at a distance, say when the instrument is towed behind an airplane, and this fact is the basis for a spectacularly successful method of prospecting for iron ore. More ordinary rocks contain only minor amounts of magnetic minerals (generally magnetite, Fe_3O_4, or the less magnetic mineral hematite, Fe_2O_3), but many have enough to deflect the needle of a nearby magnetometer. The overall magnetism of a rock specimen, which is a summation of the magnetic effects of its individual grains, generally shows a specific direction—which means that the specimen would act like a weakly magnetized compass needle if suspended so that it is free to turn. Now the important question for present purposes is the relation, if any, between this direction in the rock specimen and the direction of the earth's field.

A rock may acquire a direction of magnetization in several ways. If it contains grains that can be magnetized, and if at some time during the rock's formation the grains were free to move, the grains would tend to align themselves with the earth's field. Or the rock, after its formation, could acquire slight magnetization parallel to the earth's field simply by being struck or jostled. Particularly strong magnetization is produced if a rock is heated above its Curie point (the temperature above which it cannot be magnetized) and then slowly cooled through this temperature. Magnetization acquired in any of these ways tends to change if at some later time the earth's field changes. But the change is sluggish, and if a rock remains undisturbed at low temperatures it will retain its original direction of magnetization almost indefinitely. Thus by measuring the direction of a rock's magnetization we have a way of determining the direction of the earth's field at the time the rock was formed.

This technique is particularly effective for basalt. As basaltic lava crystallizes, tiny grains of iron-oxide minerals are among the first to form. For a time the grains are well above their Curie temperature, but as the rock slowly cools they pass this temperature and acquire a magnetic orientation parallel to the then-prevailing earth's field. As long as the solid rock is not disturbed, it preserves a record of this direction. Another rock that sometimes betrays the field direction at the time of its formation is sandstone, particularly sandstone containing abundant magnetite; if the original sand was deposited from quiet water, so that the magnetite grains as they settled could orient themselves roughly in response to weak magnetic forces, the average direction of magnetization of the aggregate will show the earth's field during sedimentation. Still other rocks may occasionally be useful, but the one that has given most information about past magnetic directions is basalt.

What you will need, then, if you set out to explore the history of the earth's magnetism, is a means of obtaining fresh specimens of basalt of various geologic ages, from lava flows that have not been strongly tilted or folded since their formation. Each specimen as it is collected must be marked so as to indicate its orientation accurately—the north-south direction and the up-and-down direction at the present time. Then the specimens must be brought to the laboratory and the direction of magnetization of each one determined with delicate instruments.

If you are fortunate enough to be working in an area where lavas of several different ages are exposed, you will find that the direction of magnetization changes as you test older and older specimens. There will be minor fluctuations in flows of nearly the same age, as you would expect from the fact that the earth's field has changed somewhat even in historic time. But superimposed on this slight variation will be a progressive and much larger change, indicating that the magnetic poles have had very different positions in the geologic past. Basalt flows from the beginning of the Cenozoic Era, for example, have magnetic orientations that point to a pole northwest of Bering Straits at the western tip of Alaska; Mesozoic lavas suggest pole positions in eastern Asia, and Paleozoic lavas various positions in the western and central Pacific Ocean.

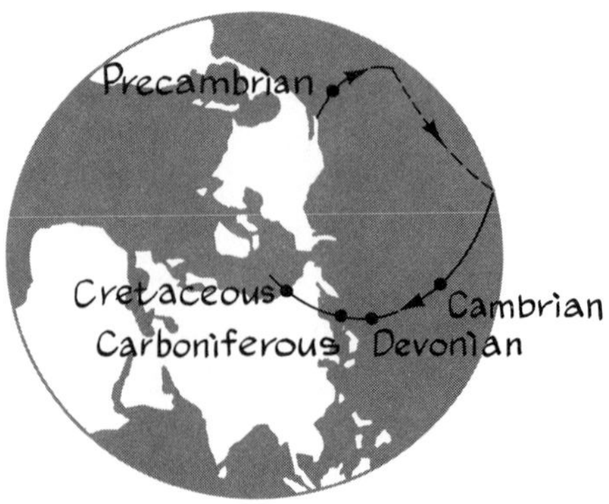

Fig. 13-11. Positions of the magnetic pole from the Precambrian to the Cenozoic, as determined from measurements on European rocks. (From Fig. 20 in *Continental Drift,* edited by S.K. Runcorn, 1962, with permission of Academic Press, Inc., New York)

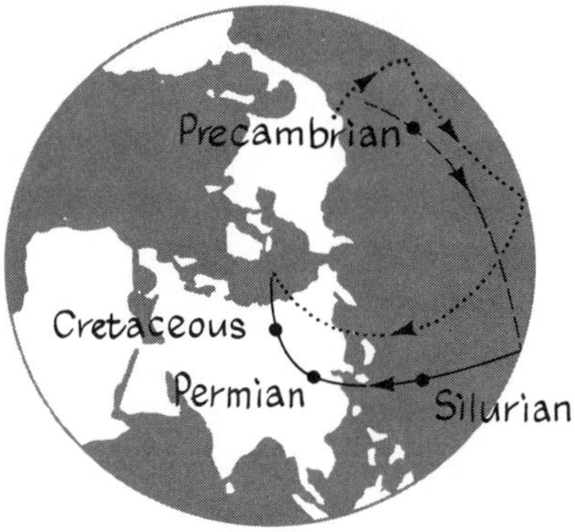

Fig. 13-12. Positions of the magnetic pole as determined from rocks in America (solid line), compared with positions obtained from European rocks (dotted line, repeated from Fig. 13-11). (From Fig. 20 in *Continental Drift,* edited by S.K. Runcorn, 1962, with permission of Academic Press, Inc., New York)

Now the startling thing about these results is that they indicate a migration not only of the magnetic poles, but of the geographic poles as well. The origin of the earth's magnetic field is still uncertain in detail, but all current hypotheses agree on a close relation of the field to the earth's rotation. The magnetic axis may deviate from the rotational axis by as much as 20°, as it does at present, but not by much more. So if the magnetic pole in the northern hemisphere has moved in the last half-billion years from the central Pacific into eastern Asia and thence into the Arctic, so also has the north pole of rotation.

What does this mean? The earth's rotational axis, from abundant astronomical evidence, is fixed in position in space (specifically, the angle between the axis and the plane of the orbit is fixed); enormous forces would be required to change the axis, and there is no sign that such forces exist. If the axis itself has not moved, then a shallow surface layer of the earth must have moved. We can think, for example, that the crust may have slid over the mantle, bringing different points on the surface directly above the constant axis on which the earth spins. Thus these simple magnetic measurements give a first hint of very large displacements of the crust (or of a shallow layer including the crust) with respect to the mantle beneath.

Another curious fact emerges from comparisons of magnetic measurements on different continents. Past positions of the magnetic pole estimated from lavas in North America do not agree with positions obtained by measurements on similar rocks in Europe, and the differences become greater for progressively older lavas. In other words, we apparently have evidence not only for a horizontal movement of the entire crust as a unit, but for differential movement of one continent with respect to another.

This striking conclusion prompted much effort to trace details of movement of the different continents by further magnetic measurements. But the effort proved unrewarding. Magnetic measurements are subject to enough uncertainty and enough difference in interpretation so that precise pole positions have become a source of endless argument. The continents almost certainly do move, and the rates of movement can be estimated, but the estimates have come from an entirely different line of reasoning.

Magnetic reversals

The most astonishing fact of all about rock magnetism is that many basalts show a direction of magnetization exactly opposite to the present field. If you are working with a lava about 800,000 years old, for example, you will find that the magnetic orientation of the rock is approximately parallel to the present-day earth's field (as it should be, for a rock this young), but that the *direction* of magnetization is reversed. At the present time the N pole of a compass is attracted to a point in the Arctic; if you could go back in time 800,000 years, you would find that the S pole would point north. Careful measurements of many rocks from all over the earth leave no doubt that the earth's field rather suddenly reversed itself about 700,000 years ago. And this is only the latest of many similar reversals. In the geologic past the magnetic field has been about as often in one direction as the other, and the intervals between reversals have had widely varying lengths (Fig. 13-13).

How can we interpret this singular behavior?

At first glance it seems completely absurd that a magnet the size of the earth should suddenly, within a few thousand years at most, change its polarity. When magnetic measurements first suggested this conclusion, it was as difficult for geologists to believe as for laymen. We still have no good explanation: the dynamo hypothesis for the origin of the earth's field can suggest possible mechanisms that might lead to reversals, but the hypothesis itself is so uncertain in detail that this is not very satisfying. As often happens in geology we must accept the facts without a complete understanding of what they mean, and hope that further investigation will gradually fill in the missing parts of the picture.

Whatever the ultimate explanation may be, the fact of reversals at definite times in the geologic past is supported by overwhelming evidence. Where basalt flows that followed one another in rapid succession can be accurately dated, a reversal is indicated by a 180-degree change in direction from one flow in the sequence to another. Presumably the field decreased to zero and then built up in the opposite direction, but the change was so rapid that intermediate stages are not recorded. The entire process must span only a few thousand years at most. Abundant measure-

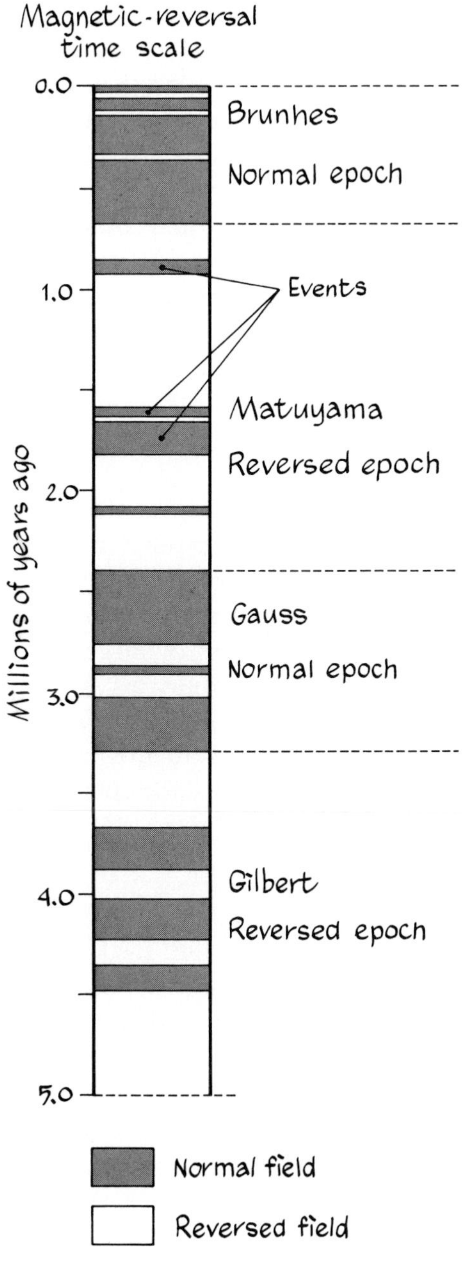

Fig. 13-13. Reversals of the earth's magnetic field. Times when the earth's field had the same orientation as at present are indicated by shaded parts of the graph, and times when the field was reversed by white parts. The major periods when the orientation was one way or the other are called *magnetic epochs,* but within each epoch there were short intervals ("events") when the field had the opposite orientation. (After Allan Cox)

ments have established that each change was worldwide in its effect—as of course it must be, if the change has its origin in the earth's core.

Because the reversals are worldwide and very abrupt, they have proved useful as fixed points in the geologic time-scale. A reversal in itself conveys no information about its age in years, but as part of a sequence of reversals its age can often be guessed by comparison with dated sequences elsewhere. If, for example, basalts in one area are interlayered with sedimentary rocks containing fossils that can be dated, their magnetic reversals can be readily recognized; in an adjacent area where sedimentary rocks are absent, basalts can be dated by matching their reversal sequence with that of the first area. So useful has this method proved, especially for rocks of the last few million years, that geologists and archeologists speak familiarly of the "Brunhes normal epoch" (going back 700,000 years), the "Matuyama reverse epoch" (2,400,000 to 700,000 years), the "Gauss normal epoch" (3,300,000 to 2,400,000 years), and so on.

Thus measurements of rock magnetism give tantalizing evidence of changes in the earth's interior. Directions of magnetization indicate apparent migration of the geographic poles and possibly migration of the continents, hence large-scale movement of the earth's crust with respect to the mantle. Reversals of magnetization suggest some sort of large-scale alteration in the liquid outer core. Mechanisms for moving the crust and for altering the core remain mysterious, but the facts are well established.

This is by no means the end of what magnetic measurements have to tell us, but to continue the story we need a background of data about the oceans.

Another look at the mantle

Before embarking on a study of the oceans, let us sit back for a moment's reflection on a seeming paradox regarding the mantle. This thick intermediate layer of the earth, we have noted, probably consists chiefly of iron and magnesium silicates, a combination that at the earth's surface we find in the heavy, dark rock peridotite. The upper part of the mantle may in fact consist largely of peridotite, as is indicated by rock outcrops of material that was probably once in the mantle. The lower mantle probably contains the same elements in different combinations, because the extreme pressures and temperatures would make the minerals of ordinary peridotite unstable; nevertheless we will not be far wrong in thinking of the entire mantle as consisting of familiar rock material, somewhat modified at great depths. To this material we have ascribed some general properties. From its

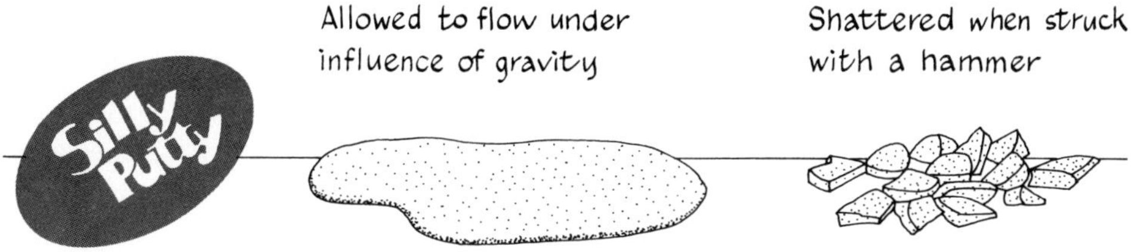

Fig. 13-14. A solid like pitch or "silly putty" shows brittle behavior when subjected to short-time forces, but is capable of flow when forces act over long periods.

ability to carry transverse seismic waves and from its small response to tidal forces we conclude that it is a brittle solid. Yet from the facts that mountains and plateaus have been raised and lowered thousands of meters, that sections of the crust have been drastically shortened by folding and faulting, and that continents have moved with respect to the lower mantle, we must imagine at least part of the mantle as capable of flowing like a viscous liquid. How can material with the composition of rock be at once solid and liquid, rigid like steel according to one set of observations and yet flowing like putty according to another?

The paradox is not as extreme as it sounds. Ordinarily we think of liquids and solids as completely distinct, the one capable of flow in response to any force, however small, and the other capable of maintaining its shape despite the action of forces. Yet for a few familiar materials the distinction is hard to maintain. Take pitch, for example, or the silicone plastic called "silly putty": if a small piece is struck with a hammer it shatters into a thousand angular fragments, just as would any brittle solid, but if another small piece is left for a few hours on a flat surface it spreads into a blob just as would a viscous liquid like honey. The difference in the two experiments is a question of *time.* When pitch is struck with a hammer a large force acts on it for a brief moment; when the same pitch is left by itself, it is acted on by the weak force of gravity over a long period. Thus a material may behave like a solid to short-term forces, a liquid to long-term forces. Even crystalline solids like rock, if subject to high pressures and to temperatures near their melting points, can be shown to act in these apparently contradictory ways. The ice of a glacier is a familiar example: it is obviously a brittle crystalline solid, yet the glacier moves slowly downhill because the ice at its base is under high pressure and can flow much like a viscous liquid.

So the properties of the mantle are not contradictory. To short-term forces—forces acting over times on the order of seconds in seismic waves, on the order of hours in tidal deformation—the rock in the mantle is a rigid, brittle solid. To the long-term forces of geologic deformation, involving times measured in hundreds of thousands or millions of years, the rock responds like putty. In response to a question, "Is the earth's interior solid or liquid?", no simple answer is possible. It is both, in a sense—solid if the forces acting are brief, liquid if they act for a long time. The time must be specified, or the question has no meaning.

What now of the distinction between mantle and core? We have said, as is customary, that the outer core is "liquid," in contrast to the mantle and inner core, which are "solid." More specifically we should describe the outer core as liquid with respect even to short-term forces (such as the forces involved in seismic waves), whereas the mantle, and presumably the inner core also, are solids unless acted on by forces over long periods.

We can likewise refine an earlier description of the asthenosphere (page 345), the layer in the upper part of the mantle between about 60 and 200 km below the surface, where seismic waves move more slowly than in layers above and below. Like the rest of the mantle, this layer shows properties of both solid and liquid, but because its material is closer to its melting point it can flow more readily—or in response to forces acting over shorter times—than the rest of the mantle. When we say that pockets and lenses in the asthenosphere may be actually molten, we mean that these parts are capable of flow even in response to very short-term forces. The distinction between liquid and solid is still valid and useful, although it is less sharp than in ordinary life. Magmas remain "liquid," in contrast to the "solid" rocks they intrude, even if the intrusion takes place at the base of the crust or below.

By way of summary, a study of earthquake vibrations that travel through the earth makes possible a division of the interior into crust, mantle, and core, with the core probably subdivided into a liquid outer part and a solid inner part. The composition of meteorites suggests metallic iron as the chief constituent of the core and silicates of iron and magnesium as the major materials of the mantle. These materials are consistent with the observed speeds of seismic waves and with calculations of densities within the earth. The mantle, making up most of the earth's volume and more than half its mass, is solid with respect to short-term forces, although much of its material is probably not far below its melting point. Heat is leaking out from the earth's interior, much of it produced by radioactive decay, but whether the earth at present is heating up or cooling off is not known. The earth's magnetism is probably a result of electric currents generated by movements in the liquid outer core, and the magnetic axis has probably always been close to the rotational axis. Measurements of rock magnetism indicate that the earth's field has reversed itself at many times in the geologic past, and that the crust has not remained fixed in position with respect to the rotational axis.

Fig. 14-1. Evidence for late Paleozoic glaciation in South Africa. The upper picture shows a rock surface striated and polished by glacial erosion, and the lower picture shows an outcrop of glacial till (recognizable by the heterogeneous mixture of fine and coarse fragments without sorting or stratification). The age of both till and glaciated surface can be determined from fossils in associated sedimentary rocks. Evidence of this sort for extensive glaciation near the end of the Paleozoic Era is found not only in Africa but also in India, Australia, and South America. The existence of glacial climates in such widely scattered regions where present climates are much milder provides strong support for the hypothesis of drifting continents. (John C. Crowell)

14

From the Bottom of the Sea

We began our long look at the sun's third planet by standing on a cliff above the seashore. At this point in our study we might well return to the cliff and gaze once more out over the blue expanse that stretches to the misty horizon.

We let our minds wander, and reflect on the role that this same expanse of restless water has played all through the planet's history. Into the water has been poured the debris of the continents, the pebbles and sand grains and mud that have been spread over shallow bottoms and sorted by shifting currents into the sediments we now so often see above sea level as beds of sandstone and shale. In the water itself has flourished a parade of always changing life, a parade of which we catch glimpses in the fossil remains preserved in some of the sedimentary rocks. We reflect further that it was in this same sea, sometime long ago when the earth was young, that life had its beginning—and that it was from this sea, uncounted millions of years later, that our own remote ancestors emerged to establish themselves in the harsher environment of the lands. Much of the earth's story has gone on beneath that blue surface, and the story is a continuing one. Even as we watch, sediments are being laid down that will be the rocks of future geologic ages, entombing a record of present-day life in the form of shells and bones like those that litter the beach below us.

What of the deeper water out beyond the horizon? About the sediments on the deep sea floor and the organisms that live in the dark, cold bottom water we know a little from dredge hauls and drill cores and photographs, but the history of the deep ocean is only scantily preserved in sedimentary rocks on land. For most of geologic time the depths of the sea have probably been much as they are today, save for the slow changes in living things.

The ocean has always had an air of mystery, in part perhaps because of its sheer size, in part because it conceals on its bottom a world that is little known, in part because it seems so alive and changeable yet so oblivious to human concerns, perhaps even in part because of our dim instinctive awareness that life itself is bound up in its origins with the moving waters. Some small part of the mystery has been dispelled in recent decades by a concerted attack with the best of modern instruments. Not only has the sea yielded up expectable information about the composition of its water, about its complicated movements, and about the shape of its floor, but quite unexpectedly it has provided a key to understanding such diverse phenomena as earthquakes, volcanoes, and the formation of mountain ranges. Many of the questions that we have left unanswered in earlier discussions can be resolved, or partly resolved, by study of the rocks and sediments hidden from view by salt water.

The essential mystery of the sea remains, of course, and always will remain, but it has been pushed a long step farther back. From the sea has come a new insight into geologic processes, an insight so profound that it is often and quite properly described as a scientific revolution.

What does the sea floor look like?

The technique of echo sounding—timing sound waves that travel from a ship down to the sea bottom and back again—has provided a quick and easy method of mapping the sea floor. As the ship moves along a carefully plotted course, the sounder automatically makes a record of the ups and downs of the bottom. Many such profiles along closely spaced lines make it possible to map the bottom in great detail.

Fig. 14-2. Profile of the Mid-Atlantic Ridge. (From *Continental Drift*, edited by S.K. Runcorn, 1962, with permission of Academic Press, Inc., New York)

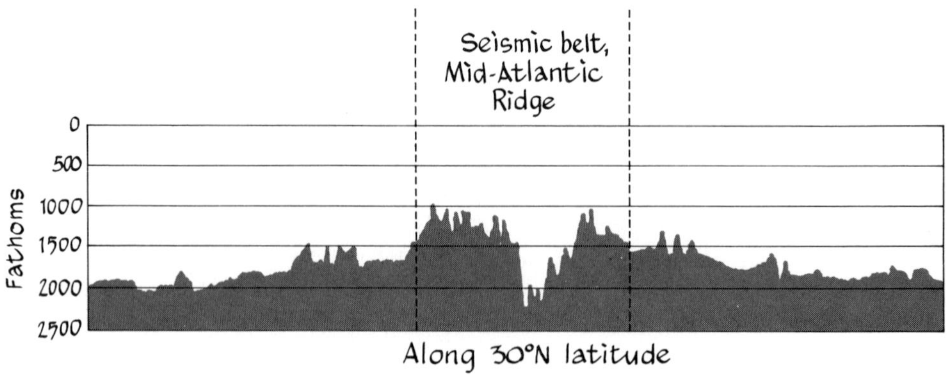

Fig. 14-3. Looking along the Mid-Atlantic Ridge in Iceland. Iceland sits astride the Ridge (Plate 27), and in parts of the country near the crest of the Ridge long straight valleys like this one are conspicuous features of the landscape. (G. Hannesson)

Sea-floor topography turns out to be almost as complex as the topography of the continents. There are broad plains, deep valleys, isolated hills, and mountain ranges comparable in size to the largest ranges on land. Particularly striking is a mountain range that can be followed from one ocean into another: part of it is a ridge that extends down the middle of the Atlantic (the *Mid-Atlantic Ridge*), then curves around the Cape of Good Hope into the Indian Ocean; here it branches, one part extending north toward India and the other eastward between Australia and Antarctica into the Pacific; the Pacific branch curves northward off the coast of South America (the *East Pacific Rise*) and can be followed until it meets the land on the coast of central Mexico (Plates 30 and 34). The existence of mid-ocean mountain ranges had been known to oceanographers since early explorations in the last century, but only with echo sounding has it been possible to establish the worldwide continuity of the range and to map its finer details.

Suppose that the water could be drained temporarily from the North

Atlantic, so that you could undertake a mountain-climbing expedition on the Mid-Atlantic Ridge. Even without much geologic background you would be struck by some remarkable features of the ridge, features unlike anything in your previous mountain-climbing experience on land. There would be no system of stream valleys carved into the sides of the ridge: this you would be prepared for. But it would come as a surprise to find that the center of the ridge, where you would expect to find the highest peaks, is instead occupied by a deep and nearly continuous valley—as if, in climbing the Colorado Rockies, you should find a valley with the dimensions of Death Valley where you expect the crest of the range. The steep-sided central valley is flanked by lesser valleys, all approximately parallel to the axis of the ridge, so that the topography of the central part of the ridge is very rugged. Then if you do some exploring along the ridge to the north or south, you would meet another surprise: the ridge is apparently offset along many deep, narrow east-west valleys, in places by many tens of kilometers, as if someone had sliced the ridge into segments with a huge knife and then had shoved the segments back and forth (Plate 27). This is indeed a strange sort of mountain range, one whose geologic history must be quite different from that of most ranges on land.

Not all the mid-ocean ridges are as rugged as the spectacular range of the central Atlantic. The East Pacific Rise, for example, over much of its length has a fairly smooth surface and lacks the deep central valley. The gash-like valleys at right angles to the ridges, along which segments of the ridges are offset, are a constant feature—in some places more closely spaced than in others, but always present in any major portion of a ridge.

A curious fact about the ridges has come to light from a study of heat flow through the ocean floor. We recall (Chapter 13) that heat is leaking out of the earth's interior everywhere, and that the flow of heat shows surprisingly little variation from one area to another. Measurements along the axes of the mid-ocean rises, however, indicate that the escape of heat here is somewhat larger than from other parts of the sea bottom. We will find this fact useful later on in trying to frame an explanation for the ridges.

Besides the ridges, the most striking elevations above the sea floor are isolated conical-shaped hills. Some of these are active volcanoes (the island of Hawaii, for example), with their tops above the sea surface, but a great many more are submerged and show no sign of present eruptive activity. The submerged ones go by the general name of *seamounts.* The shapes and the nature of rocks dredged from their flanks leave little doubt that seamounts are old volcanoes. The only difficulty with this conclusion is the presence of broad flat tops on many seamounts; it is hard to see how eruptive activity, either on land or under water, could produce flat-topped mountains. To avoid the difficulty an obvious suggestion would be that the flat surfaces were produced by wave erosion, at a time when the old volcanoes had their tops above water. The trouble here is that the flat surfaces stand at widely different depths below sea level, many at depths far too great for any imaginable process of lowering the sea's surface. If erosion is indeed the correct explanation, it must follow that the sea floor has subsided, and that the subsidence has been greater in some places than in others.

Another prominent feature of submarine topography is the great *trenches* or *deeps* near some island chains and some continental margins. In these trenches the ocean reaches its greatest depths, as much as 36,000 feet in the deep just east of the Philippines. Prominent trenches lie along the ocean side of most of the curved island chains (*island arcs*) of the western Pacific, and another follows the straight coastline of Chile (Plates 30 and 34). Trenches are conspicuously absent along most of the west coast of North America, and also from the Atlantic except for one that follows the island chain from Puerto Rico along the Lesser Antilles and a similar one in the far south. The trenches are puzzling: why have they not long since filled with sediment? Their evident persistence in some places, and their complete absence in others, demands some sort of explanation in terms of processes taking place beneath the sea.

In contrast to offshore topography, the sea bottom near the continents holds few major surprises. As we have noted earlier (Chapter 5), a nearly flat shelf flanks continental margins, wide in some places (e.g., the Atlantic coast of the United States) and narrow in others (much of the Pacific coast). One can think of the continental shelves as parts of continents, simply low-lying areas flooded by shallow seas. The true edges of the continental masses, as distinct from the deep oceans, are the continental slopes along the seaward edges of the shelves. A curious feature of the continental shelf is the presence in many places of steep-walled canyons, most of them short and localized at the brink of the continental slope, but some extending completely across the shelf to the shoreline. A few have their heads at the mouths of rivers (the canyon off the mouth of the Hudson River is a famous example), and are explainable at least in part as former stream-eroded valleys which have been flooded by a rise in sea level. This is not a satisfactory general explanation, however, because some canyons are far from land and some can be followed to depths much greater than any reasonable fluctuation in sea level would permit. A better guess is the formation of canyons by erosional processes taking place on the present sea floor, particularly by turbidity currents developed in unstable sedimentary debris (Chapter 5). The canyons, although striking features of the near-shore submarine topography, probably have little significance with respect to the processes that have shaped the major features of the sea floor.

What is the sea floor made of?

In an earlier age of oceanographic exploration, samples of the sea floor could be obtained only by lowering a dredge from a ship and bringing up whatever loose material the dredge chanced to encounter—obviously an inefficient and haphazard procedure, capable at best of sampling only the immediate surface of the bottom. A great deal of our present information about the sea we owe to the development of new instruments: drills that can bring up cores extending many meters down into the bottom material, underwater cameras capable of taking pictures even in the deepest parts of the ocean, and echo sounders that record not only waves bounced back from the bottom but also waves that penetrate into the bottom material and are reflected by deeper layers. These marvelous new tools are revolutionizing our ideas about the rocks and sediments that underlie the oceans.

Before looking at the new information, we might ask what sort of material we would expect to find—just from our general knowledge of water movements and the nature of substances being supplied to the sea.

We could certainly predict that much of the sea bottom would be covered by sediments, simply because rivers and waves have been adding debris from the land continuously for more than three billion years. Ocean currents that carry the debris, sorting and depositing it as they move, should be strongest in shallow water near the shoreline, where wave motion generates powerful undertows and longshore drifts; the currents in deeper water, like the Gulf Stream and the Japan Current, involve huge masses of water but move relatively slowly. We could expect, then, that sediments supplied to the ocean would show a rough overall sorting, from local deposits of coarse sand and gravel near the shore to finer sand and mud farther out on the continental shelf and then to very fine mud in the deep sea. Besides the fine mud particles moved away from shore by slow-moving currents, we might also expect that dust carried out to sea by winds would make up a considerable part of deep-sea sediments. From the fact that living creatures flourish abundantly in the upper few tens of meters of sea water we could anticipate that their hard parts would fall to the bottom and become part of the sediment; and in the warm water of the tropics we would expect to find much precipitated calcium carbonate in the form of limy mud and reefs built by corals and algae. Because these processes of sedimentation have been going on for a very long time, we might guess that parts of the sea floor that have not been disturbed would be covered by sediments of all geologic ages, going far back into the Precambrian—hence that these bottom deposits should provide us with a far more complete record of geologic history than anything we can find on land.

To what extent are these simple predictions fulfilled? As with most simple predictions in geology, the answer would be, "Only in part." It is true that in gross general terms sediments of the continental shelf show a progression seaward from coarse material to fine, and that sediments of the deep ocean are for the most part very fine-grained. But there are many local exceptions. The continental shelf off New England, for example, shows coarse and fine material in erratic patterns, which can be traced to advances and retreats of the Pleistocene ice-sheets. The uneven topography of the shelf off southern California gives rise to concentrations of sand and mud with little regard to distance from shore. The generally fine grain-size of deep-water sediments corroborates a guess that currents in deep water are slow, but one of the surprises in detailed study of the sea floor has been

Fig. 14-4. Ripple-marked sand at a depth of 1500 meters near Bikini Atoll, Marshall Islands. The ripples near the center are about 30 cm apart and 8 cm high, indicating rapid motion of bottom currents. (From H.W. Menard, *Journal of Sedimentary Petrology*, vol. 22, page 5, 1952; U.S. Navy photograph)

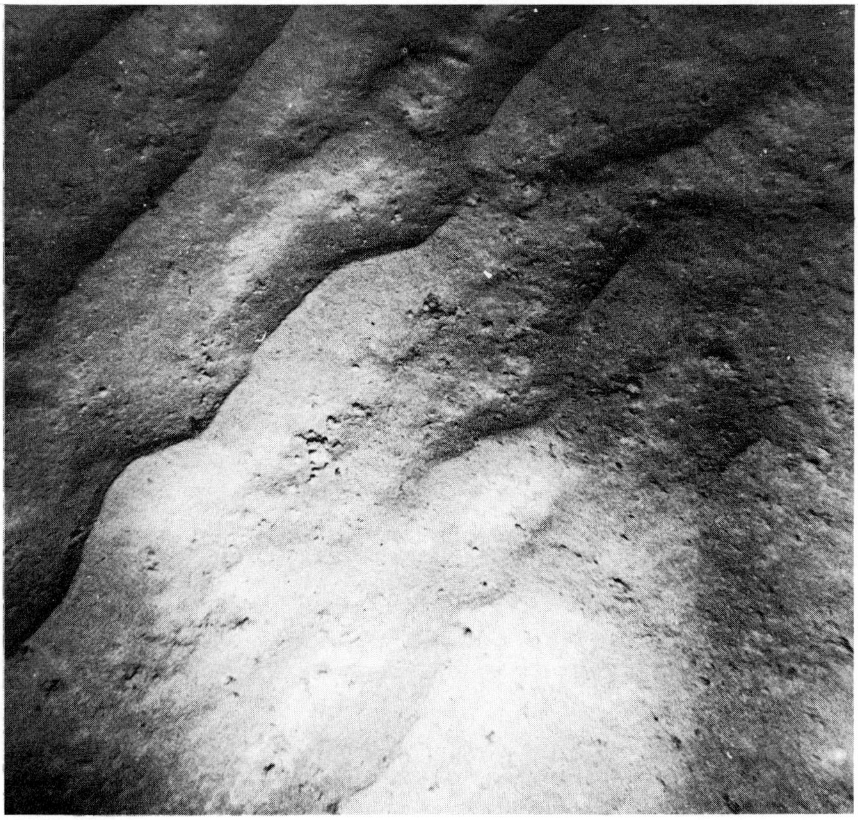

Fig. 14-5. Microfossil assemblages from deep-sea sediments in the western Pacific. Most of the shells in the left-hand picture are foraminifera, and most of those in the right-hand picture are radiolaria. Scanning electron microscope photographs, magnification about 150 times. (Stanley Kling)

the finding in a few places of fairly coarse, ripple-marked sand at great depths—which can only mean that locally the deep bottom currents move rapidly. The prediction of wind-carried dust in deep-sea sediments is abundantly fulfilled, and in places the source of the dust can be identified: for example, dust from the Sahara Desert brought by the trade winds to the central Atlantic, and dust from the Gobi Desert brought by west winds to the north Pacific.

The prediction of abundant remains of organisms in sediments of deep water is verified, but the enormous quantities of such remains could hardly be anticipated. Over great areas of the sea bottom, sediments consist almost entirely of the microscopic shells and shell fragments of one-celled plants and animals. Some organisms use calcium carbonate for their shells, and others use silica in the form of opal; since organisms of a given type flourish in particular parts of the sea, the sediments consist dominantly of either one material or the other. All such sediments go by the name of *oozes,* and are distinguished according to the principal kinds of organisms represented. The two principal calcium-carbonate sediments are foraminiferal oozes and coccolith oozes, and the major opaline-silica sediments are radiolarian oozes and diatom oozes. The existence of such marine sediments in the geologic past is proved by some well known sedimentary rocks, for example the chalk of the white cliffs of Dover in England (a coccolith ooze from the Cretaceous Period), and the pure white diatomite of part of the Monterey formation in California (a diatom ooze from the Miocene).

One of the predictions about marine sediments is emphatically *not* fulfilled: the guess that these materials should provide a complete sedimentary record for all of geologic history. Perhaps the strangest of all facts that have emerged from recent study of the sea bottom is the absence anywhere of sediments that can be dated (either by fossils or radiometrically) as earlier than late Jurassic, or about 150 million years ago. It looks as if no sediments were deposited in the ocean before the middle of the Mesozoic Era—yet this is a patently silly conclusion, because marine sedimentary rocks of far greater age are abundant on the continents. In this strange paradox lies one of the principal means we will use to unravel some of the mysterious processes that go on within the earth.

Sediments cover much of the sea floor, but not all. The sedimentary blanket is thickest near the lands and grows thin toward the middle of the ocean basins, disappearing altogether on some parts of the mid-ocean ridges. The bare rock exposed on the ridges is largely basalt, and basalt is also the chief material that underlies the sediments over all the ocean basins—as is proved by the speeds of waves set up either by artificial sources or by earthquakes. The basalt in most places is 4 or 5 kilometers thick, and is underlain by material in which seismic wave motion is faster. The precise nature of this deep material is still in dispute, but the heavy igneous rock peridotite is a widely favored guess. The contact between the basalt layer and the heavier underlying rock is the Mohorovičić discontinuity, so that the earth's crust under the oceans consists simply of a few kilometers of basalt topped by a layer of sediments ranging in thickness from zero to several kilometers.

Note the extreme contrast of this oceanic crust with the crust under the continents, where the total thickness averages 35 kilometers and consists dominantly of rocks with the general composition of granite.

Magnetic stripes

The direction of magnetization of basaltic lavas on land proved to be a very rewarding subject for scientific study, as we saw in the last chapter, and it was a natural move for geophysicists to extend this study to basalts beneath the sea. Such a study was possible because magnetometers can be made so sensitive that when towed behind a ship they respond to the slight magnetism of the lavas even under great thicknesses of salt water and sediments. The results of this investigation have been spectacular.

A magnetometer at a given point on the sea surface is affected principally by the earth's field, just as a ship's compass would be. If the nearest rocks beneath it are magnetized in the same direction as the earth's present field, the magnetometer will show a field strength slightly greater than the strength to be expected from the earth's field alone. If the rocks below are magnetized in the reverse direction, their field opposes the earth's field, and the measured field strength will thus be less than the expected figure. So if the basalts of the ocean floor are of different ages, some of them having been molten at times when the earth's field had its present direction

Fig. 14-6. Records made by a magnetometer towed across the Mid-Atlantic Ridge. Each wiggly line shows deviations of the magnetic field from its normal value for a particular traverse. The straight line in each case shows the normal value to be expected in this area; parts of the wiggly line above the straight line indicate magnetization of underlying rocks in the direction of the earth's present field, and parts below the line indicate magnetization in the reverse direction. Dotted lines connect similar magnetic features. (After J.R. Heirtzler, X. Le Pichon, and J.G. Baron, "Magnetic anomalies over the Reykjanes Ridge," *Deep Sea Research*, vol. 13, page 427, 1966, Pergamon Press)

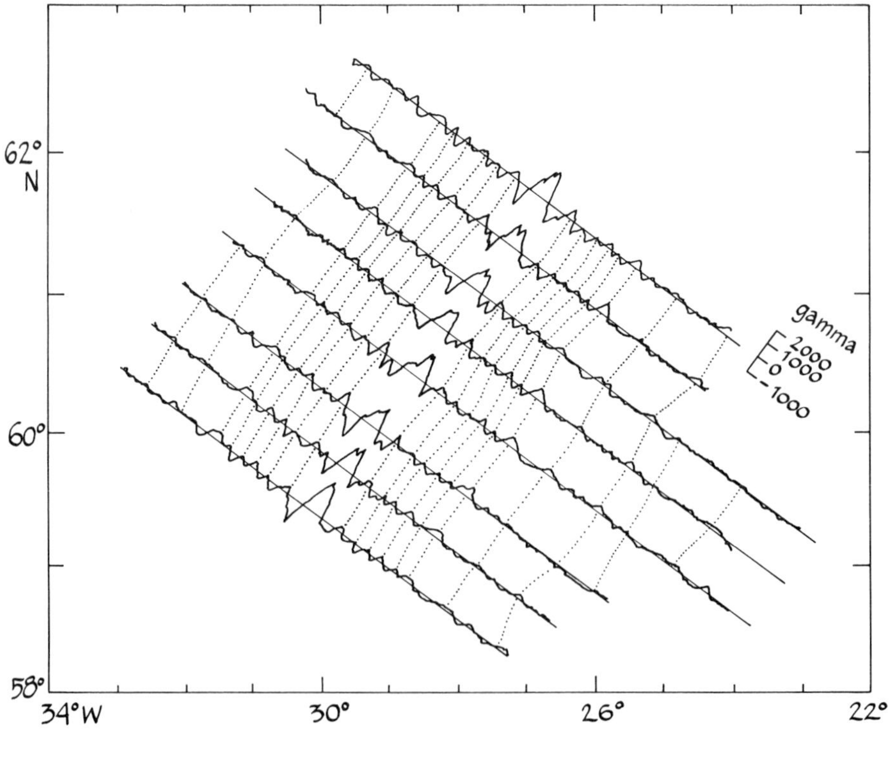

and others at times when the earth's magnetism was reversed, a magnetometer should be capable of pinpointing boundaries between flows of different ages. The ages themselves would not be known from this kind of data, but at least it should be possible to map the hidden basalt of the sea floor into areas of normal and reverse magnetization.

With this expectation, suppose we set out to map the magnetization of the basalts near one of the mid-ocean ridges, say the Mid-Atlantic Ridge. We set a course so as to cross the ridge at right angles, and we start the magnetometer record when we are still a few hundred kilometers from the ridge crest. As our ship approaches the ridge we find that the field strength shows much variation, part of the time greater than the strength to be expected for this region and part of the time less, so that the record is a wavy line crossing and re-crossing the line for the normal value. Evidently we are sailing above an area of the sea bottom where the underlying basalts have widely different ages. As we approach a point directly above the ridge crest, consistently high values of the field strength indicate that the basalt is magnetized in the normal direction, and this direction persists until we are well beyond the crest. Then comes a real surprise: as we move farther and farther from the crest, the same sort of variation shows up in the magnetic record; and if we inspect the record carefully, we find that one side of the wavy line is approximately the mirror image of the other (Fig. 14-6). In other words, the changes from one direction of magnetization to the other occur at roughly the same distances on each side of the ridge crest.

What can this mean? No answer is immediately apparent, but we obviously need more information. In great excitement we turn the ship around and set another course back across the ridge, parallel to the first but a few tens of kilometers to one side. We watch the record develop: again the wiggly line, again a broad region of high field strength over the crest, and again a mirror image of the wiggles as our distance from the crest increases. Then we compare the two complete records, and another strange fact emerges. The two lines are amazingly similar—where one goes up the other goes up, and where one goes down the other follows. The magnetic changes are not only similar on the two sides of the crest, but they occur at the same distances from the crest on two different traverses.

Now we are surely on the track of something important, so we keep the ship in motion, back and forth across the ridge on many parallel courses. As the wavy lines multiply we find them all remarkably similar—not in fine detail, but indicating clearly that magnetic reversals can be shown on a map by lines roughly parallel to the crest of the ridge. Then we can color the areas between the lines, using one color for magnetization of basalt in the normal direction and another color for the reverse direction; and our

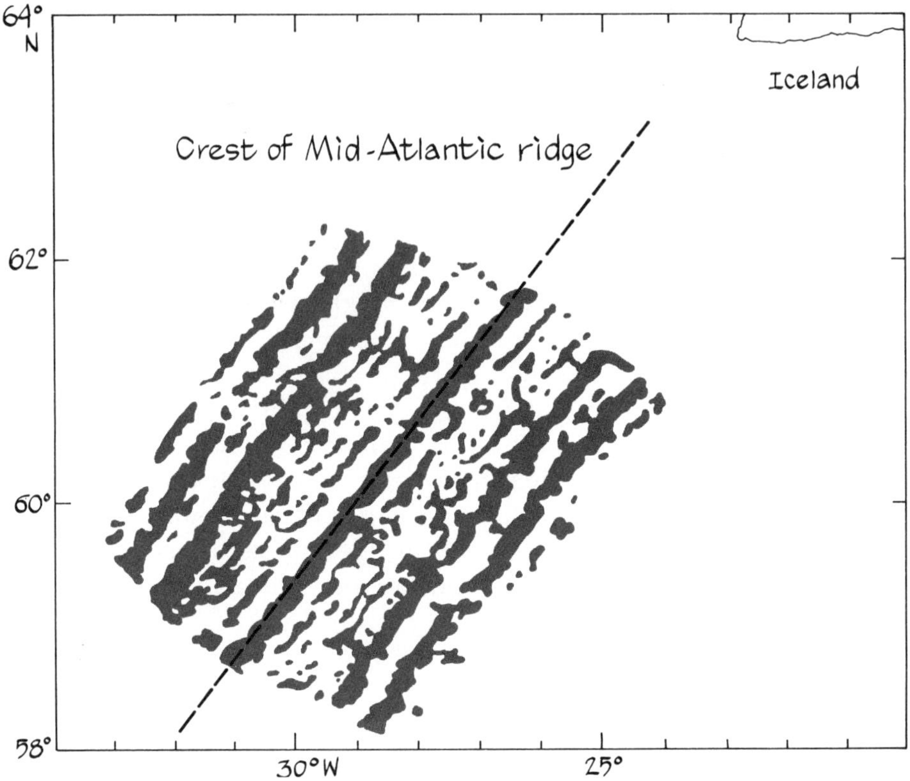

Fig. 14-7. Magnetic stripes flanking the Mid-Atlantic Ridge southwest of Iceland. Basalt on the sea floor under the dark areas is magnetized in the same direction as the earth's present field, and basalt under the light areas is magnetized in the opposite direction. (Source same as for Fig. 14-6)

map becomes a series of colored stripes, symmetrically arranged on either side of a broad central stripe of normal magnetization.

If you had been the one to make this discovery, what could you do with it? What explanation could you suggest for basalts of different magnetic orientations, hence different ages, lining up so regularly on either side of a submarine ridge? To a geologist the answer now seems evident, but it was by no means so evident when magnetic stripes first appeared in scientific publications. Many hypotheses were tried, and the now-favored hypothesis was fashioned only by putting together the mysterious stripes with many other bits of evidence.

The moving bottom

Let us try a flight of fancy. Measurements of the flow of heat from the earth's interior give slightly higher values along the Mid-Atlantic Ridge than in other parts of the ocean bottom. We know that sediment is thin or lacking on the ridge, and we know from dredge hauls that the dominant rock is fresh basalt. We know further that active basaltic volcanoes are located on or near the ridge crest, some of them forming islands—the volcanoes of Iceland in the north, St. Helena and Tristan da Cunha (among others) in the south. The epicenters of shallow-focus earthquakes, too, are concentrated along the ridge (front end paper). Evidently the ridge is a scene of considerable geologic activity. Suppose that part of the activity consists of the slow rise of liquid basalt out of the mantle, sometimes concentrated in one place and sometimes in another, but overall building up an accumulation of basalt flows and dikes all along the ridge. Together with the rise of basalt, suppose that the two sides of the ridge are slowly moving apart, impelled by some mechanism in the mantle that for the moment we will leave unspecified. The basalt as it accumulates and solidifies is then moved away from the ridge, and it is constantly renewed by fresh magma from below. If the movement has been going on for a long time, basalt at a distance from the crest may have emerged and crystallized during a period when the earth's magnetic field was reversed. Still farther

Fig. 14-8. The hypothesis of sea-floor spreading. The parts of the mantle on either side of a mid-ocean ridge are moving apart, carrying with them the basalt and sediments above. The gap as it widens is continually filled by magma rising from lower in the mantle. The magma in part freezes to the sides of the spreading blocks and in part rises nearer to the sea floor to form dikes and flows. Thus both crust and upper mantle move steadily away from the ridge, and are constantly enlarged by magma freezing to the trailing edges of the moving blocks. Symbols in the dikes indicate times of normal (+) and reverse (−) polarity of the earth's magnetic field.

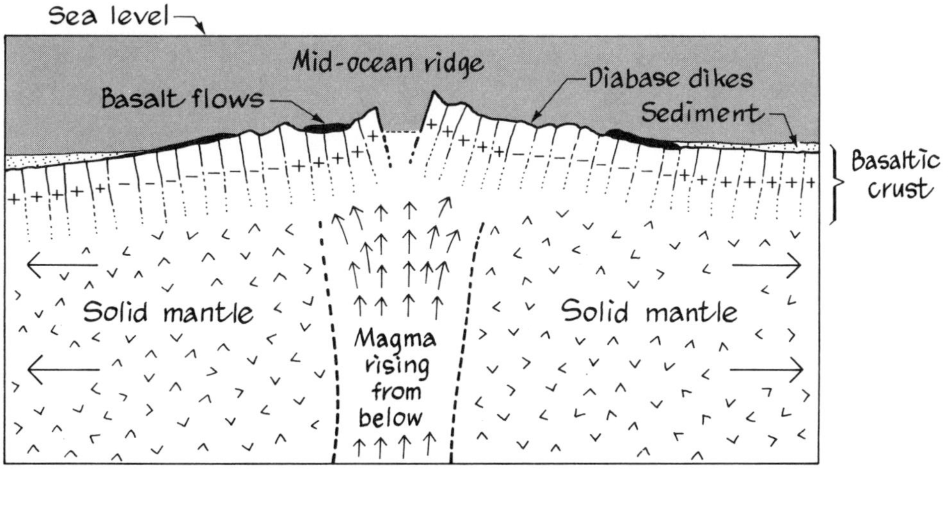

away would be basalt from an earlier period of normal field direction, and farther yet would be lava from a previous time of reversal. If the movement has been uniform along the ridge and at the same rate on the two sides, the magnetic stripes would be explained.

So far we have nothing but a wild hypothesis. It accounts for the stripes, it is consistent with heat-flow measurements, earthquakes, and volcanic activity, and it explains the peculiar longitudinal valley (and parallel smaller valleys) along the crest of the ridge. But it obviously leaves many difficult questions unanswered.

What, for example, do we imagine to be happening farther from the ridge? As long as we are spinning hypotheses we can fling caution to the winds and suppose that the whole sea floor is moving, that the material under the sediments behaves like two huge rigid blocks being carried away from the mid-ocean ridge, the widening gap being constantly filled by new basalt which then attaches itself to the trailing edge of each block. What of the continents on either side of the ocean? Losing all caution now, we imagine that the continents themselves are part of the moving blocks, huge rafts of granite resting on a layer of basalt and moving apart as the basalt moves. And we can find some justification for this outlandish guess by noting that the eastern shore of the Americas is somewhat similar in outline to the western margin of Europe and Africa, as would be expected if the two sides were once joined and have been drifting apart.

To test our hypothesis we need a figure for the rate of movement. We recall that times of reversal of the earth's magnetic field direction can be dated on land by dating basalt flows, either radiometrically or by identifying fossils in interbedded sediments. Tentatively we try matching the dates so determined with the magnetic stripes, working backward from the broad stripe of normal magnetization at the center of the Mid-Atlantic Ridge. We find that the lengths of time of the periods of "normal" and "reverse" magnetization correspond well with the relative widths of the magnetic stripes, as would be expected if the rate of movement has been constant. Thus we can assign dates to the stripes, and so estimate the speed with which the two sides of the ocean are moving apart. The rate turns out to be a few centimeters per year. During the 6000-odd years of recorded history, then, America has moved away from Europe by roughly a tenth of a mile—hardly enough to be noted by navigators or even by sophisticated surveying techniques. But in the course of geologic time, this rate of movement could produce enormous changes. The whole 3000-mile width of the Atlantic Ocean could have been opened up in something like 200,000,000 years. In other words, the New World and the Old World may have been part of a single huge continent early in the Mesozoic Era.

Then we remember that no sediment or sedimentary rock has been found in the Atlantic (or any other ocean) that is older than mid-Mesozoic. Suddenly our hypothesis takes on new meaning. *Perhaps there are no older sediments in the Atlantic simply because in earlier times the Atlantic did not exist.* The hypothesis is far from proved, but at least it provides a reasonable explanation for an otherwise mysterious fact.

The argument is straightforward for the Atlantic, but what about the other side of the planet? If the Atlantic Ocean is widening, the Pacific must be shrinking. Yet in the southeastern Pacific we find another submarine ridge (the East Pacific Rise) along which the same sort of magnetic stripes can be traced, indicating that the floor of this ocean is spreading also. Somewhere the moving blocks from the two oceans must meet each other head-on. Can we modify our hypothesis to take care of this requirement?

We reflect that the margins of the two oceans are geologically very different. Around the Pacific is a "circle of fire," a belt along which most of the world's active volcanoes are concentrated. This is also a belt of frequent earthquakes, of visible fault movements, of rugged mountains in which very recent sedimentary rocks have been steeply tilted, folded, and faulted. The Atlantic coastline, by contrast, is a region of geologic lethargy—few earthquakes, little volcanic activity, no indication that recent sedimentary layers have undergone any movement except slight regional uplift and subsidence. Could we hazard a guess that the geologically active shores of the Pacific represent places where parts of the crust moving in opposite directions are colliding?

A fact about the distribution of earthquakes gives us a possible lead. We recall from an earlier discussion (Chapter 12) that earthquake foci (the places where earthquake motion starts) are located at different depths—

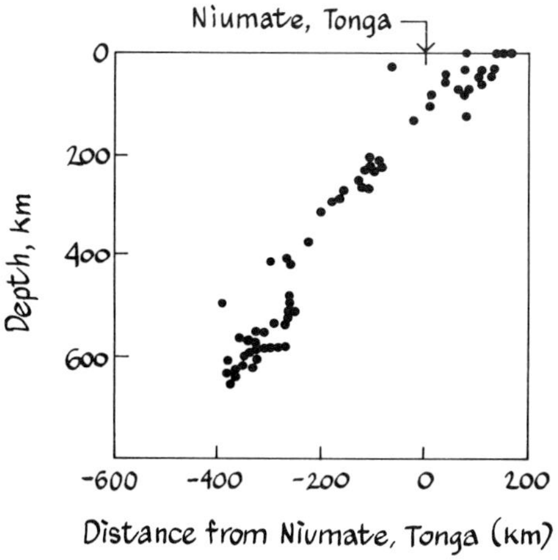

Fig. 14-9. A vertical section perpendicular to an island arc (Tonga), showing the distribution of earthquakes with respect to depth. The foci are projected onto the plane of this section from considerable distances on either side. Thus the foci are distributed roughly on a surface dipping at an angle of about 45°. (After B. Isacks, J. Oliver, and O.R. Sykes, "Seismology and the new global tectonics," *Journal of Geophysical Research*, Vol. 73, p. 5871, 1968, copyright American Geophysical Union)

most within a few kilometers or tens of kilometers of the surface, but a few at depths as great as 700 km. The deep-focus quakes (those with foci between 300 and 700 km) are restricted to a very few areas: South America east of the Andes, and the Pacific coast of Asia where the continent is fringed with curving chains of islands (*island arcs*). In the Pacific off South America and off the island arcs are some of the deepest ocean trenches. The epicenters of shallow-focus earthquakes are concentrated along the trenches and the adjacent shorelines; foci of intermediate depths lie toward the continents from the trenches, roughly on planes connecting the shallow-focus and deep-focus quakes (Fig. 14-9). One can imagine an inclined plane dipping under western South America from the offshore trench, and similar planes dipping under the island arcs of the western Pacific; these are evidently zones of weakness along which the fracturing occurs that gives rise to most of the earthquakes for which these regions are notorious.

Now again we can speculate. Suppose that the basaltic floor of the southeastern Pacific Ocean, moving eastward away from the East Pacific Rise, slides *under* the western edge of South America. The continent, riding westward on the basalt block generated at the Mid-Atlantic Ridge, moves *over* the basalt coming from the East Pacific Rise. As the eastward-moving Pacific basalt plunges downward beneath the overriding continent, local friction produces numerous earthquakes; sediment carried downward with the basalt becomes hot enough for partial melting, forming the magma that appears at the surface in volcanic eruptions; the beginning of the downward movement produces the offshore deep; and the crumpling of sedimentary layers in the Andes and lesser ranges near the coast is the result of pressure as the leading edge of the continent tries to move over the oppositely-directed basalt below. All this is highly conjectural, but it does provide a plausible way to correlate a great deal of geologic information.

The island arcs of the northern and western Pacific (the Kuriles, the islands of Japan, etc.) may be pictured likewise as places where the moving floor of the Pacific (here moving westward) ducks under the advancing front of the Asian continent. Again the offshore deeps, the earthquake foci increasing in depth toward the continent, the volcanoes, the violently disturbed sedimentary layers, all provide evidence for the movement of one part of the crust under another. A speculative zone of movement of this sort, in which oceanic crust is carried down beneath continental crust along an inclined surface, is called a *subduction zone.*

North America looks like a conspicuous exception to this grand scheme. There is no offshore deep; deep-focus earthquakes are lacking; recent volcanic activity is feeble, except in Alaska and Central America;

the East Pacific Rise is missing along part of the continental margin and close to shore along other parts. These apparent anomalies can be explained, but the explanation must wait until the next chapter.

As an attractive hypothesis, then, we can think of the mid-ocean ridges as a huge, branching, globe-encircling fissure system from which basalt wells up out of the mantle beneath. The walls of the fissures are moving apart, and the rising basalt in part plasters itself to the moving walls and in part flows out onto the ocean floor. The walls move because the ocean floor itself is moving, and where moving segments of the floor come in contact one segment may move beneath the other along a subduction zone down into the mantle. The hypothesis leaves many things unexplained, in particular the reason why such worldwide movements of crustal material should be taking place, but it is attractive because so many different kinds of geologic observations can be fitted into its framework.

Fig. 14-10. Diagram of a subduction zone. Oceanic crust and upper mantle which have been generated at a mid-ocean ridge are here consumed as they move down into deeper levels of the mantle and are ultimately remelted. The downward motion carries sedimentary beds along; the beds are intricately folded and thrust-faulted, and at deeper levels are partly melted to generate magma which feeds volcanoes at the surface and forms intrusive bodies at shallow depths. See also Fig. 15-4.

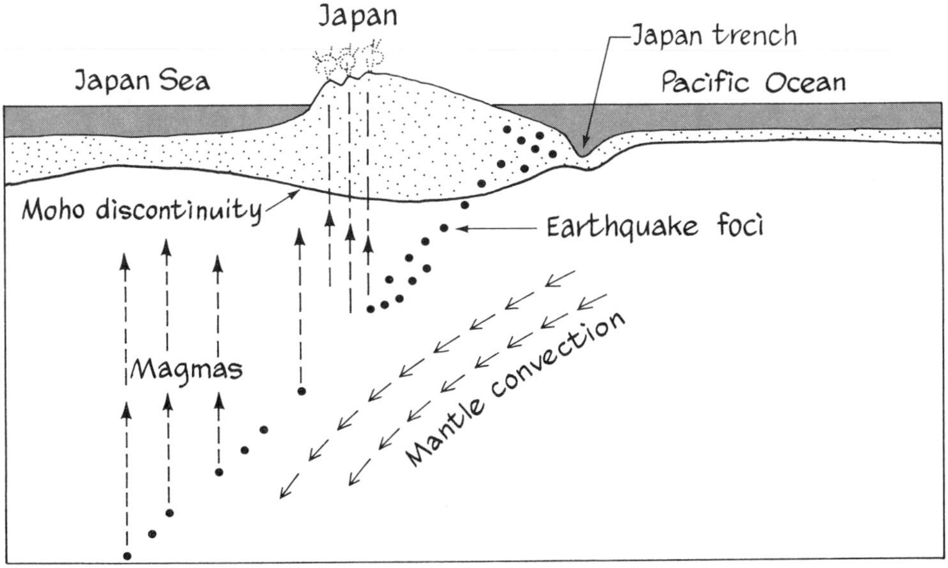

Continents: fixed or drifting?

From the hypothesis of sea-floor spreading it follows that the continents must be moving with respect to each other. Each continent is a huge raft of granitic material, carried passively on a moving layer of basalt beneath. The Americas are supposedly moving away from Europe and Africa; Australia and Antarctica are drifting away from Africa and Asia. The movement has been going on since the mid-Mesozoic, when presumably all the continents were closer together or part of a single land-mass. Earlier movements are possible, but evidence for them is still in dispute.

This picture is directly contrary to the thinking of most geologists a few decades ago. The continents were long thought to be permanently fixed in position, their boundaries established by some accident of the process by which the earth was formed. The dissimilarities of animals and plants from one continent to another, the scarcity of deep-ocean sediments on land and the lack of evidence for large areas of continental granitic rock beneath the sea, and above all the lack of a plausible mechanism to move bodies the size of continents, seemed convincing proof of the permanence and fixity of the continents and ocean basins.

Standing against the majority view, even back into the last century, was a small number of scientists who toyed with the speculative notion that the continents might move. The most prominent of these heretics in the early 1900's was a German geographer and meteorologist named Alfred Wegener, with whose name the "theory of continental drift" was for a long time associated. Wegener was impressed, as many had been before him, by the similarity in outline of the two sides of the Atlantic Ocean, which suggested that a great rift had gradually opened in a one-time single continental mass. The eastern bulge of South America supposedly had once fitted into the Gulf of Guinea (the right-angled bend in the west coast of Africa), and the western bulge of Africa once abutted the north coast of South America and the east shore of North America (back end paper). Wegener pointed out many similarities in the rocks and geologic structures on the opposite sides of the ocean, and emphasized particularly the likenesses in fossil remains in rocks of the late Paleozoic Era—at which time the continents were supposedly all joined together. He had no detailed information about the topography and materials of the ocean floor, about heat flow, or about changes in the earth's magnetic field, so that his evidence was necessarily limited to data about geologic and geographic resemblances.

The resemblances were numerous and impressive, but in the eyes of most scientists the theory foundered on one stubborn fact. No source of energy was known, either in the earth or derived from the earth's astronomical relations, that was anywhere near large enough to cause motion in masses as large as continents. The similarities in rocks and fossils could be accounted for in other ways, and numerous apparent exceptions to Wegener's broad generalizations could be pointed out. Without a plausible mechanism to make the continents drift, the theory seemed difficult to accept.

The likenesses among rocks and fossils were more striking in the southern hemisphere than in the north. In South America, southern Africa, India, Australia, and even Antarctica, the kinds of Paleozoic rocks and particularly some kinds of fossil plants (which could not easily spread from one continent to another across great expanses of ocean) were amazingly similar. In addition, the late Paleozoic rocks of all these lands contained evidence of a period of widespread glaciation—which is much easier to understand if the lands were joined together at the time of glaciation and have drifted apart to their present widely different latitudes (Fig. 14-1). It is small wonder that Wegener's ideas always had a warmer reception in southern countries than in Europe and North America.

The widespread skepticism about continental drift melted astonishingly, almost overnight, in the face of the rapidly accumulating evidence from sea-floor topography, ages of sea-bottom rocks, distribution of earthquake epicenters, and especially the bands of normal and reverse magnetism in basalts paralleling the mid-ocean rises.

It is in this sense that the new ideas constitute a scientific revolution. For many decades most geologists had simply assumed that continents were fixed in position, and speculation about such processes as folding, faulting, and mountain building was almost entirely based on this assumption. Within a few years' time this basic tenet crumbled, and now all but a handful of geologists are committed to rebuilding the foundations of their science on the radically different assumption that the continents, together with the ocean floor, have moved and are still moving with respect to the earth's interior. Seldom does a science undergo such a rapid and profound change in its fundamental ideas.

Colliding plates

North America, then, is an island of granitic rock riding westward on a layer of basalt, the basalt being constantly augmented by new lava rising from the rift at the center of the Mid-Atlantic Ridge. Each of the other continents is a similar island, carried passively on a layer of solidified lava drifting away from one or another of the globe-encircling submarine ridges. Overall one can picture the earth's crust as made up of huge segments, or

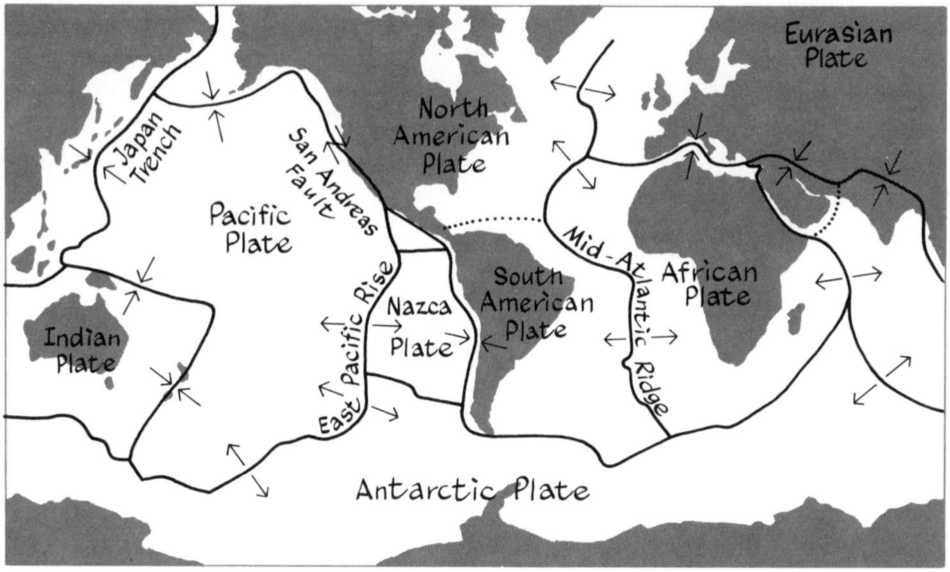

Fig. 14-11. The major plates into which the earth's crust is divided. (Redrawn from *Engineering and Science,* Nov.–Dec. 1973, published at the California Institute of Technology.)

plates, of basalt, some of the plates carrying continents on their backs, and the plates all in motion away from the ridges. From the boundaries of the magnetic stripes one can conclude further that the basalt plates are rigid, only in a few places giving evidence of bending. Many rock structures, both under the oceans and on the continents, can be accounted for by these motions of rigid plates. The idea of explaining geologic structures in this way is called the theory of *plate tectonics,* or *global tectonics.* It is obviously an extension of the theory of continental drift.

How can the plates all be in motion, when they cover the entire surface of the globe? Clearly there must be much jostling and mutual interference. For a very crude analogy, one can think of an ice sheet on the surface of a lake in winter: sometimes strong winds break the sheet into sections, which move about and pile up ridges of broken ice between them. The analogy is imperfect, because the ice segments are moving at the whim of the winds rather than steadily in a single direction, and because they are not growing continually larger by addition of new material at their trailing edges.

What happens when two crustal plates collide? One possibility, as noted in the last section, is the downward motion of one plate beneath the other, the basalt layer moving down at a 30–50° angle and carrying with it much of the sediment on its surface. This kind of relationship is especially clear around the margins of the Pacific Ocean. It may give rise either to a rugged coastline with a deep trench offshore, as in Chile, or to an island-arc-and-trench combination, as in the western Pacific. In either case the trench marks the line along which the motion of the oceanic plate changes from horizontal to obliquely downward; the mountains of the island arc or the continental margin have folded and faulted layers of geologically recent sedimentary rock, representing material crumpled and broken by the two plates moving in opposite directions; and the volcanoes a few tens of kilometers continent-ward from the trench are localized in the area where the downward-moving plate has attained a depth great enough for its material to begin to melt.

Another kind of plate collision is represented by the Himalaya and adjacent ranges of southern Asia. Here two continental plates have come together, one carrying the peninsula of India northward against the main bulk of Asia. Rather than one plate moving under the other, the result is a great piling up of continental material, with much folding and thrust-faulting, to give the highest mountains and the largest area of upland plateaus (Tibet) found anywhere on earth.

Still another kind of contact between moving plates is represented by the gash-like valleys that cut the ridge crests (page 364). These valleys, straight and steep-sided, are approximately at right angles to the crest and divide it into segments that are offset from one another. Where the ridge crest has a central rift valley, as in the Atlantic, the offsets of the rift are especially conspicuous (Plate 27). Along most of the gash-valleys the displacements are relatively small, of the order of kilometers or tens of kilometers, but a few show much larger offsets. The magnetic stripes on either side of the ridge crest are offset also, which means that the sea floor is spreading from each segment of the crest. Thus along a gash-valley between two offset segments the sea floor is moving in opposite directions (Fig. 14-12). In effect, then, two plates are in contact along the gash-valley, but their movement is sideways rather than a head-on collision.

The motion along a gash-valley is fault movement of a kind we have described previously as strike-slip (Chapter 10)—the kind that is shown so clearly by the San Andreas fault. Note, however, that the observed displacement of the ridge crest is *not* the result of the present motion—in fact, it is in the opposite direction. Why the ridge crest was broken and pulled apart we do not know, but the pulling apart had nothing to do with the present strike-slip motion of oppositely directed plates. A strike-slip fault of this sort, between two displaced segments of a mid-ocean ridge crest, is called a *transform fault.*

Thus the motion of rigid plates provides an explanation for the origin of many topographic features and geologic structures, both on the continents and in the ocean basins. The few examples we have looked at here are only the simplest and most obvious among the many applications of plate-tectonic theory that have been suggested. Some of the interpretations are highly controversial, and the arguments they call forth are among the liveliest in present-day geology.

What makes things move?

If you will sit back now and think over the reasoning of the last two sections, you will perhaps be struck by a curious inconsistency. Wegener's hypothesis of continental drift was for many decades unacceptable to many geologists chiefly because no force capable of moving the continents could be conjectured. Yet today the movement of continents, or of larger crustal segments that carry the continents with them, forms the basis of most large-scale geologic speculation—although we *still* have no clear idea as to why the plates should move. How does it happen that at one period in the development of earth science the lack of an obvious motivating force is seen

Fig. 14-12. The mechanism of a transform fault. If a mid-ocean ridge is broken, so that two segments are offset, the spreading motion away from each segment gives rise to horizontal (strike-slip) movement along the line of offset. Note that relative movement takes place only *between* the offset segments; beyond the segments the motion on either side of the line of offset is the same.

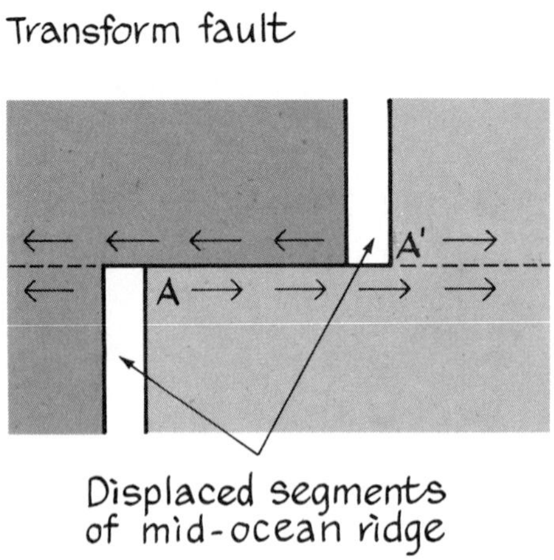

as a major obstacle to acceptance of an hypothesis, while at a later period it becomes only a minor detail to be brushed aside?

The answer probably lies in the *amount* of evidence that can be marshaled for the hypothesis at different times. As long as the idea of drifting continents could be supported only by such facts as the rough similarity of continental outlines, rather vague geologic correlations across ocean basins, distribution of fossils, and evidence of widespread ancient glaciation, the assumption of large-scale crustal motion without a propelling force seemed a needlessly fanciful sort of explanation. The situation changed completely as other kinds of evidence began to pile up—the topography of the mid-ocean ridges, the greater-than-normal heat flow from the ridges, the curiously restricted age of oceanic sediments, the distribution of volcanoes and of earthquake foci, and above all the bands of normal and reverse magnetization of oceanic basalts. Now the once-fanciful explanation seems the *only* unifying hypothesis that can account for these diverse observations, and furthermore the only one on which successful predictions about large-scale earth structures can be based. The motivating force may still be obscure, but overwhelming evidence tells us that the continents must be in motion.

What *could* the motivating force be? Several answers are possible, all highly speculative. We will look at a single one, not with the idea that it should be accepted as correct but just to show that a plausible explanation for the motion can be devised.

Fig. 14-13. A possible model of convection currents in the mantle that could account for sea-floor spreading and continental drift. (From *Debate about the Earth,* after D. H. Turcotte and E. R. Oxburgh)

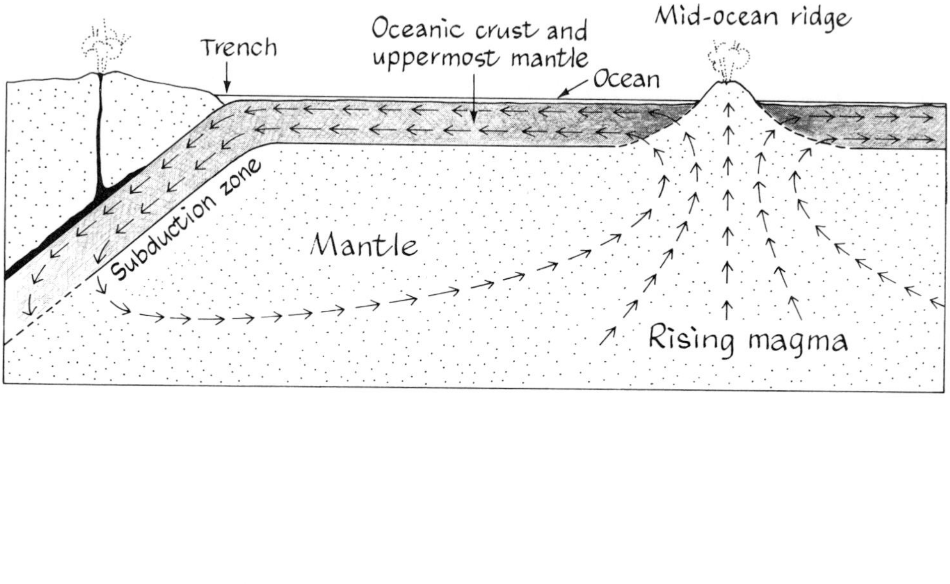

We recall that the moving, rigid plates are formed at oceanic rises, where basaltic magma comes up from the mantle; that the plates move horizontally away from the rises, as if propelled by a current beneath; and that the plates are destroyed where they move downward into the mantle under island arcs or continental margins. Let us imagine that the current on which the plates ride is in the upper mantle, perhaps in the zone of weakness, or asthenosphere (page 345). Then suppose that at greater depths there is a reverse current, moving material in the mantle back toward the oceanic ridges. In other words, we imagine the upper part of the mantle to be made up of huge cells of circulating material, moving upward at the rises, horizontally under the ocean basins, downward at continental margins, and horizontally again at great depths. This is not inconsistent with an earlier conclusion that the mantle is largely solid, because the motion is slow and evidently generated by forces that act over very long times.

What might the generating forces be? We note the similarity between moving cells of this sort and the convectional cells set up in a pan of water heated on a stove; and we recall that the deep interior of the earth is hot. Perhaps the moving cells in the mantle are convection cells, arising simply from the fact that the interior of the mantle is much warmer than the overlying crust. Heated material rises under the oceanic ridges, is cooled as it moves under the oceans, and descends as cool material under island arcs and edges of continents. The overall result is escape of heat from the earth's interior, at a much faster rate than it could escape by simple conduction through solid rock. In effect we are picturing the earth as a gigantic heat engine, propelled by the temperature difference between the interior and the crust, and the drift of the crustal plates is just a passive response to the energetic motion in the upper mantle.

This guess about the motivating force of continental drift is attractive because it appeals only to well known facts about the distribution of earth temperatures. Whether the temperature difference provides an *adequate* source of energy is less certain, and the guess can be questioned on several other grounds as well. This idea is currently the favored explanation for the mechanism of sea-floor spreading, but it is by no means established as a well verified hypothesis. Earth scientists will need much more detailed information about the mantle before convection currents can be accepted as more than a speculative guess.

For that matter, the whole theory of plate tectonics still involves a good deal of guessing, even apart from the mechanism of movement. The theory has had striking success in correlating observational facts and in making possible predictions about earth structures. But there are still some facts that seem seriously at variance with the theory, as is continually pointed out by the few brave individuals who remain skeptical. Doubtless the theory will need modification as time goes on, to accommodate these contrary observations; conceivably it may have to be abandoned completely in favor of some other hypothesis. At the moment continental drift seems so well established as a general idea that it will never be shaken from acceptance by most geologists, even though changes in detail will certainly be needed. But the evidence is not wholly one-sided, and the history of science gives warning that beautiful theories, no matter how widely accepted, have been destroyed on many occasions by new observational evidence.

Fig. 15-1. A plate junction: the San Andreas fault south of San Francisco. The zone of faulting is marked here, as in many other places (Figs. 10-15 and 10-16), by a conspicuous valley eroded in material fractured and pulverized by the fault movement. According to one version of plate-tectonic theory, the San Andreas is a transform fault separating two of the earth's major plates: the Pacific plate on the west (right half of the picture) and the North American plate on the east (left half of the picture). (R.E. Wallace, U.S. Geological Survey)

15

Our Dynamic Planet

Have you sometimes, in a reflective mood, sat before a map globe, twirling it slowly on its axis? With a little deliberate blurring of vision, a little play of imagination, you can think of yourself returning to earth from a rocket voyage to the moon. In contrast to the stark, dead world behind your rocket, the flecks of cloud and blue oceans and multicolored lands of the world ahead betoken change, movement, variety. Yet the changes and variety concern only the surface features, only the atmosphere and the living things that the rocket traveler knows he will find when he lands. The solid framework of the globe, the continents and islands that stand above the ocean surface, seem as changeless as the gray mountains and dark plains of the moon. Both planetary bodies, to all appearances, had their surface features fixed long ago, and the essential difference between them is only the air and water and life on the larger of the two.

So a geologist might have mused, turning the globe idly with his fingers—up until the middle 1960's. Now the picture has radically altered. No longer are the continents and islands fixed in position: impelled by some sort of currents beneath, they are floating about through the oceans like ice floes on a slow-moving stream. Where a moving plate carrying a continent meets resistance, the continent's edge is crumpled into mountains and perforated by volcanoes. Continents split apart, come together, and once more separate. It is true that one must watch the globe for millions of years to detect these transformations, but watching for millions of years becomes second nature to a geologist. All parts of our planet are in motion, and it is only the extreme shortness of our human time-scale that gives the illusion of fixity.

This change in view from a static earth to a dynamic earth is the real kernel of the theory of plate tectonics. Regardless of troublesome details,

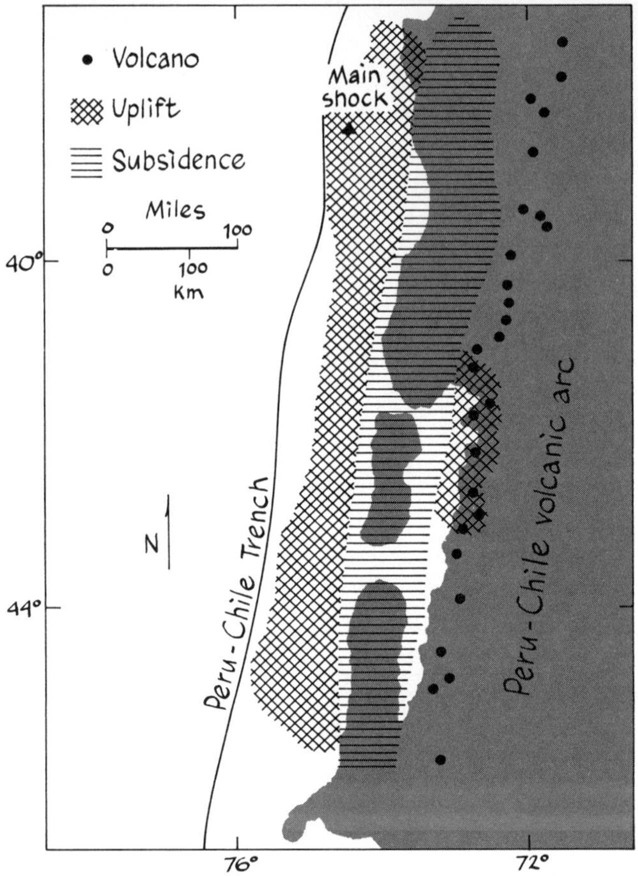

Fig. 15-2. Map of part of the south coast of Chile. (After George Plafker)

regardless of the difficulties about mechanism, the concept of an earth whose surface is racked by continuing large-scale activity of its interior has profoundly altered geologic thinking. With a few examples, let us see how some traditional geologic observations fit into the new framework of ideas.

A look at the coast of Chile

In the previous chapter we noted the west coast of South America as an example of a place where an oceanic plate is diving under a continental margin. Here the results of a plate collision are particularly clear and simple, so the example will repay a further look.

We focus attention on the island-studded coast of southern Chile, where rugged mountains rise out of the sea and where numerous islands and the valleys that deeply indent the land indicate extensive changes of elevation in the recent geologic past. We recall that a deep trench parallels the coast a short distance offshore, and that the East Pacific Rise also approximately parallels the coast a few thousand kilometers farther west (Plate 34). According to plate-tectonic theory we have interpreted these relations to mean that a rigid plate of basalt is moving eastward away from the Rise at a rate of a few centimeters per year, carrying the sea bottom on its back, and that both basalt and sea-floor sediments are plunging obliquely downward under the offshore trench.

What geologic observations might betray this submarine and subterranean activity? Certainly nothing in the placid day to day events of weathering and erosion. The wearing away of the land is rapid here, because slopes are steep and the climate is rainy; debris from the land is worked over by currents along the shore, carried out to sea, and sorted into layers of sand and mud that cover the narrow continental shelf. This is the sort of geologic activity we would expect to find on any rugged coast. Clearly it has no relation to possible movement of basalt far beneath.

But the coast of South America is notorious for severe earthquakes. A particularly damaging one struck the south coast of Chile in 1960, damaging not only because of the violent shaking but also because permanent changes of elevation with respect to sea level amounting to several meters made harbor installations inoperable. Study of post-earthquake sea levels showed that the islands farthest offshore had risen, while the coastal shoreline itself had moved downward. Now if we think of a basalt plate moving somewhere beneath, these observations seem not unreasonable. The movement is very slow, rocks can crumple and stretch to some extent, and part of the movement may be simple gliding on a fairly

Fig. 15-3. One of the deeply indented bays on the south coast of Chile, with a volcanic peak in the background. (C.F. Park, Jr.)

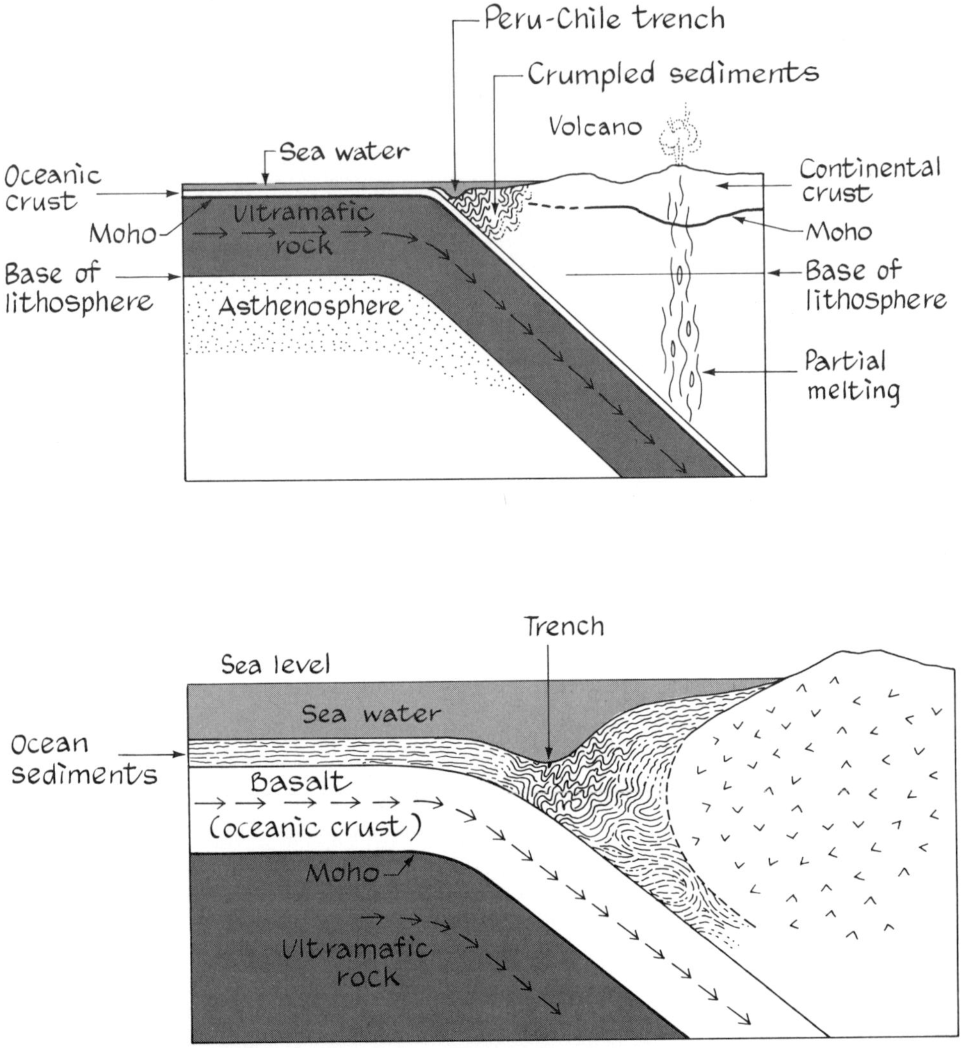

Fig. 15-4. Subduction zone under the Chile coast. The lower figure shows details near the trench.

smooth surface; so earthquakes do not occur every day, but may be expected at intervals when strain has accumulated in the rocks from several years of continuous movement. Plunging of the basalt downward at an oblique angle under the continent would plausibly explain the sea-level changes in elevation: a dragging downward of material near the submarine trench, where the downward movement of the basalt begins, a piling up of material farther east by the push against the resisting continent, and again a downward drag still farther east (Fig. 15-4). The explanation seems simple enough now, but in 1960, before plate-tectonic theory was formulated, the apparent changes in sea level posed a difficult puzzle. That similar changes have occurred here during other earthquakes in the recent geologic past is evident from many shoreline features.

If the basalt plate has been moving for many millions of years, as plate-tectonic theory suggests, one would expect sedimentary layers deposited at different times during the movement to be faulted, tilted, and folded by the continual push on the continental margin. In general older layers should be deformed more than recent ones, just because they have been subjected longer to movement, but even very young layers exposed in the islands and coastal ridges should be conspicuously folded and faulted. This expectation is abundantly confirmed by study of the meager outcrops in this heavily wooded region.

We look now a little more speculatively at what happens when a layer of basalt and an overlying layer of marine sediments move downward along a subduction zone into the higher-temperature region of the lower crust and upper mantle. The increased temperatures and pressures will presumably cause the mineral transformations that produce metamorphic rocks. While the metamorphic changes are occurring, we might expect deformation and mixing by continual movement along the contact between the descending basalt and the overriding continent. At some point temperatures will rise beyond the metamorphic range and the rock material will start to melt. The melting is a complicated process. The rock does not melt as a whole, but mixtures of minerals with low melting points form first and additional melting takes place only gradually as the temperature climbs. The nature of the first melt and the temperature of melting depend on many factors, particularly on the amount of water and salt in the interstices of the original sediment. Generally the first melt would be expected to form between 600 and 700°C, and to consist chiefly of the elements in quartz, potassium feldspar, and sodium feldspar—in other words, approximately the composition of granite or rhyolite. Continued rise in temperature would bring other materials into the melt, notably calcium, magnesium, and iron, so that eventually magmas covering a broad range of compositions might be produced.

What would happen to bodies of molten silicate material formed in this way? The melt should be lighter than its surroundings (both because silicate minerals expand on melting and because the melt contains dissolved water), hence would rise if any openings are available. Because the solid rock above the downward-moving basalt is cracked and broken, it should be weak enough to permit at least part of the magma to find its way to the surface. Thus one consequence of the moving of an oceanic plate under a continent should be the formation of volcanoes—and the great volcanic peaks of the Andes are witness to the correctness of this expectation.

Furthermore, one might anticipate that vertical movements in the crust would eventually bring some of the metamorphosed rock to levels high enough for erosion to expose it to view. Again the conclusion is verified by observation, and much of the exposed metamorphic rock has precisely the structurally complicated, stirred-up appearance that would be expected from material that has been squeezed and broken and rolled between moving slabs. Not only metamorphic rock but also bodies of granite (from magma presumably generated by partial melting, but which remained near its source rather than rising into a volcano's throat) exposed in parts of the Andes provide evidence for long-continued motion between the Pacific and South American plates.

So in this fairly simple situation a variety of geologic observations fit neatly into the framework of plate-tectonic theory. Similar clean-cut stories could be developed for other parts of the Pacific rim, particularly the south coast of Alaska and some of the island arcs fringing the coast of Asia.

(Text for Chapter 15 continues on page 402.)

Legends for Plates 27 through 34

Plate 27. The Mid-Atlantic Ridge. A view of the north Atlantic basin as it might appear without the water. The Mid-Atlantic Ridge is conspicuous curving down the center, and the continents on either side are bordered by broad continental shelves. The Ridge has a deep valley along most of its crest (the Mid-Atlantic Rift Zone), and is offset by many narrow valleys approximately at right angles. The conical hills east and west of the Ridge are in part islands (e.g., Bermuda and the Azores) and in part seamounts. (B. Heezen, M. Tharp, and the Aluminum Corporation of America)

Plate 28. A topographic map. An example of the topographic maps published by the U.S. Geological Survey, showing a corner of the Vevay Quadrangle on the Ohio River between Indiana and Kentucky. The third dimension of the topography is shown by contour lines (brown), lines of equal elevation which here are spaced 20 feet apart. Note that the lines are close together in the steep hill country of southern Indiana, but far apart on the nearly flat floodplain of the river. (U.S. Geological Survey)

(Legends for these plates continue on page 401.)

Plate 27. The Mid-Atlantic Ridge.

Plate 28. A topographic map.

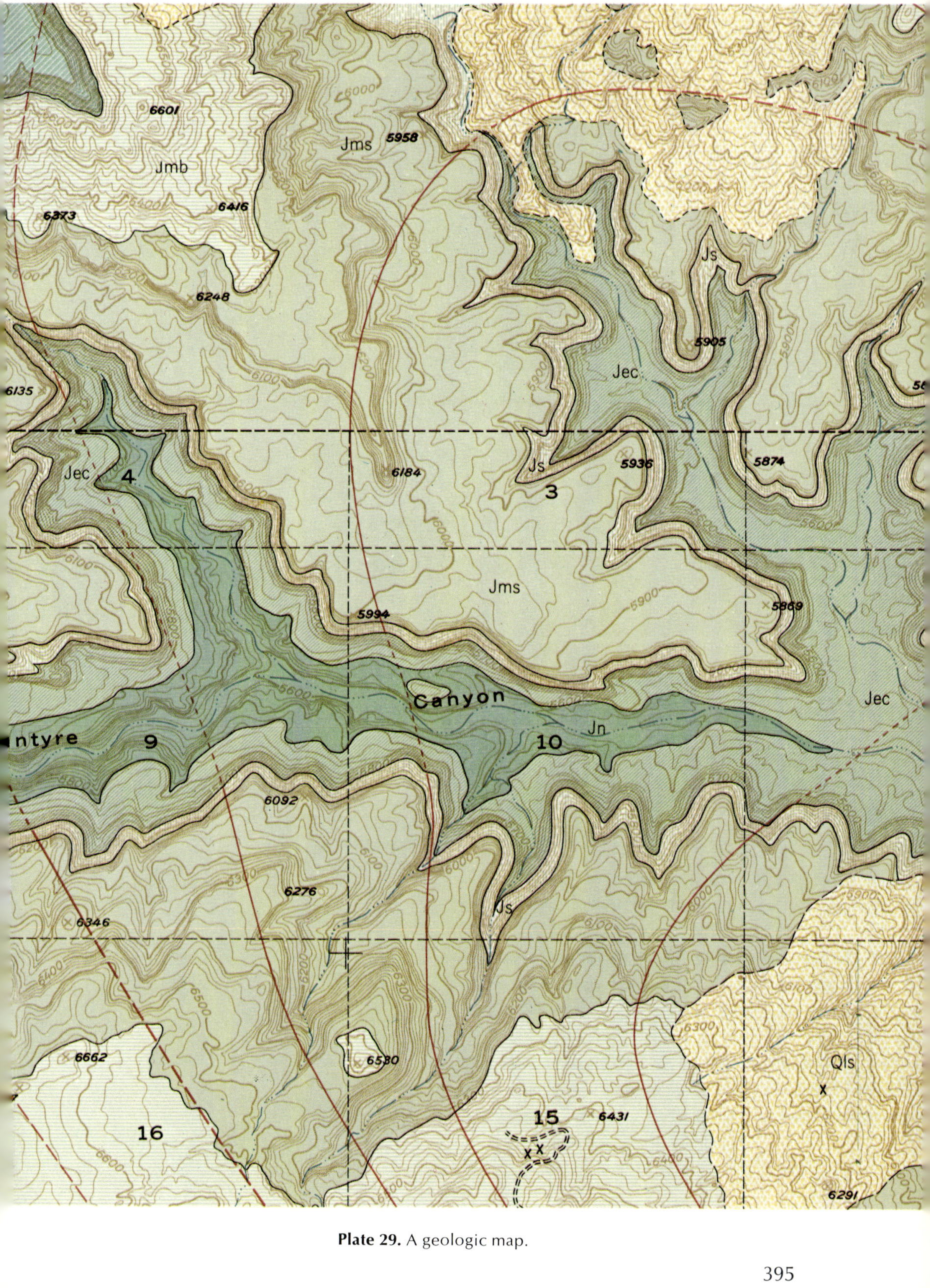

Plate 29. A geologic map.

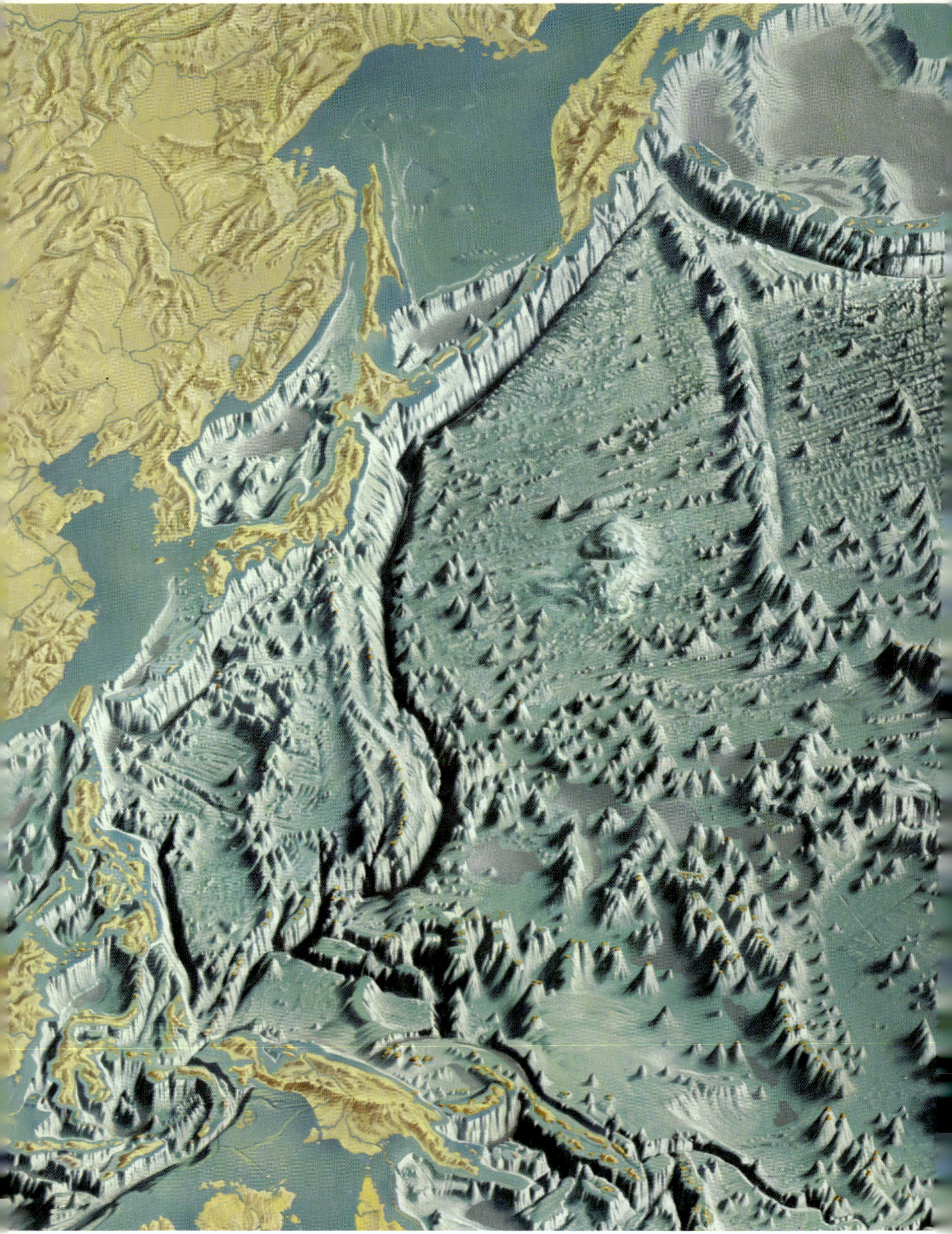

Plate 30. The north Pacific basin.

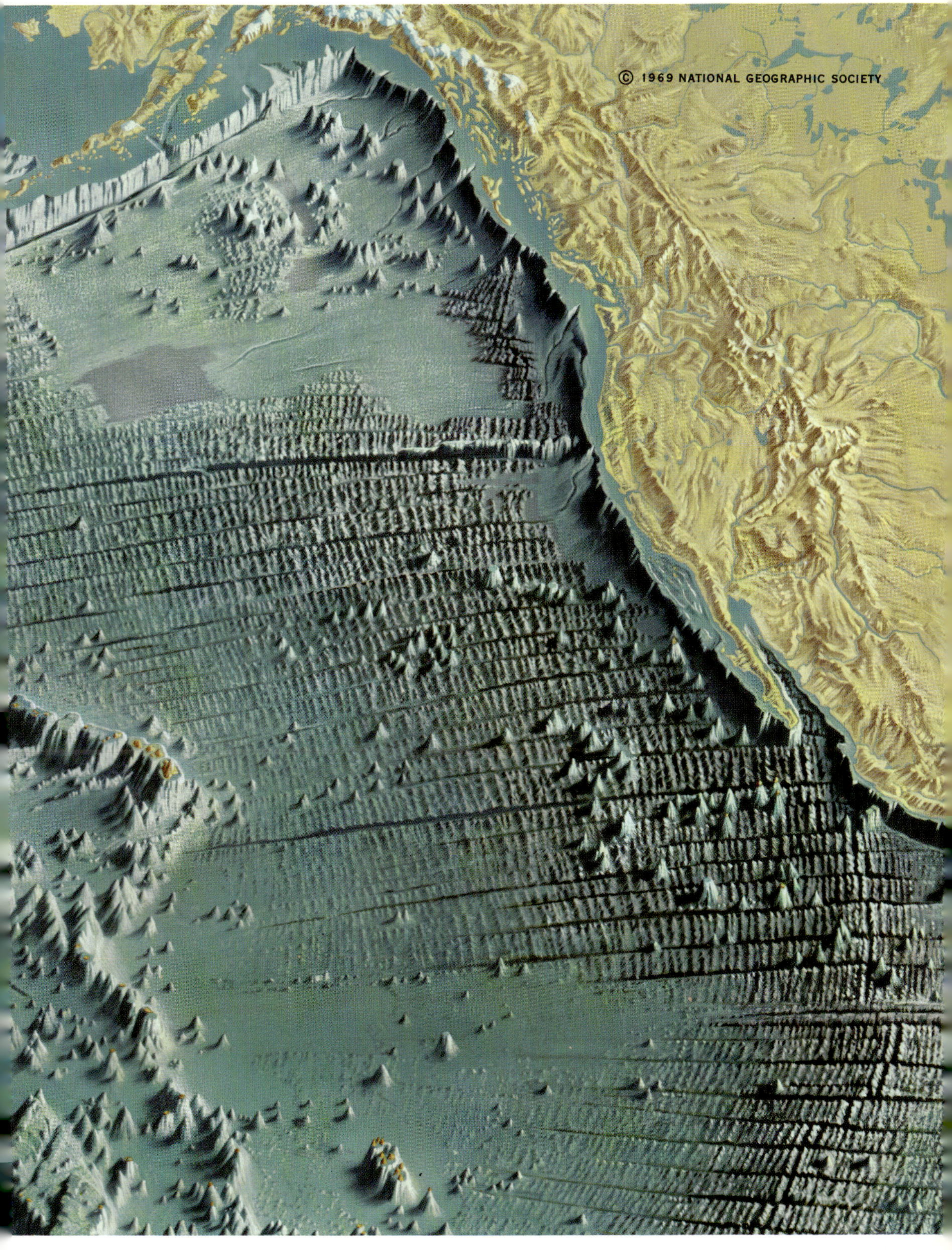
© 1969 NATIONAL GEOGRAPHIC SOCIETY

Plate 31. A uranium mine.

Plate 32. The Bingham Canyon copper mine.

Plate 33. The Sinai Peninsula from a satellite.

Plate 34. The southeast Pacific basin.

Plate 29. A geologic map. An example of the geologic maps published by the U.S. Geological Survey, showing part of the Horse Range Quadrangle in western Colorado. In addition to contour lines, this map shows by patterns and colors the kinds of rocks that crop out or are found just beneath the surface. From the fact that contacts between colored patterns are approximately parallel to contour lines, one might guess that the rocks of this area are flat-lying sedimentary beds. (U.S. Geological Survey)

Plate 30. The north Pacific basin. Conspicuous on this map are the island arcs and curving trenches of the western Pacific, the chains of islands and clusters of seamounts in many parts of the ocean, and the East Pacific Rise in the southeast corner. In contrast to the Mid-Atlantic Ridge, the Rise in this ocean is displaced far to the eastern side, and it appears to run into the continent of North America at the mouth of the Gulf of California. (B. Heezen, M. Tharp, National Geographic Society)

Plate 31. A uranium mine. The Lucky Mc mine in the Gas Hills, central Wyoming. The uranium ore here is in flat-lying sandstone beds, exposed in a large open pit. The picture shows clearly the contrast between dark unoxidized ore in the deeper levels and bright-colored oxidized ore near the surface. (Utah International, Inc.)

Plate 32. The Bingham Canyon copper mine. This huge pit in the hills southwest of Salt Lake City is one of the world's major copper producers. The colors in different parts of the pit are caused by oxidized iron and copper minerals. (Kennecott Copper Corporation)

Plate 33. The Sinai Peninsula from a satellite. The picture is taken looking northward from a point over the middle of the Red Sea. Arabia is on the right, Egypt on the left; the delta of the Nile is barely visible as a dark patch at the extreme left. From a picture like this it is easy to imagine a slow drifting apart of Egypt and Arabia, accompanied by the opening of the Gulf of Suez on one side of Sinai and the Gulf of Aqaba on the other. (NASA)

Plate 34. The southeast Pacific basin. This is a southward continuation of Plate 30, showing the East Pacific Rise off the coast of South America and the deep Peru-Chile trench near the shore. The high Andes range, dotted with volcanoes, parallels the coast a short distance inland. It seems likely that this coast represents a line of collision between the westward-moving South American plate on the east and the eastward-moving East Pacific (or Nazca) plate on the west. (B. Heezen, M. Tharp, National Geographic Society)

Complications in California

For a second try at matching geologic observations with plate-tectonic theory, we pick a more complicated example, the coast of California. Here again rugged mountains abut the sea, but there are few offshore islands, no deep trench paralleling the coast, and no submarine ridge under the ocean to the west. Layers of sandstone and shale with fossils no more than a few million years old are tilted up on end in the coastal mountains, and over large areas blocks of metamorphic rock are thrown together with sedimentary and igneous rocks in jumbles that structural geologists despair of unraveling. No volcanoes are active at present, but fresh lavas and areas of hot springs are sure signs of activity in the not very distant past. These characteristics of the geology point clearly to a collision of crustal plates, but the submarine evidence of a plate moving toward the continent is completely absent. How can we reconcile this seeming contradiction?

A further complication is the great fault, the San Andreas, which is a major feature of California geology. Skirting the coast from northern California to San Francisco, then bending inland slightly and crossing the Coast Ranges to the south end of the central valley, then again turning southward across the desert to the head of the Gulf of California, this line of movement along much of its course separates rocks of very different ages and compositions. The movement is mostly strike-slip, with the west side moving northward. Correlating rocks and structures across the fault indicates that it has been active for at least all of the Cenozoic Era, and that the total movement of one side relative to the other has been more than 300 miles. Any reconstruction of California geologic history must include an interpretation of this impressive strike-slip fault.

What happens to the East Pacific Rise in this part of the world? If you will again look at a globe, you can follow the Rise northward from its position off the coast of Chile. The line of the Rise may be marked—it is shown on many modern globes—but if not, you can imagine it trending northward, a little west of the Galápagos Islands, offset by many of the transform faults that are so common along all the mid-ocean rises, and at length running into the land at the mouth of the Gulf of California (Plates 30 and 34). From there on the Rise apparently vanishes, but a possible continuation reappears a few hundred kilometers off the coast at about the California-Oregon border. Segments of the Rise, again offset by transform faults, can be followed northward along the Canadian coast.

What can be made of these confusing relations? Suppose we try projecting ourselves into the past, back to a time 100 million years ago when North America had not moved so far westward from the Mid-Atlantic Rise. Perhaps in those days the East Pacific Rise was continuous off the west coast of the continent, and the basalt plate moving eastward from the Rise was descending along a subduction zone under the continental margin, just as it is today under the coast of Chile. The deformation of sedimentary rocks, and the formation of metamorphic and igneous rocks at depth, would have been natural consequences. Then imagine the continent, continually pushed from the Atlantic side, moving until its western margin touched the ridge—and even farther, until the bulge in the margin that is now California and northwestern Mexico actually overrode the ridge. In this imaginative reconstruction, the reason that no submarine ridge appears off the southern California coast is just that this part of the ridge lies beneath the continent; and the rocks and structures that so strongly suggest a collision of two plates are left from an earlier time when such a collision was in fact taking place.

What, then, of the San Andreas fault? Recall that the East Pacific Rise runs approximately northward into the coast of central Mexico, then is apparently broken off and reappears near the Oregon border. In effect, the broken ends have been displaced by something more than a thousand miles. Now the San Andreas—if the geometry is interpreted generously—lies roughly along the line of displacement. Hence it is possible that the great fault is simply one of the transform faults that cut the Rise in many places,

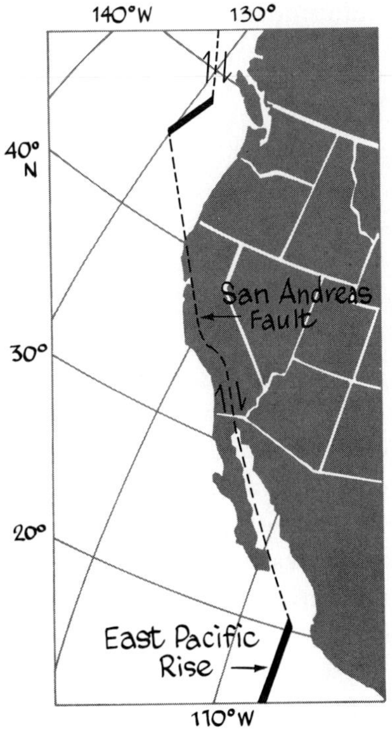

Fig. 15-5. The San Andreas fault as a transform fault. This oversimplified diagram shows the fault as connecting two widely offset segments of the East Pacific Rise. Actually the situation is much more complicated, for the Rise has many additional offset segments in the Gulf of California. (After J.T. Wilson, "A new class of faults and their bearing on continental drift," *Nature*, vol. 207, page 343, 1965)

perhaps elongated by the westward push of the continent against the northern part of the Rise.

This, in broad general terms, is a possible way to reconcile the anomalous geology of the western United States with plate-tectonic theory. It is only one of several ways, however, in which a plausible reconciliation can be made. It should be taken only as an hypothesis, with none of the assurance we can give to our reconstruction of geologic happenings in southern Chile. The hypothesis has many weaknesses. The geometry is not quite right, for example: the San Andreas fault lies at an oblique angle to the East Pacific Rise, not at right angles as do nearly all the other transform faults. The hypothesis does not explain geologic peculiarities of the western states inland from California—the many desert ranges outlined by normal faults in Nevada and Utah, the high Colorado Plateau, the frequent earthquakes in western Nevada and western Montana. The northern termination of the San Andreas fault is poorly understood. In part these deficiencies in the hypothesis can be patched up by further plausible assumptions, but too many assumptions make an hypothesis suspect.

We are looking here at an area about which geologic controversy is still very active, where observational evidence is still insufficient to permit a clean decision among rival interpretations. Such a situation is a joy to geologists. One of the great attractions of plate-tectonic theory is precisely the fact that it leads to controversy about major geologic problems, and therefore stimulates examination and re-examination of rocks and structures to find the critical evidence that will favor one speculative hypothesis over another.

The other side of the continent

California and Chile are examples of advancing continental margins—places where continents are moving, or have moved in the past, over oppositely directed oceanic plates that are descending into the mantle. Such margins are scenes of intense geologic activity. Where sedimentary rocks are forming, they are laid down on narrow continental shelves or in small basins like the present central valley of California; individual layers can be followed only short distances, and change rapidly from one kind of sediment to another; tilting and folding are so general that it is rare to find sedimentary layers in their original horizontal position; volcanic material is commonly interbedded with the sedimentary rocks. Where older rocks are exposed, they commonly include patches of metamorphic rock and plutonic igneous rock. All these observations along active margins, as well as the frequent earthquakes, give evidence of the disturbance created when a continental plate bumps into an oceanic plate at depth.

What about the other kind of continental margin, the "trailing edge," represented by the east coast of the United States? The situation is entirely different, because here there is no motion between the continent and its underpinnings. According to plate tectonics the continent behaves as a huge slab of relatively light rock, carried westward by the moving basalt underneath, like a pile of rock traveling on a conveyor belt. This eastern edge of North America was once joined to Europe and Africa; after the rift between the continents opened, the basalt plate under the western Atlantic has grown by additions to its eastern edge, and at the same time has moved steadily westward with North America riding passively on its top. Thus the continent remains on the same part of the basalt plate as when its travels started. There is no geologic activity except the usual processes of erosion, transportation of debris, and laying down of sedimentary strata on the coastal plain or in shallow water offshore.

By contrast with the Pacific side of the continent, then, Atlantic sediments are better sorted, they are deposited in layers that can be followed for tens or hundreds of miles, and they remain, for the most part, nearly as flat as when they were laid down. (This applies only to sediments formed since westward movement began in the middle of the Mesozoic Era, of course; older sedimentary rocks of the Atlantic states show more complex structures, but these belong to an earlier period of earth history.) Because sediment has accumulated undisturbed since the mid-Mesozoic, the continental shelf is broad and gently sloping. Volcanic materials are absent, except for debris eroded from much older rocks; very few intrusive or metamorphic rocks of later than Triassic age are known from the eastern states. Earthquakes are not completely absent—the quake that struck Charleston, South Carolina, in 1886 was highly destructive—but are few in comparison with the west.

The passive accumulation of sediment on a continental margin where there is little motion of adjacent plates is especially striking in the Gulf of Mexico. The supply of sand and mud here is enormous, especially from the Mississippi River, which drains much of the interior of the continent. Since the middle of the Mesozoic, some 150 million years ago, the great river has been pouring sediment into the Gulf, building its delta southward from an original shoreline at the southern tip of Illinois. The lowlands along the river in Arkansas, Mississippi, and Louisiana are a product of the river's activity; the slow migration of the present delta out into the Gulf is only a continuation of a process that has been going on since the heyday of the dinosaurs. In the Gulf itself, waves and currents have transported and deposited debris not caught on the delta, forming beds of sandstone and shale with a total thickness of several thousand meters—some now underlying coastal portions of Texas and eastern Mexico, some still wholly

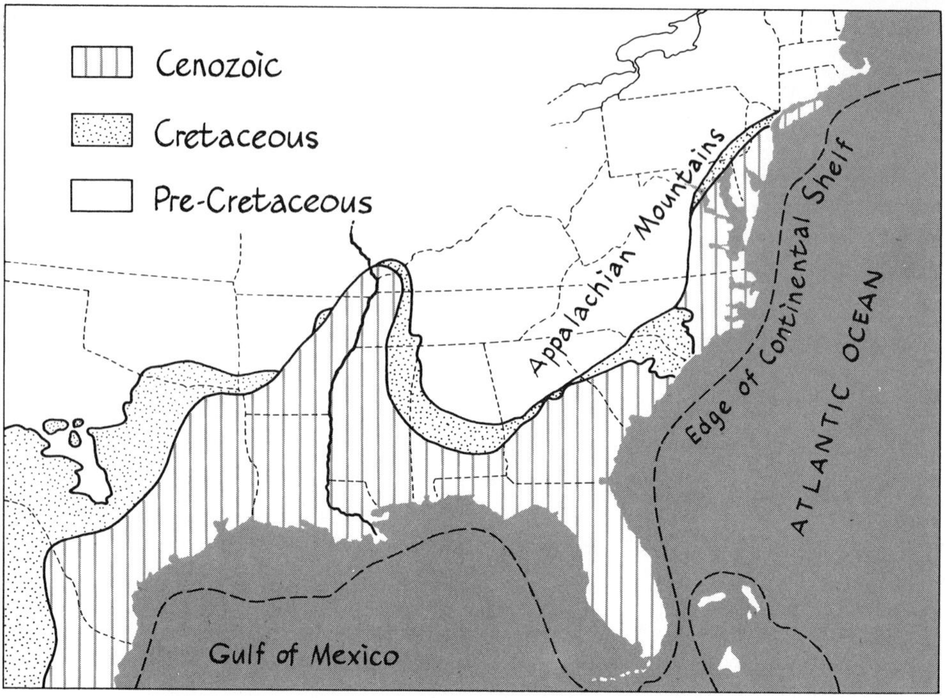

Fig. 15-6. The growth of the Mississippi delta. The map shows sediments along the Gulf and Atlantic coasts that have accumulated during the past 150 million years. The central part of this sedimentary pile, from southern Illinois to Louisiana, consists largely of material deposited in a succession of deltas formed by the Mississippi River during this long time interval. (Redrawn from Thomas H. Clark and Colin W. Stearn, *Geological Evolution of North America*, second edition, copyright 1968, The Ronald Press Co., New York)

submerged beneath the Gulf. Parts of this huge accumulation have been arched into broad anticlines and synclines by minor crustal movements, and parts have been more sharply deformed very locally by plugs of salt pushed upwards into the sediments from thick salt beds far below. Many of the anticlines and broken strata near salt domes have provided traps for petroleum, resulting in some of North America's largest oil fields.

What was happening in the middle of the continent, while the western edge was being crumpled and the eastern and southern margins were being buried by sediment? From plate-tectonic theory, one might guess that the interior would be relatively inactive, because it is just the broad surface of an inert continental mass drifting slowly westward. For the region east of the Rocky Mountains, this guess is fairly accurate. River and lake sediments were deposited at times over parts of the high western plains, and the Pleistocene ice sheets dropped debris from Canada in the northern states, but otherwise the continental interior was largely a scene of long-continued erosion.

In the states from Colorado west, however, the geologic record is much more complex. Toward the end of the Mesozoic a shallow sea covered much of the mountain states, extending eastward for a time even across the Dakotas into western Minnesota; and the start of the Cenozoic was punctuated by the formation of many mountain ranges and much volcanic activity in the western states. These events are not easily predicted from plate tectonics, and in fact constitute one of the most troublesome parts of American geology to bring within the theory's framework.

Before the Mesozoic

Look once more at your globe, and try to imagine how it might have appeared some 250 million years ago. This was the end of the Paleozoic Era, before the dinosaurs, before there were birds or flowering plants—and before an ocean separated the Old World from the New. All the continents were clustered in a single landmass, the Americas joined to Africa and Europe, Australia and Antarctica connected with India and Africa. One side of your late Paleozoic globe would have been land, the other side water. Very probably the south pole would have been on land, somewhere

Fig. 15-7. Sketchmap of the southern hemisphere as it might have appeared at the end of the Paleozoic Era, with all the continents clustered into a single landmass. The shaded portion shows the area of the late Paleozoic glaciation. (After A. Holmes, *Principles of Physical Geology,* 1965, Ronald Press Co., and Thomas Nelson and Sons, Ltd.)

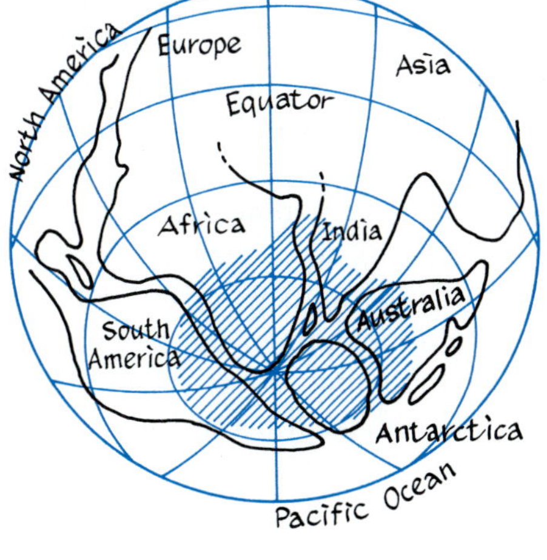

in the part of the landmass that is now fragmented into South America, southern Africa, India, Australia, and Antarctica. Glaciers would have covered the polar area, leaving records of their movement that we can still see in the rocks of these now widely scattered continents (Fig. 14-1). There would be little about the Paleozoic world represented by your globe that you would find at all familiar.

If your globe could be equipped with a mechanism to show changes through geologic time, you could follow the break-up of the single huge continent and the scattering of its fragments. A rift would have opened in the western part of the landmass, probably first in the south and gradually spreading toward the north, and the sea would have invaded to form a narrow strait. Similar straits would have formed between Australia and Asia and between Africa and Antarctica, and over millions of years the straits would have steadily widened (back end paper).

If your clairvoyant globe could show submarine movement as well as the changing outlines of continents, you would see molten lava rising from rifts at the centerline of each strait, freezing to the walls of the rifts and becoming part of basaltic plates carrying the continents away from the straits. In places where the continents were actively overriding the ocean floor, you would see chains of volcanoes appearing and new mountains formed by the crumpling of sedimentary layers. As the mechanism of your globe grinds on, with millions of years compressed into a few minutes, you would see the continents move and turn and gradually assume the familiar positions of your present-day globe.

These positions, of course, represent only a fleeting moment in a continuing process of change. If your animated globe could show the future as well as the past, you would see the continents spreading farther apart, the Atlantic growing ever wider and the Pacific smaller, toward a new arrangement of lands that no one can accurately predict.

So we read the long history of the Mesozoic and Cenozoic Eras. For the details we rely on study of rocks and fossils, for the overall framework on the theory of plate tectonics. Here the theory serves us well despite a few weak spots, for it shows in a very plausible way how mountain-building, igneous activity, sedimentation, and climatic change can be fitted into a grand scheme of continental movement. But what of earlier times? We recall that the 200-odd million years of the two most recent eras are only a small fragment of the earth's much longer history. Can the same theory of plate tectonics be extended back into the Paleozoic Era, or perhaps even farther back into the Precambrian?

This is a difficult question, one to which so far there is no firm answer. We no longer have actively spreading submarine ridges and precisely known continental outlines to guide us. If plate tectonics did play an earlier role, presumably the single large landmass of late Paleozoic time must once before have been separated into scattered continents, but the shapes of these ancient lands we can only conjecture from the dim record preserved in the rocks. The ocean can no longer help us. It was study of the sea floor that led to plate tectonics, and further study that provided much of the corroborating detail, but materials found on the sea floor are nowhere older than mid-Mesozoic. All the earlier part of the earth's history must be read from rocks of the continents.

What sort of story can these rocks tell? We have seen glimpses of it in previous chapters, and now we are in a position to put it together in more connected fashion.

We look first at the magnificent display of Paleozoic sedimentary rocks that cover much of the central and eastern United States. In the Appalachian Mountains of the eastern states, and again in the Ozark and Ouachita Mountains of Arkansas and Oklahoma, the layers of sandstone, shale, and limestone are strongly folded and cut by thrust faults. In the north-central states these same beds are monotonously flat-lying, with only the gentlest of dips to indicate broad regional domes and shallow basins. A given bed can be followed from outcrop to outcrop for hundreds of miles; it may retain the same composition, or the composition may change

Fig. 15-8. Horizontal beds of limestone and shale in Kansas. (W.G. Pierce, U.S. Geological Survey)

gradually, say from limestone into shale or shale into sandstone. Where continuity cannot be established by outcrops, fossils permit correlation of beds over huge distances. The fossils for the most part represent marine organisms that lived in shallow water, and shallowness is indicated also in many beds by sedimentary structures. If the thickness of beds dated by fossils is measured in the mountains and in areas where the layers are flat, a given bed generally turns out to be thicker in the mountains. From all these facts we can reconstruct the geologic history, as we have done once before (page 309): a broad area of what is now North America was covered during much of the Paleozoic by a shallow sea; as sedimentation went on, the land subsided slowly and erratically, in general at a faster rate in the areas that were eventually to become mountains.

The thick, far-spreading sedimentary strata are the most conspicuous part of the Paleozoic record in the eastern half of the country, but by no means the whole of it. In New England, and farther south in the Blue Ridge of Virginia and the Piedmont that borders the Appalachians from New Jersey south to Georgia, are Paleozoic rocks of a different stamp. These are schists and gneisses, produced by metamorphism of sedimentary and volcanic rocks, invaded here and there by masses of granite and other intrusive rocks. The famous garnet-mica schist of Manhattan, the gneiss that crops out in the islands and rapids of the Potomac River above Washington, the granite dome of Stone Mountain in Georgia, all belong to this belt of rocks.

These rocks of the east and middle west, we recall, are part of the stable continental mass that broke away from Europe and Africa sometime in the early Mesozoic. Their deposition, their structures, their metamorphism all date from a far earlier time. Through all the plate-tectonic history we have been talking about, these rocks remained inert—moved up and down from time to time with respect to sea level, but not pushed into folds or broken by thrust faults. If plate tectonics played any role in their formation, it was long before the present episode of sea-floor spreading. Can we see any indication of such earlier continental movement?

The answer must be somewhat equivocal. Certainly part of the picture does not fit our first expectations from plate tectonics: there is nothing in the theory to account for the wide spreading of shallow seas over the continent, which is such a prominent part of Paleozoic history. These shallow seas of the past remain mysterious. They reappeared, on a reduced scale, in the western part of the continent during the Mesozoic Era, and at one time in the Cretaceous Period stretched from the Arctic Ocean to the Gulf of Mexico. There is nothing comparable in the present-day world, or in the rock record of the Cenozoic. Since the end of the Mesozoic, shallow seas have been restricted to continental margins. Such incursions of marine water over the land require only vertical subsidence of the entire continent, or major parts of it, and have no apparent relation to plate-tectonic theory. Yet they remain important facts of geologic history, which someday must be accounted for in a truly comprehensive theory of the earth's behavior.

Leaving aside this question of broad continental submergence, we find some features of the Paleozoic rocks along the eastern seaboard that do suggest possible plate-tectonic activity of the sort we recognize today around the Pacific. The sharp folding and thrust-faulting in the Appalachians, the metamorphic and intrusive rocks nearer the coast, the presence of volcanic material both in the metamorphic rocks and in some of the sedimentary layers, all indicate that we may be looking at a belt where long ago an ocean plate dived under a continent. How big the ocean was, where the shoreline and the offshore deep and the mid-ocean ridge of that time were located, we can only conjecture.

The conjecturing is fun, and geologists have spent much time in trying to reconstruct plausible patterns of pre-Mesozoic oceans and continents that will fit the known structures of Paleozoic rocks throughout the world. It seems likely that continental masses were separated from one another before their aggregation into a single landmass at the start of the Mesozoic, and that continents moved so as to collide and to override oceanic plates just as they do today, but there is no agreement as to the forms or paths of movement of these ancient masses. The theory of plate tectonics can be projected into the distant past, but the projection becomes more and more uncertain as the time increases.

Before the Paleozoic

The early Paleozoic rocks of western New York state can be followed across the Niagara River into Ontario, one particularly resistant limestone bed forming the cliffs over which the river tumbles to make the famous Falls. Traveling northward in Canada, one finds scanty outcrops of successively older Paleozoic beds, until the oldest is observed to lie unconformably on rocks whose complex structure and metamorphism suggest still greater antiquity. These are Precambrian rocks, and if you have the fortitude to continue traveling north into the wilderness you will see nothing but Precambrian rocks until you reach the straits separating the Canadian mainland from the Arctic islands. This huge area of Precambrian rocks that makes up most of eastern Canada is often called the *Precambrian Shield*, "shield" in reference to its shape on a map (Fig. 11-12).

Now we repeat our earlier question: Does plate-tectonic theory give us any guide to interpreting these most ancient rocks? Were there ocean basins and drifting continents even back before the Paleozoic? Just from the standpoint of lengths of time involved, this is a major question of earth history—for we recall (page 308) that out of the nearly four billion years of history recorded in the rocks, fully five-sixths belongs to the Precambrian.

The question is important, but not easily answered. Precambrian rocks are a polyglot assemblage, ranging from sedimentary layers as fresh and beautifully preserved as if they had been deposited only yesterday to enigmatic gneisses so profoundly altered that one cannot say with assurance even whether the original rock was igneous or sedimentary. Layering in some Precambrian rocks is as regular and undisturbed as the layering in sediments of a modern lake, but elsewhere the rocks are so contorted that one can imagine them having been stirred with a giant spoon. Fossils with hard parts—shells or bones—are completely absent, so that geologists must work without the usual tool that enables them to arrange rocks in a time sequence and to distinguish clearly between beds laid down in salt water and those from fresh-water lakes. The widespread alteration, the structural complexity, and the difficulty of establishing rock sequences make applications of plate tectonics even more questionable than for Paleozoic rocks.

A few facts about the Precambrian stand out clearly. Sandstones,

Fig. 15-9. Precambrian conglomerate and sandstone near Eagle Harbor on the Keweenaw Peninsula, Michigan. (Carl Dutton)

Fig. 15-10. Precambrian shale, south shore of Lake Superior. (Michigan Department of Natural Resources)

shales, and limestones formed in that remote time are similar to rocks of later periods. Some show the cross-bedding, channeling, and crude sorting characteristic of stream deposits; some have remarkably preserved ripple marks and mudcracks that betoken deposition in shallow ponds ruffled by the wind and subject to periodic drying; some suggest by their thickness, uniformity, and wide areal extent that they were laid down in shallow seas spreading over the continent like the later seas of the Paleozoic. Volcanic rocks also show no general differences in composition or structure from the familiar volcanic materials of the present time. If you could have explored the Precambrian lands, the barrenness of the landscape would have seemed strange; but you would have found the familiar processes of erosion, sedimentation, and volcanic eruption going on just as you see them in the modern world.

Actually this statement is too sweeping. If your Precambrian exploration were extended, and if it could include some oceanographic observation

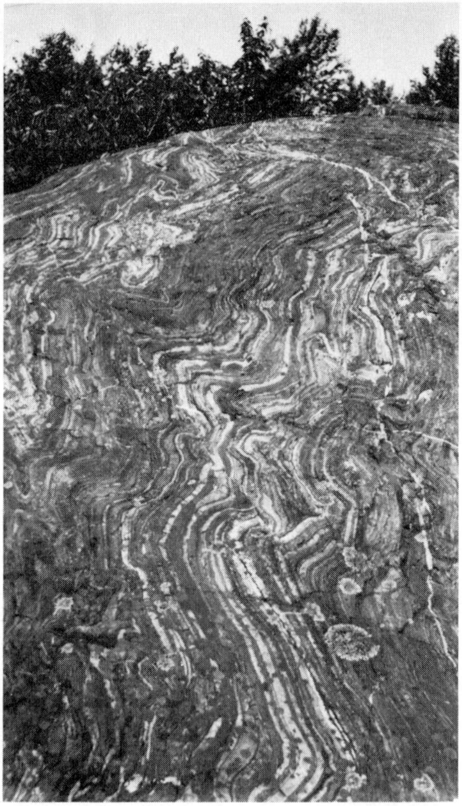

Fig. 15-11. Contorted Precambrian sedimentary rocks ("iron formation") near Soudan, northern Minnesota. (Carl Dutton)

as well as study of the land surface, you would note a few differences from later times. You would see no salt beds in process of formation, for example, in contrast to the extensive salt deposits of more recent ages. The limy muds that harden eventually into limestone you might find accumulating abundantly in the later Precambrian, but they would be scarce in the older seas. One strange kind of sediment you would find accumulating (perhaps in large lakes rather than the sea—the evidence is ambiguous) in amounts far greater than in any later era: a mud consisting largely of iron minerals, mixed with or alternating with beds of gelatinous silica that would ultimately harden into chert. These iron-rich beds (often called *iron formation*), cropping out today in most of the major Precambrian areas of the world, often enriched by later weathering to form huge masses of the red iron-oxide mineral hematite, are the source of much of the iron on which modern industry depends (Fig. 15-11).

These peculiarities of Precambrian sedimentation, together with generally more intense and widespread volcanic activity, would convince you before your exploration had progressed very far that the Precambrian world, despite much similarity in basic geologic processes, was in some ways a very different kind of environment from the world of later times.

All this has little to do with plate tectonics. If we look for structures in

Precambrian rocks that might give a record of continental movement, we have no difficulty in locating belts of folded and faulted sedimentary rocks, or metamorphic rocks that once were sedimentary, which suggest the existence of ancient mountain ranges now worn down to their roots by long erosion. These highly folded beds, together with the volcanic rock that accompanies them and granitic rocks that intrude them, suggest places where ancient crustal plates may have collided. But to outline the Precambrian plates, and to reconcile their movements with other kinds of Precambrian structures, involves much speculation. We are left with the same conclusion as for Paleozoic rocks, except that now it is even less sure. Fragmentary evidence suggests that the earth's crust may have been broken into rigid, moving plates during at least part of the Precambrian, but how important such movement was in Precambrian geologic history as a whole we simply do not know.

Back to the beginning

Through a geologist's eyes we see the earth as a huge, active machine, its solid outermost part segmented into plates that are moved by slow currents in the semi-solid material beneath, the energy for the motion still shrouded in mystery but perhaps explainable by the difference in temperature between the core and the upper part of the mantle. The movement that is going on today has been continuous since early in the Mesozoic.

Fig. 15-12. Early Precambrian metamorphic rocks invaded by granite. The granite has been dated radiometrically at 2.5 billion years, and the metamorphic rocks must be still older. Northern Minnesota. (Carl Dutton)

Earlier movement of the same sort may have existed; certainly forces of sufficient magnitude to produce mountain ranges were active even far back in the Precambrian, although the pattern and mechanism of crustal movement may have been very different. While these enormous forces, about which we actually know very little, were operating in the interior, the familiar processes of weathering, erosion, and sedimentation were busy on the surface, and their effects in the remote past were little different from their effects today. As a grandiose generalization, we can say that all geologic history is a record of the interplay between movements generated by the interior forces and the continual work of surface agencies. The single strangest thing about the record is how constant the interplay has been, how uniformly geologic processes have operated during the nearly four billion years of history that we can read from the rocks.

Yet sometime, if in imagination we take the record back before the earliest known rocks, the world must have been very different. There must have been a time before erosion was active, before an ocean existed to receive sediments, before forces generated by interior movement could fold sedimentary beds into mountain ranges, before there were beds to be folded. Probably there was even a time before the earth existed as a planet. What can we say of these very earliest days of earth history?

Such a question is really more astronomical than geological, but it is nonetheless fascinating to geologists. In attempting an answer we no longer have rocks to provide us with clues, for rocks more than four billion years old simply do not exist on earth—or at least have not yet been found. We depend on what astronomers can tell us about the earth's relations to its sister planets, about the meaning of meteorites, and about the original composition of the material from which planets might be formed. Then we look carefully at the compositions of materials now at the earth's surface, and speculate as to how these compositions might have evolved from primitive planetary material. We can have no assurance that our answer will be right, and we can expect the answer to change from year to year as new facts come to light. Despite the difficulties and despite the assurance that any answer favored today may look silly tomorrow, the question about the origin of things has long intrigued both scientists and laymen.

From the radiometric age of meteorites, and from the astronomers' conclusion that earth and meteorites were formed at roughly the same time, we date the earth's origin at about 4.6 billion years ago. How did the solar system and the earth form? Again we turn to astronomy, and learn that the most plausible current hypothesis begins with a cloud of gas, small particles, and solid chunks with sizes up to meters or tens of meters in diameter, scattered thinly through a volume much larger than that of the present solar system. Gravitation would slowly pull the mass together, and any slight tendency to rotate would be exaggerated as the diameter shrank. In this huge whirling cloud small aggregates of matter would collect and start rotating on their own. These would ultimately become the planets and satellites, and the great central mass of the cloud would shrink to become the sun. To account for the slowness of the sun's rotation compared with the speed of the planets' motion in their orbits, one must imagine that during the condensation great quantities of matter escaped from the cloud into outer space. The earth and its sister planets, then, are the residue of a very wasteful process of aggregation of matter that originally had about the average composition of the universe, but that probably changed its composition as the condensation went on. The four-and-a-half-billion-year "age of the earth" refers to the time when the aggregation into a planet was approximately complete.

In a geologist's eyes the most important aspect of this hypothesis—and the one most strikingly different from previous hypotheses—is the idea that the earth was put together from cold, solid chunks of matter rather than by condensation of a hot gas. Heat would be generated as the chunks fell together, and additional heat would accumulate from radioactive decay, so that the interior of the growing planet would gradually heat up; but the surface need never have been very hot. Just how hot the surface became is a matter of conjecture. The temperature may have risen enough to cause widespread melting of the rock material, but only briefly and never approaching the high temperatures of the interior.

During the original aggregation of solid fragments the planet would have lost most of the gas originally present, especially the lighter gases, simply because the gravitational attraction would have been too small to hold rapidly moving gas molecules until most of the solid mass had accumulated. Once the planet approached its present size, however, heating of the interior would cause the solid chunks to lose gases adsorbed or frozen on their surfaces. The released gases would find their way to the surface, and now the planet would be large enough so that its gravity could hold them. Thus the atmosphere accumulated, very little of it from the original gas of the cloud but mostly as a secondary product sweated out of the earth's interior.

One of the major gases to be sweated out was water, which would remain as a vapor as long as the surface was hot but would collect to form the primordial ocean when the surface cooled. Other important gases would presumably be nitrogen, carbon dioxide, and compounds containing sulfur and chlorine—precisely the assemblage of gases that we now find escaping from volcanic fumaroles. And we might well wonder: does this mean that gases are *still* escaping from the deep interior, that some of the gases from present-day volcanoes represent material that was imprisoned when the earth was formed? In particular, is water still escaping, so that the oceans are growing larger? To such questions there is no sure answer, but the best current evidence suggests that the amount of primitive gas now escaping is small, that the major part of the atmosphere and ocean came to the surface far back in Precambrian time.

We turn now from the air and water above the solid surface to the interior, and the obvious first question is how the massive part of the earth developed its layered structure—core, mantle, and crust. If the solid chunks that formed the earth had the average composition of meteorites, so that the metallic iron now concentrated in the core was once spread through the whole body of the planet, some sort of partial melting would be necessary to separate the iron from the silicate minerals. Reasonable estimates of internal temperatures and pressures indicate that iron in the interior would have melted before the silicates, and because of its high density would have drained downward toward the earth's center. Partial melting of some of the silicates might also have permitted the lighter material to rise and form the crust. Alternatively, one might suppose that in the initial aggregation of solid chunks to make the earth, some separation of light and heavy material took place: perhaps heavy iron-rich chunks like present-day iron meteorites were the first to accumulate, and silicates of varying density were added only later, so that the earth's layering was a direct consequence of its mode of formation. The relative merits of these two hypotheses are a subject of lively current argument.

Whatever our guesses may be about the tumultuous early days of earth history, the general segregation of material into atmosphere, hydrosphere, crust, mantle, and core was probably accomplished within a short time—say a few million years—after the planet's birth. Some additional sorting out of heavy and light material probably went on for a long time, and may even be continuing today, but the overall layering is very ancient. Certainly by the time the oldest known rocks were formed, a half-billion years or so after the earth's material accumulated, an ocean much like the present one lapped against rocks of the same sort that now form the earth's crust. The surface had long since cooled, the crust had separated from the mantle, and presumably most of the earth's iron had collected into the core.

The scene was apparently set for crustal plates to move, impelled by currents beneath, but we know so little about the precise interior conditions necessary for crustal movement that this conclusion is shaky. Plate tectonics may have governed the development of major earth structures from earliest times, but we have no sure evidence. The only certain conclusion is that crustal rocks have been actively deformed, whether by plate tectonics or some other mechanism, from the time when the geologic record begins.

An observer who could go back to stand on a seashore of three or four billion years ago would recognize the rocks beneath his feet, and would find the ocean about as salty as the present one. But the atmosphere would be very different, as he would quickly discover from the difficulty in breathing. There would be no oxygen, or no more than a trace. The chief gases would be nitrogen, carbon dioxide, and water vapor, with perhaps a little methane, ammonia, hydrogen cyanide, hydrogen sulfide—not a very healthy mixture for a modern land-dwelling organism. If the observer could revisit the seashore at intervals over the subsequent three billion years, he would find the composition of the air slowly changing. The carbon dioxide would decrease, the oxygen would increase—and at the same time he would note that the sea was becoming populated by primitive organisms, which would change in character from one observation to the next. At long last, about half a billion years ago, he would find that the atmosphere had evolved to nearly its present composition, and that the sea was crawling with larger creatures who had somehow mastered the secret of making hard parts to support their bodies. He would wonder, doubtlessly, whether the change in atmosphere during three billion years was somehow related to the changes in living things.

His wonder is shared by geologists and biochemists the world over. In all science there is no more fascinating question than the origin and development of early life, and the corresponding changes in the atmosphere. Somehow, during those hidden first few hundred million years of earth history, the first organisms appeared in the sea. How did it happen? No one knows, but we can make a plausible surmise. Take a mixture of carbon dioxide and nitrogen and water, or of methane and ammonia and water, whatever be your assumption about the composition of the early atmosphere. Put the mixture in a flask, subject it to an electric discharge, to ultraviolet light, to high temperatures—by way of simulating thunderstorms and bright sunlight and lava flows on the primitive earth. Analyze the brownish droplets that collect on the sides of the flask, and you will find a surprising variety of complex organic compounds, many of them similar to substances in living organisms. Such compounds, then, must have been present in the primeval ocean. Does it take a great flight of imagination to think of some of these molecules joining together into larger ones, and ultimately a combination being formed capable of ordering the molecules around it into a replica of itself?

Once life began there would be no stopping it. The first living things must have been like the anaerobic organisms of the present day, capable of growing in water without dissolved oxygen. Later, perhaps much later, some variant developed with a new talent: the ability to utilize the energy of sunlight in making carbon dioxide react with water, a reaction that produces organic materials directly, under conditions that the organism can control, and gives off oxygen as a byproduct. So began the slow conversion of carbon dioxide in the atmosphere to oxygen, and eventually a different kind of organism evolved that could take advantage of this new gas as a source of energy. The two great branches of the living world—plants that convert carbon dioxide to oxygen, and animals that turn oxygen again to carbon dioxide—were established well back in Precambrian time. The balance between their activities gradually increased oxygen at the expense of carbon dioxide, until at the beginning of the Paleozoic Era the ratio of the two gases in the air had reached something like its present value. Why the approach to this value was accompanied by the rapid evolution of animals with preservable hard parts remains one of the great geologic mysteries.

Thus the history of the Precambrian time, both the internal history of the planet and the external history of the surface and the beginnings of life, can be pieced together in rough outline from a few facts and a generous use of imagination. The meager detail contrasts sharply with the history recorded abundantly in the fossils and rock structures of later eras.

Fig. 16-1. A limestone quarry in Michigan. The thick horizontal beds are typical of the Paleozoic marine sedimentary rocks that cover large parts of the northern Middle West. Limestones like this are one of the most valuable of nonmetallic mineral resources. (Michigan Department of Natural Resources)

Fig. 16-2. An open-pit mine in a small body of iron ore. The ore forms a tabular mass standing approximately vertical (the dark material in the center of the picture). It consists of iron oxides (hematite and magnetite) deposited by hot solutions that rose through a thick deposit of volcanic rock, presumably from a magma somewhere below. The volcanic rock near the ore has been bleached and altered by the solutions. (Hercules mine, Coahuila, Mexico, courtesy of Ing. Enrique Prado Ruiz)

16

To Live with the Earth: Material Resources

In the language of evolution, man is an animal well adapted to his environment. The "environment" of this statement means most of the land surface of the earth, for man has shown a capacity remarkable among animals for living in a variety of habitats—from the bleak coasts of the Arctic Ocean to the rain forests of Brazil and central Africa, from the high plateaus of Tibet to the hot, arid wastes of Arabia. These extremes of adaptation are possible because of man's unique ability to *use* his environment, to find and modify the materials he needs for food and tools and protection. The materials he seeks have been fashioned for him by the geological and biological processes that have shaped the surface of the planet. From a strictly human standpoint, then, the most important of all geologic questions is the efficacy with which useful substances have been produced and concentrated over long ages by erosion, sedimentation, volcanism, and tectonic activity.

How effectively have useful materials been concentrated in nature? One can ask such a question on two levels: the obvious query about the local availability of needed materials to a community or a nation, and the broader problem of the sufficiency of earth materials to satisfy the long-time needs of the race as a whole. The second question has become important only recently. For most of human history the ultimate adequacy of resources seemed obvious. If in time of famine the soil or water supply of a given area was insufficient, there were always new lands to be occupied; and the quantity of metals needed was so small that trade with foreign lands could make up for local deficiencies. It is only in modern times, with the explosive increases in population and the ever-growing need for raw materials in huge quantities to maintain an industrial civilization, that a query about overall planetary resources has become pertinent.

To change your sights from the urgent problems of day-to-day living to a concern about the future of humanity is a big jump. Too often the jump is largely emotional. You see pictures of swarming humanity juxtaposed with pictures of oil tankers or exhausted mines or eroded soil, and you read frightening statistics comparing population trends with rates of depletion of resources; and suddenly the prospect of your grandchildren eking out a savage existence on an exhausted and crowded planet seems very real. Radical solutions sound attractive: halt industry, close down mines, recycle everything possible, go back to simpler ways of life. A more sober second thought suggests that such a solution would be monstrous, for just the simple maintenance of the world's population depends on the intricate network of production, fabrication, transportation, and communication to which we are accustomed. To abandon technology overnight on any sizable scale would be to condemn much of the world's population to starvation. We are truly caught in a sorry plight: if increase of population and exhaustion of resources continue at present rates, humanity is headed for disaster within a few decades; yet any major interference with the present order would mean another sort of disaster in the nearer future.

So say the pessimists. If we put emotion aside and try to look at the problem analytically, is the situation really so desperate? If the increase in population can be controlled, if the industrial use of raw materials can be regulated, if other sources of raw materials can be found and developed, might it not be possible to avert a serious calamity and leave a habitable world for our grandchildren? These are complicated questions of sociology, economics, and international politics. Only one aspect is related to geology, the availability of useful materials and the search for additional supplies. Yet this one aspect is basic to all others. Without detailed knowledge of what materials the planet can supply, no social scientist can make reliable predictions or suggest reasonable changes in current trends.

In the final two chapters, then, we will explore this question of planetary resources, using as a guide what we have learned about the geology of our restless earth. Where do we find useful materials, why are they concentrated in certain geologic situations, how do we go about finding new supplies, what estimates can we make about total amounts available? These are the questions that will now concern us—very practical questions, in contrast to the theoretical problems that have been the focus of the last few chapters. In the present chapter we will concentrate on sources of materials, in the next chapter on sources of energy.

Renewable versus nonrenewable resources

What earth materials are essential for human existence? We could start a list with food, air, and water. Then would follow a series of substances not required for bare maintenance of life, but certainly needed for existence in civilized society: coal, gas, and oil for fuel, wood for fuel and construction, metals for tools and gadgets and construction, clay for bricks and chinaware—the list could go on indefinitely. Some of these materials, like air, are renewed by natural processes as fast as we use them; we take oxygen from air as we breathe, but oxygen is continually supplied to air by photosynthesis in the leaves of green plants. We complain about pollution of air in our immediate vicinity, but on a worldwide scale pollution does not appreciably affect the constant renewal of the essential gas oxygen. Other earth materials, by contrast, are not being renewed by any visible process. As gold is mined from a quartz vein, no gold is being formed to replace it; once the vein is mined out, the gold is gone forever. The supply of renewable resources is scarcely a problem, because we can always trust nature to produce more, but for a nonrenewable metal like gold a continuous supply for the future depends entirely on our ability to find new gold-bearing quartz veins (or other kinds of deposit) as the known ones are depleted.

The contrast between renewable resources like air or water and nonrenewable resources like gold and other metals is obvious enough, but some substances would occupy an intermediate position. Take wood, for example. Certainly the earth's wood supply is being constantly renewed by new growth of trees, but whether the renewal will be fast enough to keep pace with the ever-increasing use of wood for lumber, paper, and fuel is an open question. Over wide areas in China and the eastern Mediterranean forests have disappeared during historic time simply from over-use, and the same fate is not inconceivable for forests on a much larger scale. Even water may someday be only marginally renewable at a rate fast enough for human needs. At present the world's rainfall is more than sufficient for adequate supply even to water-deficient areas, provided the technical problems of storage and transportation can be solved, but increasing population and industrial consumption could well put an end to this happy state.

Renewability for any substance, as a theoretical concept, must be a relative term. Gold is nonrenewable, in the sense that a gold deposit once exhausted is not replenished, but presumably other gold deposits are in process of formation somewhere in the earth. The renewal is very slow and localized, in contrast to the rapid and widespread renewal of atmospheric gases, but strictly speaking gold is nonrenewable only in a relative sense. A major conclusion from geologic study is the similarity of processes taking place in the world today to processes that have gone on in the past; it must follow from this conclusion that the processes by which ores were formed, or by which petroleum accumulated, almost certainly have their counterparts in the present world. The processes today may be very slow and very restricted in area, but if they are going on at all the resources they produce are not in an absolute sense nonrenewable.

Every kind of natural resource is probably somewhere, to some degree, in process of renewal, but unless the renewal is fast it will be of little consequence for human needs. The renewal time of oxygen in air is so short that we can assume it available for an indefinite future; fresh water is almost in this category, but a limit is conceivable; wood is renewed more slowly, over times measured in decades or centuries, so that some effort at conservation is needed to ensure a continuing supply; soil, with a time of formation measured in hundreds or thousands of years, is only marginally renewable for human purposes. The metals, with renewal times in the millions of years, on a human scale are completely exhaustible. For any natural resource, then, an important question is its rate of replenishment. If the time is short in comparison with human life, maintaining an adequate supply is chiefly a matter of conservation practices; but if the time is very long, so that the resource is effectively nonrenewable, insuring a supply for the future is a much more difficult problem. In the following sections our concern will be chiefly with the nonrenewable resources.

Resources without glamor: the nonmetallic materials

A friend who is building a house asks you where he can find some gravel to serve as foundation material for walks and driveways. What can you suggest? One possibility, certainly the easiest, is to take him to the nearest dealer in construction materials and let him order gravel by the truckload. But suppose your friend is trying to cut expenses: can you help him find a natural deposit?

A pretty dull and trivial geologic question, you would say. After all, gravel is abundant. All he needs to do is go to the nearest stream and find a gravel bar, or to a lake shore or ocean shore where gravel has been concentrated by wave action. In some parts of the country this answer would suffice, but many communities are distant from large bodies of water and no gravel is visible at the surface. Then the problem becomes a matter of hunting for older deposits that may be hidden by soil and vegetation. In this form the question is anything but dull; you may need all your geologic acumen to solve it. You would try to recognize from landscape forms, for example, the presence of ancient stream channels, or beach deposits raised above present water levels, or outwash deposits associated with moraines of an ancient glacier. When you have located a deposit of old sediment, your geologic background would help you further to predict where the best gravel would be found, and where the material would be only fine sand and mud. If, in helping your friend, you become expert in locating gravel deposits, your advice would thereafter be widely sought, for in the construction industry gravel and sand are needed in huge amounts. The locating and exploiting of good deposits constitute a major business.

Despite your best efforts, in some areas gravel would be impossible to find in quantities sufficient for the needs of local industry. For many purposes a good substitute is crushed rock, and your problem then would be to locate a place where rock of suitable characteristics could be easily quarried. What kind of rock would you look for? If you need rock simply for fill or foundation material, there are many possibilities. Basalt and limestone are widely used, but almost any rock that is reasonably hard and durable would be a good candidate. With a little knowledge of local geology you should be able to pick out rocks with the right properties and to suggest places where the rocks would be available in large amounts just under the surface cover.

Finding nonmetallic materials suitable for special uses is a more difficult problem. Quartz sand of sufficient purity for glass making, for example, or limestone for the manufacture of cement, or clays that can be used in fine chinaware, are found in only a few places. They have obviously required special geologic conditions for their formation; ordinary processes of sedimentation do not give materials of such purity. To locate deposits of these substances, or to trace extensions of known deposits, may tax a geologist's skill to the utmost.

The list of nonmetallic materials needed in large quantity by modern industry could be extended almost indefinitely. Sulfur, salt, gypsum, asbestos, phosphate rock, stone for building ("dimension stone"), are only a few examples. Coal and petroleum might be included here, but the special problems of these organic materials will be deferred until the next chapter. Deposits of the important nonmetallic substances are often found by accident, or by old-fashioned prospecting methods, but geologic knowledge about the way these substances have formed and about their relations to other rocks is often a valuable guide in exploring and evaluating possible places of accumulation.

To what extent are resources of nonmetallic materials renewable? The answer, of course, will be different for different materials. Sand and gravel, at least in small quantities, may be replenished in river bars and beaches during every major storm. A layer of salt may be deposited in arid basins in every dry season, although the layer formed in any one season is generally thin. Limestone is being renewed in the sense that limy mud is accumulating in tropical seas and corals are building reefs along tropical shores, but these materials hardly constitute replacements of the old limestone beds being quarried in the interior of continents (Fig. 16-1). Thus a few of the nonmetallic materials are being renewed in small quantities by geologic processes of the present day, but in the amounts used by modern industrial civilization most would be properly grouped with the nonrenewable resources. Except for a very few, however, existing deposits are so large and numerous that exhaustion will not be a serious threat for a long time in the future.

"Gold is where you find it"

The metals pose a more glamorous geologic problem, perhaps because their deposits are scarcer and the chance of making a fortune from a single deposit is greater. The legendary fortunes that have been won from a few mines, however, should not obscure the fact that in sheer dollar value the nonmetallic materials used by modern industry outstrip the metals. You would be richer if you owned all the operating limestone quarries in North America than if you owned all the gold mines.

To a geologist, metallic ores have an additional reason for glamor: their origin is one of the great unsolved geologic problems. For most of the nonmetallic resources, the geologic processes responsible for their formation are reasonably clear. We can watch sand and clay being deposited in the world around us, or salt crystallizing from brine, or crusts of sulfur growing at the mouths of fumaroles, and we can readily imagine similar environments that existed when older deposits were in process of formation. But places where metals are visibly accumulating in any quantity today are extremely few, so that in accounting for ore deposits we must rely much more heavily on imagination. The geologic clues that we can use to guide imagination are generally ambiguous, permitting a number of hypotheses to be formulated regarding any one deposit. Controversies among rival hypotheses are long continued and often bitter. This means that the search for new bodies of ore has little established theory to guide it, and discoveries are often made more by accident than by application of geology. If you ask a geologist where to look for gold, his most honest answer will be a shrug of the shoulders and repetition of the old maxim, "Gold is where you find it."

Why should the origin of ore deposits be such a peculiarly stubborn problem? We are asking, in essence, about a process of concentration. The metals of ore deposits are widely distributed in trace amounts in ordinary rock and sediments, and somehow from these ordinary materials the different metals have become concentrated into small regions of the earth's crust. A million grams of ordinary granite, for example, contain roughly 10 grams of copper, 20 grams of lead, and 5 hundredths of a gram of gold. To form *ores* of these metals—meaning materials from which the metals can be profitably extracted—requires that some geologic process must be capable of concentrating each one by hundreds or thousands of times over these quantities. One can suggest several possibilities, but all demand rather special geologic conditions and none are demonstrably operating on more than a trivial scale in the world today. Familiar geologic processes, in fact, for the most part lead to mixing and dispersion of materials rather than to concentration. It is this combination of failure to find adequate concentration processes in the world around us and difficulty of getting clear evidence for the operation on a large scale of unusual processes in the past that makes the problem of ore deposition so troublesome.

In very general terms, what sorts of geologic circumstances might lead to metal concentration? The simplest example that comes to mind is the accumulation of gold in stream gravels (*placer* deposits). In placers, metallic gold is found in particles and nuggets mixed with sand and gravel, mostly near the base of gravel deposits and often in pockets eroded into the stream bed. The concentration is evidently a purely mechanical process, the gold being carried along by the stream with other fragmental material and accumulating at the base because of its much greater density. Placer deposits formed in the same manner are important sources of a few other metals, for example much of the tin of Malaysia and platinum in the Ural Mountains of the USSR. Here there is no mystery about the concentration process, but an awkward question remains. The gold (or platinum, or tin

Ore concentration

Average content of metals in rocks of the earth's crust, and degree of concentration needed to form economically exploitable ore deposits.

Metal	*Average amount in crust*	*Enrichment needed to form ore*
Iron	56,000 grams/ton	5 times
Copper	55	140
Zinc	70	500
Lead	13	2500
Silver	0.1	5000

oxide) must have been present in unusual amounts and in unusually large crystals in rocks somewhere upstream; how did this earlier partial concentration come about? Placers do not form everywhere or even commonly, as any frustrated prospector can tell you, so the original concentration process that makes the gold available to stream erosion in a few localities remains to be explained.

For another example, many of the world's major iron deposits consist of ferric oxide (hematite, Fe_2O_3) in layers interbedded with chert, limestone, or other sedimentary rocks. The iron mineral was deposited as a sediment, and the process of concentration was somehow related to erosion and sedimentation. This much is obvious, but why was *so much* iron concentrated in these few places, and during deposition of any one hematite layer why was it laid down in such pure form, with very little other sedimentary material? Ferric oxide is common in modern sediments, of course, minor amounts forming the red, brown, and yellow coloring matter of many sands and shales. The source of iron is the weathering of iron-containing silicate minerals like biotite and hornblende, and the processes by which it gets into solution and is carried to the site of deposition are well understood. But in modern sediments the iron oxide is a minor constituent, always mixed with other material. The mechanics of the depositional process of iron ore are clear enough, but the problem of *concentration* remains: why so much, and why in such pure form, in only a few places and at only a few times in the geologic past?

A third example of the kind of situation in which metallic ores are concentrated is provided by the Bingham Canyon mine in the hills a few miles southwest of Salt Lake City. If you visit this mine, one of the world's great copper producers, you will be guided to an observation point high on a ridge from which you overlook a huge pit carved in colorful rock. The pit is a scene of much activity, with huge trucks and mechanical shovels operating at various levels on its sides. Perhaps you will be shown some of the ore that comes from the pit: a rock that you will recognize as granite from its abundant feldspar and quartz, but not normal granite because it shows much fracturing, some of its feldspar is cloudy and soft, the dark minerals are replaced by dull greenish material, and scattered through the rock you will see brassy glints of the ore minerals (chiefly sulfides of copper and iron). In some ore specimens you will note an abundance of quartz in small pockets and veinlets, often associated with crystals of the shiny sulfides. The widespread alteration of the granite and the dissemination of copper minerals through it suggest that the entire mass was fractured and then permeated by solutions that carried copper, iron, and silica.

If you explore the Bingham Canyon district away from the main pit you will find several other mines, most of them now inactive, consisting of

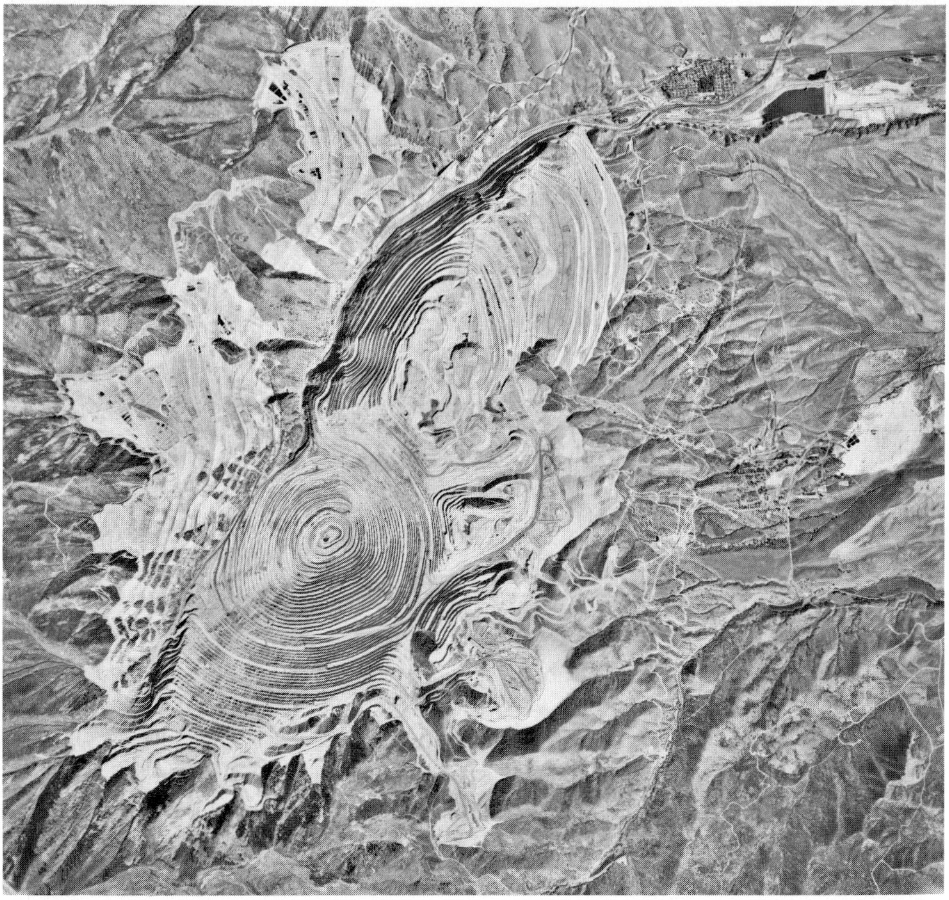

Fig. 16-3. Vertical air photo of the huge open-pit copper mine at Bingham Canyon, Utah. A view from the edge of the pit is shown in Plate 32. (Aero-graphics Inc.)

shafts and tunnels into the mountainside. If you can find a guide to take you into the old workings, you will be shown the kind of ore that was obtained—quartz veins showing up as white streaks on the tunnel walls, cutting across layers of various kinds of rocks that include lavas and slightly metamorphosed sedimentary material. In the quartz you may be lucky enough to find a few pockets of large, well formed crystals of sulfide minerals, including as a prominent variety bright silvery cubes of the lead sulfide, galena (Plate 14). This kind of material, which makes handsome museum specimens, fits the popular conception of an "ore" better than any other we have discussed.

Bingham Canyon and its surroundings have produced metals in extraordinary amount and variety—not only the obvious copper from the open pit and lead from smaller mines on the periphery, but much gold, silver, and molybdenum as well. What sort of guess can we make about the process that concentrated these metals? We note first the evidence for solutions, which permeated the granite exposed in the pit and evidently moved through fissures to deposit the quartz-sulfide veins of other mines. We can reasonably infer that the solutions were warm, because both quartz and sulfide minerals are very insoluble in cold water but become more soluble as the temperature rises. We note the evidence for igneous activity—the granitic body exposed in and near the pit, the abundant volcanic rocks in the surroundings. We recall that volcanic eruptions are accompanied by much water vapor released from solution in the ascending magma, and that hot springs and fumarolic gases in volcanic areas often contain detectable amounts of some of the common ore metals. Putting these bits of evidence together, we guess that the metals were originally contained in a silicate magma at great depth, and that as the magma moved to regions of lower pressure near the surface its dissolved water was liberated, carrying the metals with it. In part the metals were deposited in cracks in areas of the granite that had already cooled, in part they were carried away from the magma in solution and deposited along with abundant quartz in fissures cutting the surrounding rocks. Many details are uncertain in this reconstruction of past geologic events, but the general idea of metals being carried in hot solutions, and of the solutions being somehow associated with igneous activity, seems the simplest way to account for the relations observed at Bingham Canyon.

In our three examples we have looked at some common methods of metal concentration—mechanical processes of sedimentation, chemical processes of sedimentation, and derivation of metal-bearing fluids from a cooling magma. Note that each explanation is incomplete, for each avoids the deeper question of how the metals came to be available in such large amounts to the agents of sedimentation and igneous activity. Yet as explanations of the final deposition of ores as we see them, the three mechanisms seem beyond dispute. One can find details to argue about, but the kinds of processes involved are clear. For many less simple examples, however, even the general mechanisms of concentration remain open to controversy.

Take the lead-zinc deposits of the upper Mississippi River Valley, for example. The geologic relations are disarmingly simple: the sulfide minerals of lead and zinc (galena and sphalerite), together with some quartz and pyrite, are found in limestones that are part of the sequence of flat-lying Paleozoic beds which cover much of the Middle West. The ore minerals are concentrated in large part along the bedding planes of the limestone and to some extent in steep fracture zones that cut the beds. Now how did the metals get here? The fact that so much of the ore lies parallel with the beds suggests that it was deposited chemically during sedimentation, like the iron ore of a previous paragraph; yet this seems impossible because the sulfides of lead and zinc are so insoluble in cold water. The greater solubility in hot water, together with the distribution of some ore along fracture zones, suggests the alternative possibility that lead and zinc were deposited by hot solutions from a cooling magma, like the ores of Bingham Canyon; yet there is no sign of igneous activity anywhere near these deposits, except in the ancient Precambrian rocks far beneath. Could hot water be produced in any other way? Could the lead and zinc be somehow concentrated from the minute amounts present in the limestone or adjacent sedimentary rocks? Such questions have been a fertile source of argument among geologists for a long time, and they remain without completely satisfactory answers.

With this look into the mysteries of ore deposition we have barely opened the door on an exceedingly complex subject, but perhaps it is enough to show why theory is still so inadequate either to explain the concentration of metals or to give much aid in the search for new deposits.

Ore deposits are nonrenewable

The ways in which metals have been concentrated geologically may remain mysterious, but one fact stands out in stark clarity: all the suggested processes are so slow that the formation of new deposits in the present world can have no human importance. Ore deposits are nonrenewable, period. Great concentrations of ore have been built up slowly over geologic time, and these we can find and exploit. When they are gone we cannot expect new ones to take their place.

If you try hard, you can think of trivial exceptions. Iron oxide in some places accumulates in swamps, at a rate fast enough so that small amounts can be "harvested" at intervals. Stream placers and beach placers may be stirred up enough during storms to re-sort the sand and gravel and generate minor new concentrations of gold or other minerals. But the very effort required to think of *any* exceptions only underlines the extreme nonrenewability of metal resources.

This simple conclusion from elementary geology is all too easily forgotten. We have been exploiting our metal resources, especially in North America and western Europe, at a steadily increasing rate for many centuries. We are accustomed to having all the metal we need, in our homes and in industry, at prices that not only are reasonable but for many metals have actually declined as mining techniques have improved. If we need more metal, we have only to work existing deposits more intensively or discover new ones. So long has this pattern persisted that we scarcely dream of a day when mines will be exhausted and new ones cannot be found. Yet ore deposits are patently exhaustible, as is dramatically illustrated by the "bonanza" deposits of precious metals in the western states that lasted only a few decades, and by the more recent exhaustion of rich iron ore in the fabulously productive mines of northern Minnesota. The number of new deposits that await discovery is necessarily limited. Someday mankind must reckon with the fact that metal resources are finite and are not being renewed by any known geologic processes.

The important word in the last sentence is the adverb "someday." What day does this mean? Are metal resources so greatly depleted that we must listen to the criers of doom and take drastic measures immediately? Or are we talking about a future so remote that neither we nor our children will have to worry about it?

To such questions there are no sure answers. From a geologic standpoint—and we should remind ourselves that the questions are only in part geologic—we can note with alarm the limited life that is predicted for many of the world's great mines, and the increasing difficulty of finding new deposits despite the expenditure of huge sums for exploration. The alarm, however, is tempered by the fact that enough new deposits, or extensions of old deposits, have been discovered in recent years to forestall actual shortages—sometimes seemingly just in the nick of time, when reserves of one or another metal were dangerously low. Either a pessimist or an optimist can find ample reason for his belief, either that we are skating near the brink of disaster or that we will continue to be saved by new discoveries.

Suppose that the pessimist is right, and that within a few years the minable deposits of some important metal are completely worked out. What might we expect to happen? One answer is that we can always fall back on lower-grade ores. The profitable mining of any metal depends on its price: if the price is raised, the more expensive exploitation of low-grade material can be undertaken. A really confirmed optimist can carry this thought to extremes, by suggesting that when we get desperate enough for copper, say, we can start mining granite and recover the minute amount of copper it contains. In other words, the supply of metal is practically infinite because we can find any metal we wish in ordinary rocks. This far-out line of reasoning neglects the huge amounts of energy that would be required and the problem of disposing of the vast piles of waste that would be generated.

In any event, the treatment of low-grade ore would make the price of a metal higher. When we look into our crystal ball and try to read the future, then, we can predict that metals will become harder to get and more expensive, but that the world is in no danger of waking up some morning to find that one of the familiar metals is suddenly no longer obtainable. Very likely the prices of all metals will rise, some much faster than others, and the increasing scarcity may well have drastic effects on the world's economy. The scarcity will be alleviated by recycling of existing supplies and by substitution of other substances for a few of the metals. Nevertheless, it is hard to see how other materials can ever take the place of some metals for vitally important purposes. Plastics are amazingly versatile, for example, but it is hardly likely that they can be made strong enough to replace the iron of structural steel. We can perhaps take comfort from the reflection that the changes resulting from increasing scarcity of metals will probably be slow enough so that adjustment will not be an insurmountable problem.

Two morals can be drawn from such a discussion. One is that humanity's general standard of living, insofar as it depends on the ready availability of metals, will decrease over the next century or so. The second and more important moral is that we can postpone or slow down the decrease if we recognize that metal resources are exhaustible and nonrenewable, and hence should be used with less profligate waste than we are accustomed to.

A third moral will suggest itself if you have an apocalyptic turn of mind. Our present industrial civilization, it can be claimed with much cogency, developed primarily because mankind found readily available metallic ores in concentrated form that could be worked without much effort—first the copper, zinc, tin, and lead of the Bronze Age, then iron, then the great variety of other metals that have become common since the Industrial Revolution. Within a few decades or a few centuries, depending on how optimistic you are, the rich and readily discoverable deposits of these metals will be gone, over the entire surface of the earth. If a cataclysm should destroy most of humanity—global war, disease, an astronomic event—so that the miserable remainder would have to start again from a savage beginning to build civilization, no repetition of history would be possible. Metals would not be available in quantities sufficient for more than a simple hunting or agricultural existence. The human race, so to speak, has had one chance to build civilized life on this planet; it will never have another chance, because the metals that were concentrated into ore deposits through long geologic ages have been dug from the ground and widely dispersed.

Capricious distribution

We have been considering metal resources from a whole-earth standpoint, as if the metals of an ore deposit regardless of location were available to all who might need them. On such a basis we can talk in simple terms about the problems of maintaining future supplies of metals for the entire planet. But with the present political organization of humanity, such a picture is grossly oversimplified. Ore deposits of any given metal are found in some countries and not in others, so that the availability of metals depends on politics as well as geology. Without stepping off into the murky water of politics, what geologic data can we find about the distribution of metals?

It is easy enough to compile the facts of distribution, but exceedingly difficult to find an explanation for the facts that will permit predictions about metal distribution in little-explored regions. Take the useful metal tin, for example. The greater part of the world's tin comes from southeastern Asia, where the principal ore mineral (cassiterite, an oxide of tin) is mined from quartz veins in granite and from placers in the beds of streams that drain the granite areas. Other important tin deposits are located in Bolivia, Nigeria, and eastern Siberia. Older mines, now largely exhausted, were once famous tin-producers in southeastern Germany and southwestern England. Recent discoveries have established Australia as a potentially major source of the metal. Tin mines, present or past, exist in all the continents except North America—but in North America, despite extensive searching, the only workable deposits that have been found are very tiny ones in Mexico and western Alaska. Why should the distribution be so exceedingly spotty, and why should an entire continent be almost devoid of tin concentrations?

Questions like this are embarrassing to geologists, because answers so far are little more than conjectures. Study of tin deposits shows that the ore minerals are nearly always found in or near granite, and granite of a rather special kind—characterized mineralogically by an abundance of biotite (black mica), and chemically by a high content of potassium. No reason is known for this preference of tin for biotite granites, but the simple fact of association suggests that tin deposits should be looked for in areas where such granites crop out. The trouble is that granites of appropriate composition are by no means uncommon in North America—but they are not accompanied by tin concentrations. Now what has gone wrong? Does this mean that the process of granite formation in North America was in some obscure way different from that in southeastern Asia? Or does it mean that the original materials of the earth, the materials from which the granite magmas were generated, contained more tin in one place than in the other? We simply do not know. Recent geologic opinion tends to favor the second guess—that capriciousness of ore distribution is inherited from original differences in the materials that formed the earth—but evidence is very scanty.

Tin is the outstanding example of a metal abundantly concentrated in only a few widely scattered regions, but other metals show similar anomalies of distribution. Copper, for example, is being mined in many parts of the United States, but rich deposits show an extraordinary clustering in southern Arizona; nearly three-fourths of the country's total production has come from this one area. Geologically copper concentrations are found with many kinds of rock—with basaltic lavas and conglomerate in northern Michigan, with sedimentary rocks in Oklahoma, with mafic igneous intrusives in southern Canada, with granitic rocks in Arizona and at Bingham Canyon—suggesting that the metal can accumulate in a variety of geologic environments. This is an interesting fact that can give some guidance in the search for new deposits, but it falls far short of explaining the abundance of copper in Arizona.

Iron, the metal on which industrial development primarily depends, has a similar erratic distribution. Unlike copper and tin, iron is one of the

Fig. 16-4. Sources of iron-ore concentrates in 1970. (C.F. Park, Jr., *Earthbound,* 1974, Freeman, Cooper & Company, and U.S. Bureau of Mines)

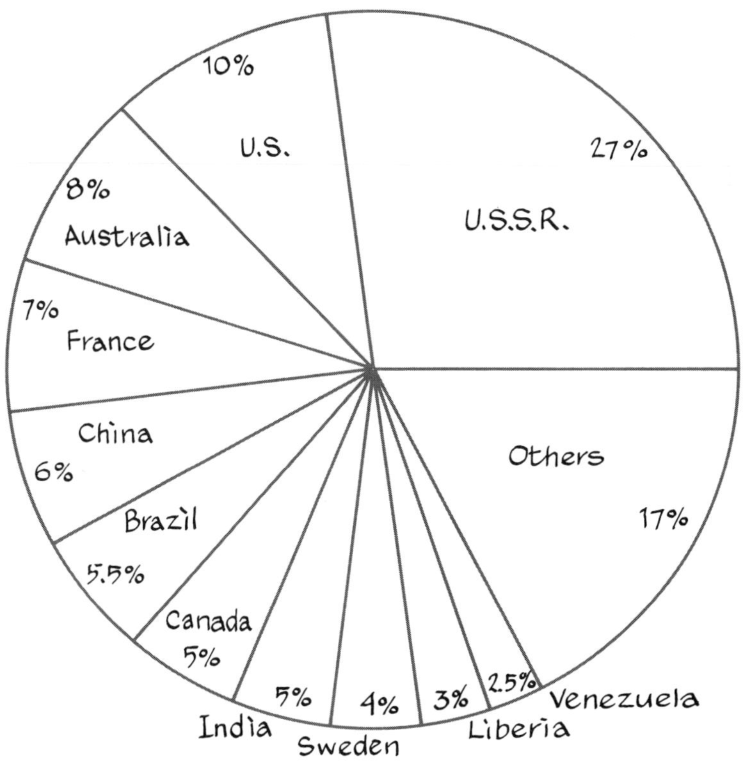

earth's very abundant metals, making up about five percent of the weight of the crust, so that iron minerals are important constituents of most ordinary rocks. Hence iron could theoretically be obtained from a great variety of materials, but in an economic sense a rock is usable as iron ore only if its iron content is greater than about thirty percent, and only if the iron is present in a mineral that can be easily smelted (most commonly one of the oxide minerals, hematite or magnetite). The United States was fortunate in having enormous quantities of easily worked hematite ore readily available in the Precambrian sedimentary rocks around Lake Superior, as well as smaller deposits of Paleozoic age in the Appalachians. The very rich ore of the Lake Superior region is now largely worked out, but technological developments have made it possible to handle the much larger volumes of lower-grade ore in the same area, and exploration has revealed other similar Precambrian ores in Canada and on other continents. Besides the sedimentary concentrations of iron ore, large deposits have formed also by igneous processes at the contacts of granitic rocks, for example in Pennsylvania and Utah. Again we know from geologic study the general way in which the different kinds of iron ore have been concentrated, but the explanations offer little help with questions about why the concentrations

Fig. 16-5. Iron ore produced and imported in the United States between 1945 and 1972. The amount of ore imported has increased in three decades from zero to one-third of the total quantity used. (*Minerals Yearbook*, 1972, U.S. Bureau of Mines)

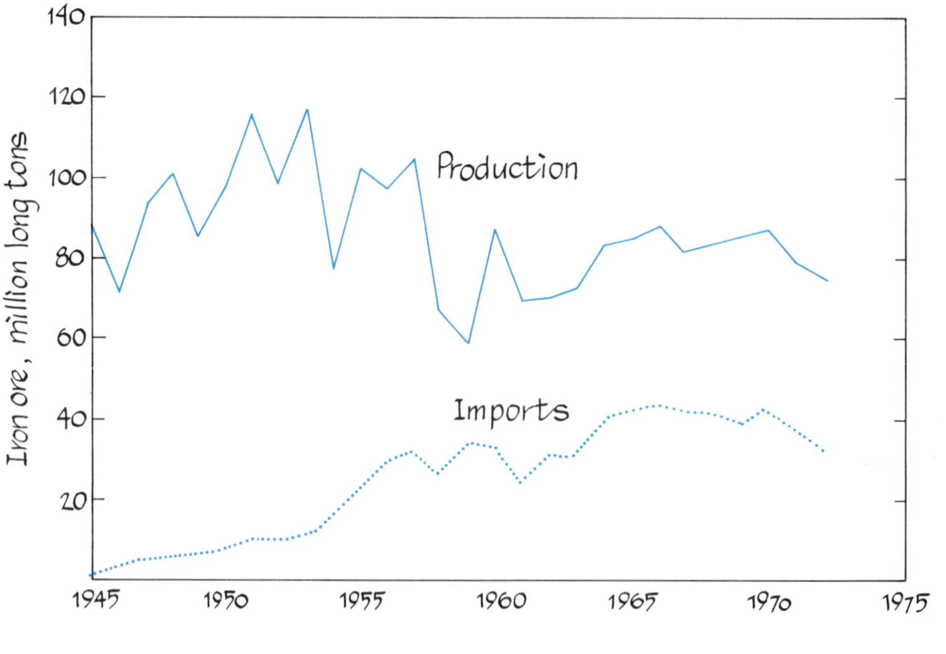

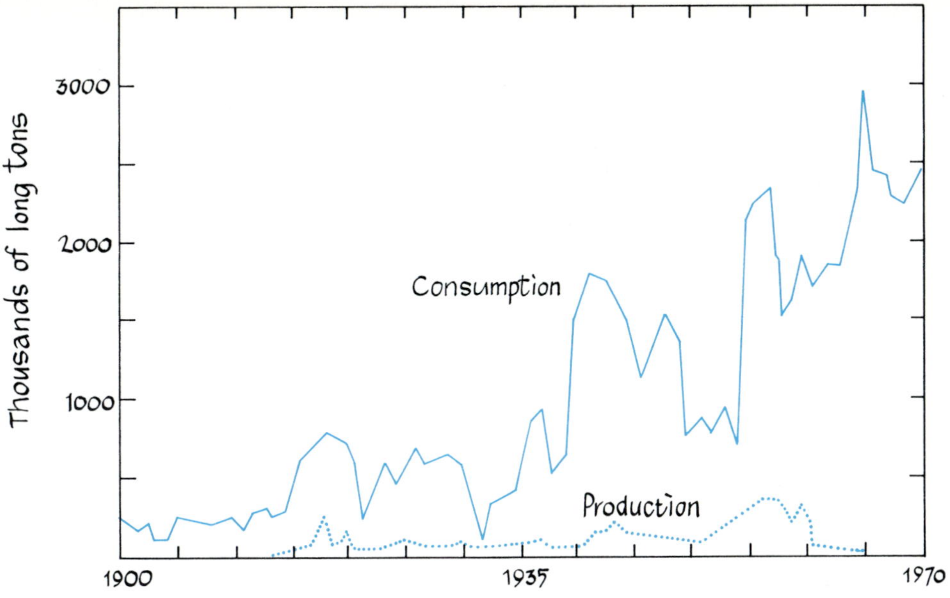

Fig. 16-6. Manganese ore produced and consumed in the United States between 1900 and 1970. Manganese is one of the metals needed for alloying with iron to make steel. The United States has only a few small deposits of manganese ore, and is almost completely dependent on imports. (C.F. Park, Jr., *Earthbound*, 1974, Freeman, Cooper & Company, and U.S. Bureau of Mines)

occurred in particular places and at particular times in the geologic past.

One fascinating aspect of metal distribution is the effect that the location of ore deposits has had on world history. In ancient times, for example, the dominance of Athens among the Greek city-states depended in large measure on the rich silver mines at Laurium, only a few kilometers east of the city. The Spanish conquests following the discovery of America ensured a steady flow of precious metals from the rich deposits of Mexico and the Andes, giving Spain a position of power in European politics. Later, when a process for making steel by using coke was discovered, and large-scale production of steel became important for industrial development, the availability of iron ore and coking coal in close proximity made England, France, and Germany the dominant European powers. In more recent times, much of the twentieth-century history of the Far East can be related to Japan's need for overseas metal (and fuel) supplies to supplement her own meager resources. In the same vein one can look into the future and find cause for worry about even so richly endowed a nation as the United States. For some metals the deposits in this country are small or nonexistent, and our need to obtain elsewhere increasing amounts of tin, aluminum, tungsten, chromium, zinc, and manganese may well have a strong influence on our relations with other countries.

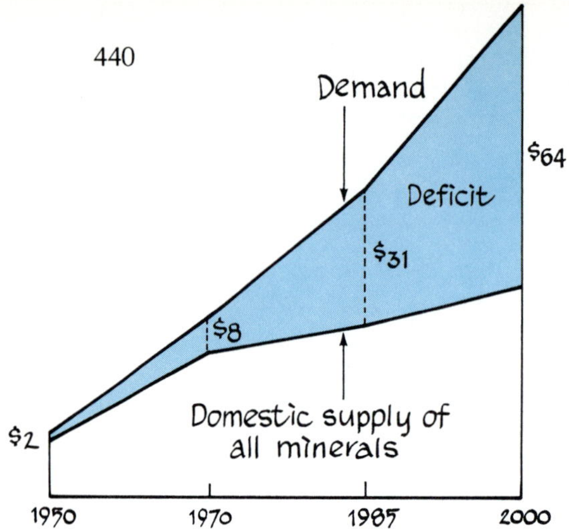

Fig. 16-7. Developing deficits: mineral demand and supply in the United States for 1950 and 1970, and projections to 1985 and 2000. The deficits are given in billions of dollars at 1970 value. (C.F. Park, Jr., and U.S. Department of the Interior)

Many facts about the concentration of metals into ores remain mysterious to geologists, but none more so than the facts of distribution over the earth. Perhaps one can hope that the new ideas associated with the hypothesis of plate tectonics will someday provide the needed tool for unraveling this mystery.

Prospecting

Because metallic ores are a nonrenewable resource and because known ore deposits are being rapidly depleted, the search for new supplies is a matter of major importance. Geologists, in their role as experts on all kinds of earth materials and processes, should seemingly be able to aid this search. As we have seen, however, geologic knowledge about ore deposits is far from satisfactory; little is known for sure as to how ore deposits have formed or why deposits occur in one place rather than another. It is hardly cause for wonder that most of the world's major deposits have been located

Scarce metals

Examples of imports in 1970 of ores of metals vital to the United States. For each metal the figure is the percent of the total amount used that was obtained from imports.

Metal	%	Metal	%
Manganese	100%	Aluminum	87%
Chromium	100	Mercury	66
Cobalt	97	Zinc	66
Antimony	91	Lead	58
Nickel	90	Tungsten	52

more by the intuition and luck of prospectors than by the application of geologic theory. Yet despite this gap in basic theoretical knowledge, geologists play an important part in ensuring a continuous supply of metals to industry.

Suppose, for example, that you have bought stock in a small copper mine. You know that copper ore is present, but you don't know how much. Should you increase your investment, on a hunch that the ore will expand into a large body as it is followed downward? Or should you sell your stock immediately, on the grounds that the ore you can see looks low-grade and discontinuous? On such questions a geologist should be able to give you sound advice. Study of rocks exposed in the mine would tell him whether the copper ore occurs in a quartz vein, along a granite contact, disseminated through altered granite, or in a sedimentary bed; and mapping of rocks around the mine would indicate whether the vein or the granite or the bed extends for a reasonable distance. Perhaps at one end of the mine workings the ore is cut off abruptly by a fault, and you need to know where the displaced ore might be found on the other side of the fault. Again you would call on a geologist, because his training should enable him to work out the geometry of the fault movement from relations in the surrounding rocks. With this sort of problem—evaluating the possibilities of a poorly exposed ore body, predicting the direction in which an ore body can be most profitably followed, finding extensions of ore bodies cut by faults or pinched off by folding—geologists daily render important service to the mining industry.

Suppose that you are more of a gambler and would like to invest money in hunting for a completely new deposit rather than in looking for extensions of a known ore body. Would you be wise in again seeking advice from a geologist? Or to make the question more personal: if you decide to try your hand at prospecting yourself, rather than finance someone else, could you use your geologic background in deciding where to start your search?

You would recall some of the ways in which metals have apparently been concentrated. Gold, it is well known, accumulates in placers, so a possible procedure is simply to start panning stream gravels. This would get dull very quickly, unless you have enormous patience, and it would be discouraging to reflect that practically all accessible stream gravels have been worked over by many prospectors before you. So you would turn your attention to other kinds of deposits, and recall that metals may be concentrated by igneous processes. In particular, metals accumulate in fluids that escape from a magma as it crystallizes, and are deposited by these fluids as they move into cooler rocks in the outer part of the igneous body or in the metamorphic rocks adjacent to it. So you guess that the contact of a body of granite might be a good place to look for ore. If you find outcrops of granite on a hillside, and outcrops of schist and marble nearby, you would do well to restrict your prospecting to a belt a few tens of meters on either side of the contact—especially if you find pieces of white vein-quartz near the contact, and areas where the surface rock is heavily stained with brown and yellow from the oxidation of iron-containing sulfide minerals. Finding ore along such a contact is by no means guaranteed, but at least you will have a good chance of picking up rock fragments with unusual minerals.

To recommend granite contacts as likely places for prospecting requires no very profound geologic knowledge. The frequent association of ore with granite margins has long been known to prospectors as an empirical rule of thumb, and in many parts of the western states granite contacts can be recognized simply from the number of shallow trenches and pits dug by prospectors of the last century. A little geologic background, however, should enable you to see what part of a contact is worth intensive study, and what part is likely to be barren. Particularly if you enlist the help of a trained geologist, he should be able to focus attention on the most promising areas. The geologist may not know for sure just how metals are deposited at an igneous contact, but from long experience he knows enough about the behavior of fluids in different kinds of material and about the appearance of rocks in and near ore deposits to judge where prospecting has most chance of success.

Geology applied to the search for ore depends in very large measure on experience. The abstract theory of ore deposition is not very helpful, but a geologist who has worked with many deposits comes to know the kinds of situation in which ore occurs, and develops skill in recognizing similar situations in new areas. Basically his reasoning is not far different from that of the old-time prospector, but his greater knowledge of rocks and geologic processes enables him to choose likely areas and eliminate other areas with much more assurance.

In recent decades the search for metal concentrations has become more sophisticated. Now that most deposits exposed in surface outcrops have been discovered, new methods are necessary for locating deposits hidden beneath surface cover. Iron ore, for example, can be detected by a sensitive magnetometer carried in an airplane, so that huge areas of inaccessible jungle or tundra can be surveyed rapidly. Measurements of resistivity or electric potential in the ground give indications of buried sulfide deposits. Soils above metal deposits generally contain small amounts of the metal or metals, and analytical methods have been found for detecting these traces rapidly. Even distant observation from satellites, using photographs taken

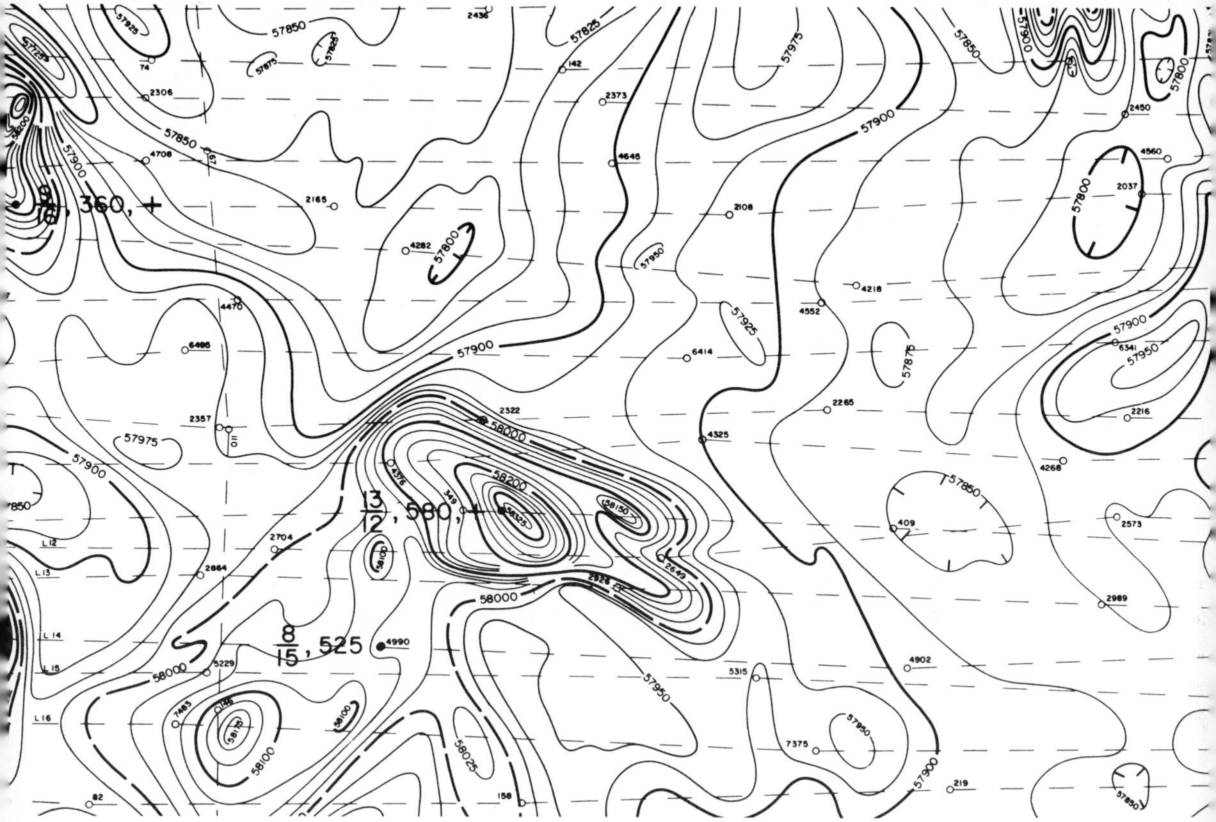

Fig. 16-8. A magnetic anomaly map for an area in northern Washington, drawn from a survey by airborne magnetometer. The lines are "magnetic contours" drawn through points of equal magnetic field strength. The area of high values near the center of the map (a "positive magnetic anomaly") suggests that a body of iron ore may be present a short distance beneath the surface. Scale: 1 inch = 1 mile. (Washington Division of Mines and Geology)

at different wavelengths ("remote sensing"), can aid in locating areas of ore concentration. The need for new supplies of metal is stimulating the steady development of such instruments and techniques to aid the prospector.

The use of new techniques and the discoveries of new ore deposits should not obscure the fact that prospecting is still a largely empirical activity. Geologists have long played an important role, but their contribution to maintaining our metal supply will be far more significant if future work can lead to a theory of ore formation from which predictions about the location of new deposits will be possible.

Fig. 17-1. Geothermal power—The Geysers, northern California. This is an area where volcanoes were active during the Pleistocene, and rock beneath the surface is still hot enough to convert much groundwater into stream. The steam is being used to generate electricity. A few similar areas elsewhere may be exploitable, but it seems unlikely that geothermal power can serve as a major source of future energy. (Pacific Gas and Electric Co.)

Fig. 17-2. Strip-mining of coal. The coal is a thick bed interstratified with other sedimentary rocks, chiefly shale. As petroleum and natural gas become scarcer, coal will be increasingly important as a source of energy. (Department of the Interior, U.S. Bureau of Mines)

17

To Live with the Earth: Energy Resources

Metals are a key to man's future welfare, but a continuing supply of metals demands energy to win them from their ores. Without energy to crush the ore, to separate the metallic minerals from waste, to reduce the metals from their compounds, and to transport the finished product, even the richest metal deposit would be of little use. Conversely, if enough cheap energy were readily available the problem of metal resources would be solved for centuries. Given unlimited energy, metals could be obtained from low-grade ores, or even from ordinary rocks, in any amount needed. There would remain the problem of disposing of the waste from mining and smelting operations, a problem that necessarily grows more serious as the grade of ore diminishes. But waste disposal also is largely a question of energy, for if enough energy is available waste in any quantity can be transported or treated so that it is no longer a nuisance.

The energy needed for metal recovery and waste handling is only a small part of the amount required by industrial civilization. It is often said that the single most striking difference between life today and life in the ancient world is the amount of energy available to the individual—the energy of electricity and gasoline contrasted with the energy of slaves and domestic animals. The quantity of energy used in daily life by individuals, enormous though it has become, is still far less than the energy consumed by industry. The awesome engineering achievements of this century, from the mechanization of warfare to the sending of rockets to the moon, have been possible only because of our ability to concentrate energy in huge amounts.

So far the energy needed by industry has come almost entirely from materials in the earth. These materials must be quarried or mined or pumped like other mineral resources, and new deposits must be sought by prospecting. Thus the problem of ensuring adequate energy for the future is in large part a geologic concern, very similar to the problem of ensuring adequate supplies of materials.

What are the chief sources of energy? How have they formed, and to what extent are they being renewed by natural processes? Where do we look for additional supplies? Is there really an "energy crisis," as the pessimists tell us, because the world's available energy resources are being rapidly depleted? How are known energy resources distributed over the earth, and how does the distribution affect politics, now and in the future? These are the questions that face us in this chapter.

Where does energy come from?

The word "energy," a physicist would tell us, means the ability to do work, in the sense of making things move. In ordinary speech we often give the term a wider meaning, referring for example to spiritual energy or moral energy or psychic energy. But for our purposes here we restrict ourselves to the scientific definition, to the energy an object or material possesses that enables it to produce motion.

Energy takes many forms. The food you eat has chemical energy, because by chemical processes during digestion it has the ability to make your body move. An electric current has energy because it can set a motor to turning. A moving billiard ball has mechanical energy because it can cause another ball to move. Heat energy supplied by burning coal or oil can turn the wheels of a locomotive. Radiant energy from the sun evaporates water from the oceans and moves the water upward into clouds. Thus energy comes in many guises, and one form can change into another. An energy source is simply a material which by reason of its position, its motion, its temperature, its chemical make-up, or its electric and magnetic properties can cause something to move.

Now what sort of natural materials in our planet have this capacity to get things moving? We think first of the fuels, coal and petroleum and natural gas, that are used in so many ways to keep the wheels of industry turning. We think of uranium, the widely publicized element that serves as a source of power in nuclear reactors. We note that water in a reservoir, simply by virtue of its elevated position, has the ability to turn the turbines of a hydroelectric plant in the valley below. We might list the familiar active agents of nature—water in waves pounding against a coast, water lifted twice a day in tides, water in a mountain torrent, the moving air of wind. We recall that rocks a scant few kilometers below the earth's surface are hot enough to boil water, and wonder whether this heat might be

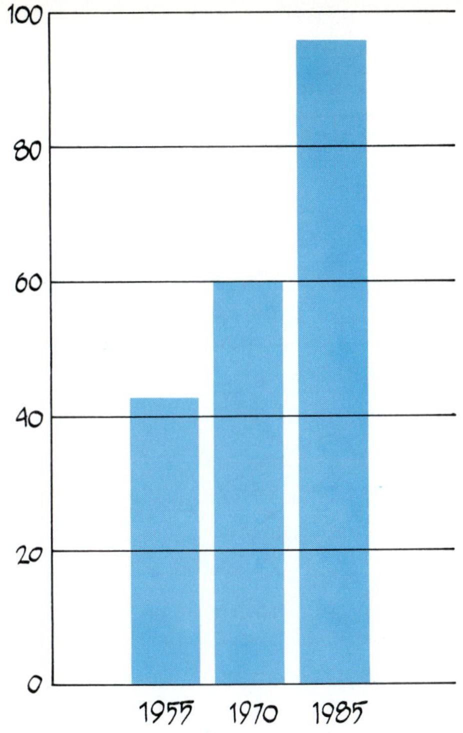

Fig. 17-3. Per capita use of energy in the United States in 1955 and 1970, and a projection to 1985. The energy is expressed in terms of barrels of petroleum. (The Chase Manhattan Bank)

harnessed as a source of energy for machines. Then perhaps we look wistfully at the sun, which bathes the earth continually with energy in the form of light; sunlight is not strictly energy of earth materials, but it is a possible energy source if we had the know-how to use it.

From this list, which items would you select as the most important ones for human purposes? Which are the materials that will be necessary in large amounts in the future if our industrial civilization is to prosper? A question like this cannot be answered with complete assurance, for technological progress may enable us to use tomorrow an energy source that seems unimportant today. But if we look at the future conservatively, assuming that our principal needs will be more of the same kinds of energy we are using at present, the four items that stand out are coal, petroleum, natural gas, and uranium.

Coal is the historic source of industrial energy, the source that made possible the Industrial Revolution, and is still a major contributor to the world's energy needs. Petroleum and natural gas have supplanted coal for many purposes because of the greater convenience of handling fluids and avoiding problems of ash disposal. Uranium, a newcomer to the list of energy sources, has a bright future in the development of nuclear power. These four are the important suppliers of energy in today's world, and

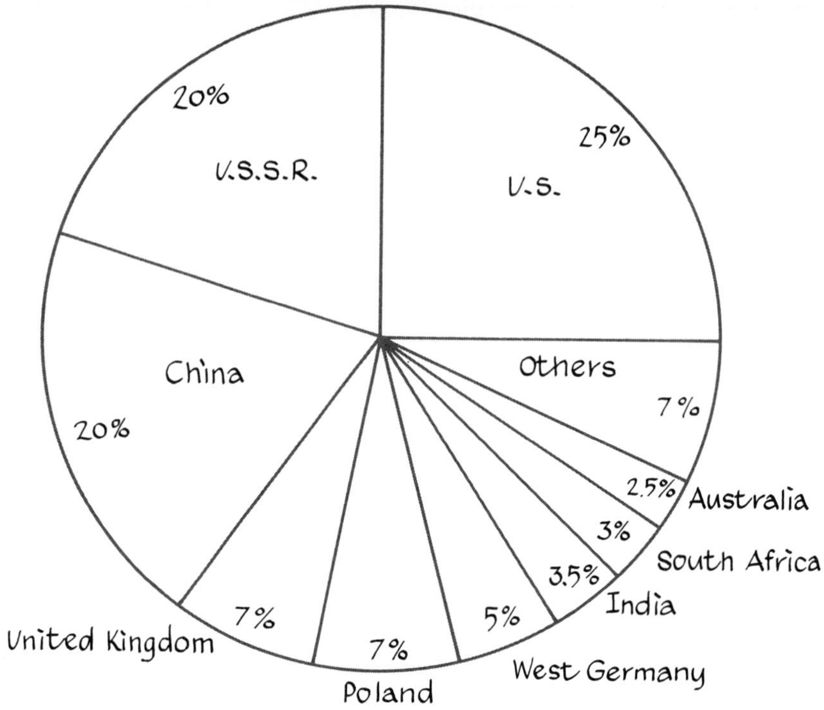

Fig. 17-4. World production of coal in 1970. (C.F. Park, Jr., *Earthbound,* 1974, Freeman, Cooper & Company, and U.S. Bureau of Mines)

questions about adequacy of energy resources for the future are largely concerned with their reserves and with projections of their rate of utilization and exhaustion.

What about other possible energy sources? Is there a chance that another might be developed, so that supplies of coal, petroleum, gas, and uranium would no longer be so critical?

Water power we can dismiss at once, simply because amounts are grossly insufficient. This is not to say that hydroelectric power is unimportant; it accounts for something like five percent of the energy produced in the United States, and is especially valued as an energy source because it puts no pollutants into the environment. But the number of rivers that can be dammed is limited, and even if it were esthetically desirable to construct dams wherever feasible the amount of energy obtained would be no more than a few percent of future needs.

Immense quantities of energy go to waste in wave motion on the ocean surface, in tidal currents, and in wind, but it is energy dispersed over huge areas and highly variable with time. Except for minor uses like windmills, the problems of concentrating and controlling the energy from such sources have proved insuperable. The same might be said for direct use of energy from the sun, although possibilities here are somewhat brighter. Large reflectors or photoelectric cells in desert areas are demonstrably capable of concentrating enough radiant energy to generate electricity in commercial amounts, but the cost of constructing and maintaining the surfaces has so far made such a project uneconomical, and the problem of storing electricity during periods when the sun is low or obscured is formidable. Energy from the earth's internal heat is currently being exploited on a small scale in a few volcanic regions, where groundwater in large amounts is converted to steam by hot rocks not far below the surface ("geothermal power"). Similar development will be possible in a few other especially favorable areas, but widespread use of the internal heat would require fracturing rock at great depth and maintaining rapid circulation of large volumes of water, operations which at present are uneconomical.

Of the alternate sources of energy, enough is known about water power, wave motion, tides, and wind to say confidently that these can never serve as major contributors to the world's industrial needs. Sunlight and geothermal power have much more promise as possible major energy sources for a remote future, but both will require solution of some very stubborn technical problems. In the years immediately ahead no other source is likely to be competitive with energy from coal, petroleum, natural gas, and uranium.

Energy as a solid fuel: coal

In several parts of the United States—notably a belt stretching from Pennsylvania and West Virginia westward into southern Illinois, and a large area of the western states reaching from Montana and the Dakotas south into New Mexico—a traveler sees occasional layers of black rock exposed in cliffs and roadcuts. The black rock forms beds interlayered with shale and sandstone, and in places at its contacts the black material and sediments are intermixed. The black substance is evidently a sedimentary rock, but it differs from the other layers in its ability to burn. Why should it have this ability? A chemist would explain, because it consists largely of the elements carbon and hydrogen, both of which combine readily with oxygen from the air and give out much heat in the process. If you ask about the particular form of these elements in coal, the chemist will not be very helpful. He will mutter something about complicated organic compounds containing also some oxygen and nitrogen, and about differences in composition from one kind of coal to another, but beyond this he would be intelligible only if you are fluent in the language of organic chemistry.

Coal is a sedimentary rock that will burn. From an energy standpoint this is more important than the details of its composition. In burning it provides a concentrated source of heat energy—much more concentrated, for example, than wood, which has a somewhat similar chemical make-up. Coal is chiefly prized as a cheap source of energy, which may be used directly as heat or may be converted to the mechanical energy of a steam engine or to electrical energy for transmission to homes and factories. But coal is a versatile substance, also needed in large amounts for smelting metallic ores and for providing raw materials for the organic chemical industry.

If we look at coal now with the eye of a geologist rather than that of a chemist or a consumer of energy, how did this strange kind of sedimentary rock come to be formed? As always, we bring out our hammers and handlenses and look carefully at a coal outcrop, or a pile of coal brought from a mine. If we are lucky, the answer will be immediately evident. Some pieces of coal are full of the carbonized remains of plants—chunks with the recognizable structures of wood or stems, and sometimes entire leaves with details marvelously preserved (Fig. 6-5). Coal is plant material that has been buried and somewhat altered. This much is clear from casual inspection. If we pursue the matter further and ask what kinds of plants are involved, we find that many of the fossil leaves and stems can be identified with species that lived in the moist environment of swamps.

So we picture the origin of a coal bed as a swamp, say on a river floodplain, with abundant plant growth whose remains accumulated in stagnant water and were thus protected from immediate decay. Eventually a change in climate, or of elevation, or of the channels of rivers over the plain, brought mud and sand to bury the plant debris under layers of sediment. Slow chemical decay aided by bacteria, plus pressure from the weight of accumulating sediment, over geologic ages changed the material to coal. Confirmation of this postulated origin of coal comes from observations of organic matter in various stages of alteration: the thick accumulation of dead organic matter in present-day swamps; the ancient plant remains of peat bogs, compressed and somewhat altered by partial decay over thousands of years; the soft, light, coaly material called lignite, often with abundant plant forms still clearly recognizable; and finally the jet black, highly compressed, often shiny layers of bituminous coal.

Thus the geologic interpretation of coal seems reasonably simple. The rock bears within itself traces of its origin, and we can find in the world around us precursors of future coal beds apparently in different stages of formation. Much mystery still remains about details of the chemical conversion and about the distribution of coal beds in the geologic past. One interesting question, for example, concerns the abundant coal of the Appalachians and the Midwest: the coal here comes from late Paleozoic rocks, and coal of this age is particularly common elsewhere in the world also; why should huge swamps have been so numerous in many places at this one time in geologic history?

Just as the origin of coal seems easy to understand, so the search for new supplies holds less difficulty than the search for most of the important mineral raw materials. If coal beds are present in a sedimentary sequence, outcrops can generally be found—or if not outcrops of coal itself, perhaps exposures of sandstone and shale containing abundant organic matter which suggest that coal beds may be found by drilling. Because coal beds are simply layers in a sedimentary pile, they can be expected to show reasonable continuity and can be followed by working out the geologic structure of the enclosing beds. Thus in any area where coal is encountered, determining the amount present is a fairly simple matter of geologic mapping and exploration underground by means of drillholes.

Reserves of coal for the future, as shown by exploration of this sort in many parts of the world, are fortunately very large. Like metallic ores, coal is distributed unequally over the world, so that some countries (for example, most of South America) have very little while others are well supplied. A problem in areas where coal has been mined for decades or centuries, as in

Fig. 17-5. Strip-mining of lignite, North Dakota. When material overlying the coal is only a few feet or tens of feet thick, this method of mining is far more efficient than an underground operation. The ridges on the right are piles of waste produced by the mining. (U.S. Department of Agriculture)

Fig. 17-6. Waste produced by strip-mining after a few years of exposure to erosion. One of the problems of strip-mining is how to return land like this to productive use. (U.S. Department of Agriculture)

the eastern United States and western Europe, is the fact that the coal of best quality and most accessible to mining has been exhausted; reserves are still large, but as time goes on recovery will become increasingly expensive. The uneven distribution of coal and the economic problems of exploitation may lead to local political difficulties, but as a worldwide resource the amount of coal is sufficient to satisfy man's needs for at least another century or two.

In view of this happy conclusion about coal reserves, why is there such alarm over the rapid exhaustion of energy supplies? The answer lies in some obvious disadvantages of coal as an energy source: for many purposes a solid fuel is just not convenient. In large fixed installations like factories or electric power plants, or in large machinery like a steam locomotive, coal serves well enough. But for running an automobile or airplane or small motorboat coal would hardly be suitable. Even for fixed installations coal has the handicap that it produces abundant ash which must be disposed of, and in urban areas it is anathema to environmentalists because of the black smoke and noxious gases it forms on burning. In addition, coal has become increasingly unpopular because much of it is recovered by "strip-mining"—removing soil, vegetation, and underlying rock over wide areas so as to expose a coal bed to large-scale quarrying operations. If not done with proper precautions, strip-mining can leave the landscape denuded and unsightly for generations. It is small wonder that the cleaner, more easily handled liquid and gaseous fuels have been steadily displacing coal as an energy source in recent decades.

But the story of coal is almost certainly not yet ended. With improved technology it may well be possible to remove the polluting substances from coal and even to convert it economically to gaseous and liquid fuels. If this can be accomplished, the enormous reserves of coal still in the earth will be a major energy source for the future.

Fluids as energy sources: petroleum and natural gas

The chemical compounds in petroleum, for the most part, are simpler than those in coal. The major ones, called *hydrocarbons,* contain only the two elements carbon and hydrogen. Petroleum has little ash, and only minor amounts of compounds containing nitrogen, oxygen, and sulfur. The hydrocarbons come in bewildering variety, the differences depending on the size, structure, and carbon-hydrogen ratio of their molecules. Those with small molecules and much hydrogen in relation to carbon (for example methane, CH_4, and propane, C_3H_8) are gaseous at ordinary temperatures, and are the principal constituents of *natural gas.* Mixtures of hydrocarbons with somewhat larger molecules make up volatile liquids like gasoline and kerosene, and those with still larger molecules and low hydrogen-to-carbon ratios are the heavier liquids of fuel oil and lubricating oil.

Separating these different kinds of fluid mixtures from crude petroleum is accomplished by fractional distillation—by heating petroleum gradually and collecting the materials that vaporize in different temperature ranges. The amount of any one fraction can be somewhat increased or decreased by chemically altering the hydrocarbon molecules. Thus the yield of gasoline can be improved by joining together ("polymerizing") the small molecules of natural gas, or by breaking ("cracking") and adding hydrogen to the larger molecules of heavy oil. All these characteristics—the simple fact that petroleum and natural gas are fluids, the small amount of impurities, the ease of separating fuels adapted for special purposes, and the possibility of increasing amounts of one fuel or another as the market requires—give the petroleum hydrocarbons a tremendous advantage for many domestic and industrial uses. They are needed in ever-increasing amounts, and their continued availability constitutes the most critical question about energy supplies for the near future.

Like coal, petroleum is almost certainly a product of the slow partial decay of organic matter from dead organisms. For petroleum the proof is not quite complete, because a liquid cannot preserve the fossils which make the origin of coal so obvious. The proof of organic origin is incomplete also because stages in the conversion of dead organisms to oil are not found, and because the conversion has not been duplicated in the laboratory at reasonable temperatures. The argument for an organic origin rests chiefly on the almost universal occurrence of petroleum with marine sedimentary rocks. Such rocks generally contain a percent or so of organic matter, often much more, and were deposited on shallow sea bottoms, an environment where living things are especially abundant. Their association with petroleum seems a clear indication that hydrocarbons have formed from part of the organic debris laid down with the original sediment. The particular kind of organic matter is not certain, but it was probably chiefly the remains of one-celled plants and animals that accumulated in enormous amounts on the floor of ancient seas and were buried by later sediments.

Petroleum and natural gas presumably form at the same time, sometimes more of one and sometimes more of the other, depending on the

Fig. 17-7. Diagrams to show some of the ways in which petroleum and natural gas may be trapped.

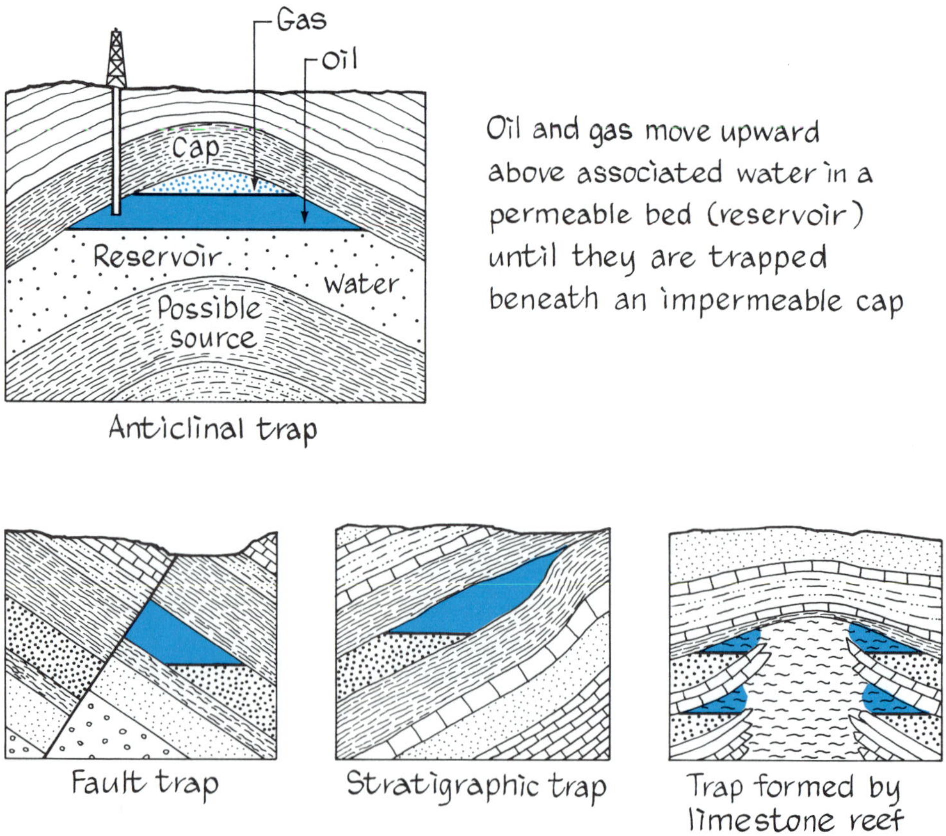

Fig. 17-8. An anticline in shale, with an oil well drilled near its crest. Presumably a layer of more permeable rock like sandstone, in which oil may have accumulated, is present in this same structure at depth. (M.N. Bramlette, U.S. Geological Survey)

nature of the organic source material and on the temperature and pressure during its partial decay. Once formed, the two fluids may remain as part of the sediment, particularly if it is fine-grained and impermeable, producing an organic-rich, usually dark-colored shale. More commonly the fluids would be expected to migrate, particularly if the enclosing sediment is permeable or if rock movement squeezes them out of the layer in which they formed. If all the sediment above them is permeable, or if they move into rock broken by faults and fissures, they may escape to the surface and be lost by evaporation or slow oxidation—and most of the petroleum produced during geologic time has probably suffered this fate.

Only when the fluids move into some sort of trap can they be preserved in large quantity underground. Possible traps include a variety of rock structures, as illustrated in Fig. 17-7: a dome or closed anticline, with oil and gas accumulating in permeable beds under a layer of impermeable shale; a fault that blocks off the upper end of a tilted sandstone bed; a sandstone layer that changes gradually to shale, so that fluids are trapped in the more permeable part. In any of these traps the fluids are commonly arranged according to density—gas at the top, petroleum under it, and salt water at the bottom.

Armed with this kind of information, how might you set about looking for oil? You could hardly expect to find hydrocarbons in surface outcrops to serve as a guide for your search; oil seeps do exist, but they are very rare. Drilling would be necessary, and your problem is to find the right place to drill. Since drilling is expensive, you would have to be prepared to gamble a good deal of money. To reduce the gamble as far as possible, you would call on all the geologic information at your command. You would concentrate on places where outcrops indicated a considerable thickness of marine sedimentary layers, and avoid areas with outcrops of igneous and metamorphic rocks. You would watch particularly for layers of sandstone or conglomerate that might serve as permeable reservoirs for oil and gas accumulation under the surface. You would look for a trap by carefully mapping outcrops to see whether the strata could be fitted into an anticline or were cut by faults. Then on the most promising structure you would put your drill to work. Even with all this geologic sophistication you might well end up with a dry hole, because all too often a seemingly perfect trap turns out to be empty. But success in applying this kind of reasoning has been frequent enough so that geologists have long been in great demand for guiding oil exploration. In the practical world this is where geology has had its greatest triumphs.

If you should set out now, as a newcomer, to try your hand in the oil business by looking for structural traps on the basis of geologic mapping at the surface, you would be frustrated to find that the obvious traps have already been thoroughly explored, most of them several times over. Petroleum and natural gas are needed in such huge quantities, and successful exploration is so very lucrative, that both wildcatters and large oil companies have combed the country from one end to the other for rocks and structures that might contain the precious fluids. The search has gone far beyond the simple technique of working out structures from the mapping of surface outcrops. In areas where bedrock is hidden by soil and alluvium, for example, the presence of folded and faulted strata can often be detected by the seismic waves from artificial earthquakes generated by explosions; or exploratory holes can be drilled through strata at depth, and characteristics of the strata can be mapped either by examination of drill cores or by lowering sensitive instruments down the holes. Modern petroleum exploration requires that geologists have at their command a great variety of such techniques for detecting minor structural details and changes in rock textures even at great depths. The methods have become bewilderingly complex, but the essential nature of the search remains the same: one must look not directly for oil, but for structures and changes in rock characteristics that might permit oil to accumulate. The search is exciting and the rewards are very great, both in money and in the satisfaction that comes from using geology to work out some of nature's most complicated puzzles.

Why is geology so much more successful in aiding the search for oil than in exploration for metallic ores? Neither for oil nor for metals do we have a completely satisfactory theory to explain the origin of large accumulations. But once oil has been formed, we do know a good deal about how it moves and how it may be trapped; our deductions about the migration of oil have been tested in so many oil fields that the general pattern seems well established. For metals the situation is much more complicated. The fluids that deposit them must not only carry them to the site of accumulation, but by complicated chemical reactions must precipitate them; the paths of migration are not limited to one kind of rock or one kind of structure; sometimes there is evidence that the solutions were trapped by impermeable material, much as petroleum is trapped, but the evidence is less clear and less consistent. Another handicap of the metal prospector is the fact that he is generally looking for a small target, an ore body with dimensions in the range of hundreds of meters or at most a few kilometers, while an oil field may cover an area ten to a hundred times greater.

Thus the greater success of geologists in oil exploration is explainable on the basis of a better established underlying theory, restriction of oil and gas to particular kinds of rock, fewer complications in the process of accumulation, and greater size of the target sought. A geologist interested in ore deposits has an important role in exploration, but it will be a long time before he can be as confident a guide as his counterpart in the search for oil, simply because his problem is more difficult.

Oil and gas for the future

From a conservationist's perspective, the future of petroleum and natural gas looks bleak. On the one hand he knows that these fuels are a nonrenewable resource; they may well be in process of formation in present-day marine sediments, but the process is too slow for any human usefulness. Sometime in the future a drill-rig will penetrate the last oil field and the last gas reservoir, and from the present declining rate of discovery of new fields the conservationist estimates that this time is not far off. On the other hand he sees the consumption of gas and petroleum products rapidly increasing, as population grows and the superiority of fluid fuels for one purpose after another is demonstrated. In his view much of the use is little better than profligate waste, as these precious resources go into unneeded travel, excessive air-conditioning, and generation of electricity for trivial purposes. The use is often inefficient, spewing into the atmosphere the products of incomplete combustion and generating the smog that shrouds our cities. Between dwindling resources and over-use of its best fuels the world spins gaily toward a major calamity. The "energy crisis"

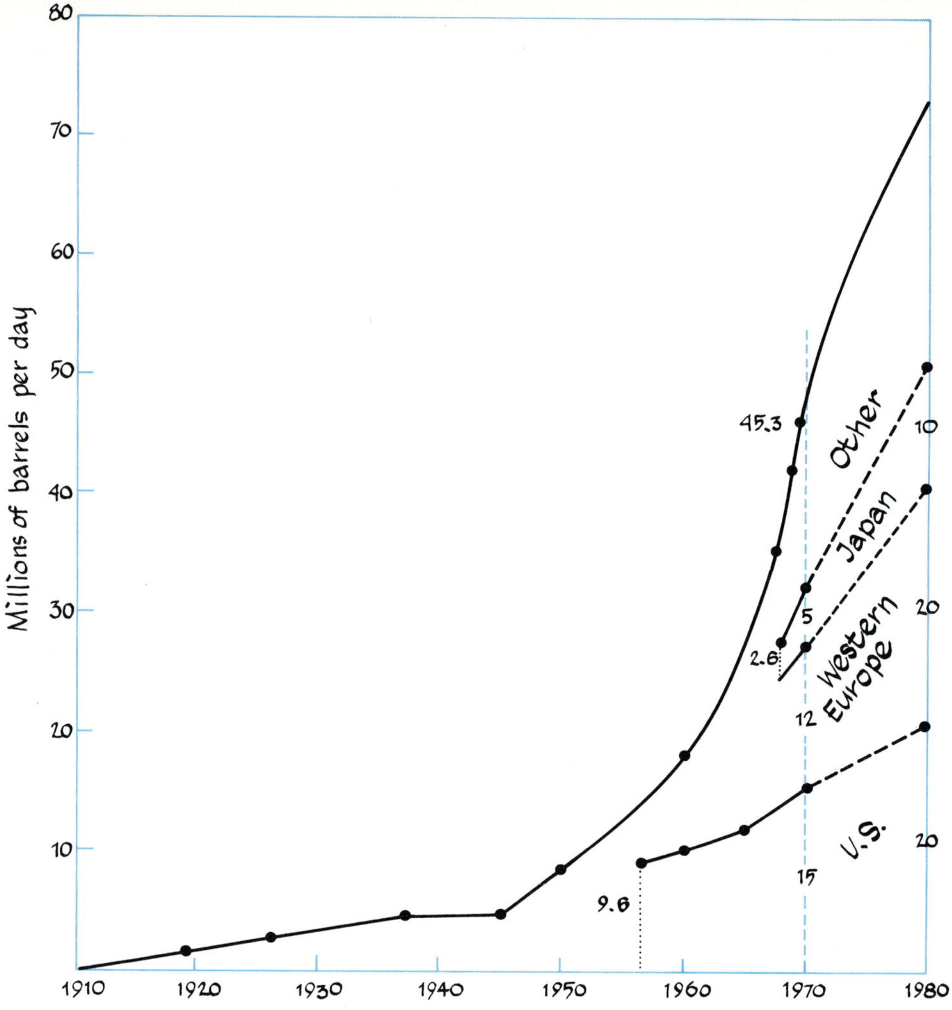

Fig. 17-9. World demand for petroleum, 1910–1970 and projection to 1980 (Marshall Ayres)

proclaimed by the more pessimistic prophets is in large part a prediction of imminent scarcity of these fluid fuels.

How accurate is this dismal picture? No one really knows, of course. One hears all shades of opinion, from the deepest gloom to blithe optimism. A rational stance, presumably, lies somewhere between.

It is an incontestable fact that the world's petroleum supplies have a finite limit. Where the limit is cannot be certainly known, because not all areas of marine sedimentary rock have been thoroughly explored, but any reasonable extrapolation from available data indicates that the ultimate quantity is not nearly as large as that of coal, and that acute scarcity of oil and gas is probable in decades rather than centuries. It is sobering to note that in the United States, where exploration has been particularly intense for many years, the rate of new discoveries fell off sharply several years ago, and has shown no sign of probable recovery. Much oil will still be produced from known fields, but the amount will decline as the fields approach exhaustion. Even at present this country would have difficulty in supplying from domestic sources all the petroleum it uses; and the supply of natural gas from domestic fields is even more precarious. One need only project the experience of the United States to other countries to be convinced that sometime in the not distant future, certainly within the next century, the worldwide rate of new discoveries will be inadequate to maintain reserves.

On the other hand, an optimist can note that several major new discoveries have been announced in the last few years (the north slope of Alaska, the Arctic Islands of Canada, the North Sea), and that a continuation of such discoveries will postpone the evil day significantly. He can also take comfort in the fact that exhaustion of oil and gas will not

Fig. 17-10. The distribution of petroleum reserves. The map is distorted so that each oil-producing country has a size roughly proportional to its known reserves of petroleum. The huge reserves of some of the countries in the Middle East are especially striking. (Exxon and *Oil and Gas Journal.* Copyrighted and used with permission.)

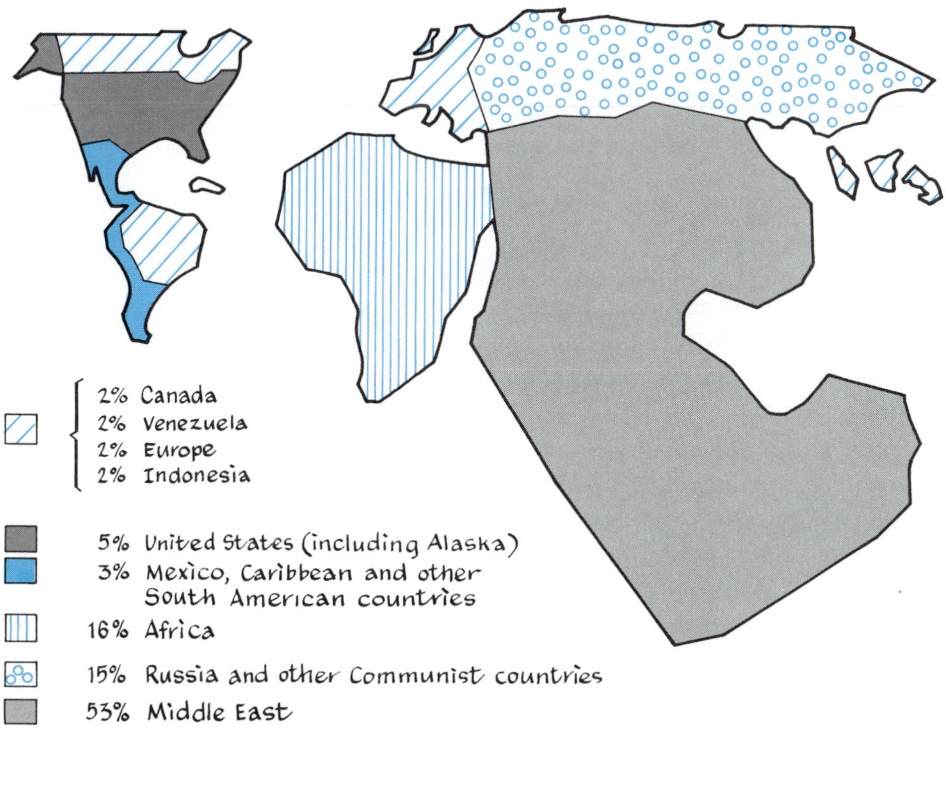

come suddenly, but will be a long decline over many decades. Prices of gas and petroleum products will rise, but men will have ample time either to find substitutes or to adapt themselves to scarcity.

Worldwide resources of petroleum pose a long-time problem for man's continued occupancy of the planet, but a more urgent immediate problem arises from the geographic distribution of petroleum accumulations. Like the metals, petroleum is widely distributed over the continents, but in grossly unequal amounts in different areas. The uneven distribution is very marked within the United States: one thinks of the fabulous accumulations in the thick Cenozoic and Mesozoic beds along the Gulf Coast of Texas and Louisiana, the many fields in Paleozoic rocks from Pennsylvania to the midcontinent, the oil of many small sedimentary basins in the much folded and faulted rocks of California—and of the complete absence of oil in the New England states and the northwest (except possibly offshore). Elsewhere in the world the greatest accumulation of all is in the lands around the Persian Gulf, where thick Cenozoic sediments contain roughly two-thirds of the world's known reserves. The United States, facing a time of growing dependence on foreign sources of oil, is necessarily concerned about political relations with other oil-producing countries, particularly with those of the Middle East.

Suppose that oil and gas become unacceptably expensive, or that international accidents greatly impede the flow of oil from other countries. What recourse do we have for maintaining supplies of liquid and gaseous fuels? Many alternatives are possible, none of them economically competitive at the moment but some very promising if technological difficulties can be overcome.

One frequently in the news is the distillation of petroleum hydrocarbons from *oil shale.* Huge deposits of this rock—a shale formed from mud containing much organic matter, but from which the products of partial decay did not migrate—are found near the junction of Colorado, Utah, and Wyoming, and the amount of hydrocarbon that could be obtained by heating it is estimated to equal at least a large fraction of the country's known oil and gas reserves. The technical problems of efficient heating of the rock and disposal of waste have prevented more than token development, but in the future the shale will almost certainly become an important source of energy in fluid form. Another promising material whose exploitation involves similar problems is *tar sand,* a rock found in beds that cover wide areas in west central Canada.

Still another possibility is the preparation of liquid and gaseous fuels from waste organic matter. Almost any kind could be used—wood, grass, paper, sewage—and the resulting fuel might consist largely of compounds other than hydrocarbons, for example alcohol. The chemistry of such processes is well known, but the fuel obtained would be more expensive than those in common use. A similar possibility that is closer to economic realization is the conversion of coal to gaseous and liquid compounds. Coal has long been used to prepare domestic gas in small quantities by heating it with water (to produce "water gas"), but the process is too expensive for large-scale application and one of its products is the poisonous gas carbon monoxide. Better fuels can be prepared by letting coal react with hydrogen, and if technical problems can be overcome, a successful process of this sort would make possible efficient use of the world's ample resources of coal.

Diminishing supplies of natural gas and petroleum products, with consequent increasing prices, are a safe prediction for the future. We must hope that as these materials become scarce, other sources of fluid fuels can be developed without too greatly increased costs and without an inordinate increase in the amount of energy required to produce new energy. Although our grandchildren will probably not know the cheap and abundant fuels that we have enjoyed, there is little risk that they will lack fluid fuels entirely. But what the cost will be in quality of life we cannot tell.

Nuclear energy: salvation or menace?

Coal, petroleum, and natural gas have this in common: they are materials that once were part of living organisms. Because of this relation to past life, they are often referred to as *fossil fuels.* Their origin in living substance means that the energy they contain came ultimately from sunlight—light energy absorbed by leaves of green plants in the reaction of photosynthesis, and stored both in the plants and in animals that used the plants for food. When we burn the fossil fuels, in a very real sense we are using the energy of sunlight that came to the earth in the long distant past. Most of the other kinds of energy that man has learned to use likewise have their source in the sun: the energy of domestic animals, which comes from the food they eat and hence ultimately from photosynthesis; the energy of moving water, or water stored in high reservoirs, which depends on evaporation of water by the sun's heat; the energy of wind, generated by the sun's warming of the atmosphere. While we have not yet mastered the trick of using energy from the sun directly, we have taken advantage of many ways in which solar energy is stored in natural materials.

With the development of the atomic bomb during the Second World War, a possibility emerged of deriving energy in huge amounts from an entirely different source. The energy released by atomic bombs and nuclear reactors has nothing to do with the sun, but comes from the nuclei of heavy atoms. It is energy that was stored in rocks at the time the earth was formed, energy whose history goes back to the mysterious astronomic event in which the atoms of the solar system were fabricated.

Nuclear energy is liberated when the nuclei of the heavy atoms are split apart by bombarding them with neutrons. Because the splitting generates more neutrons, the reaction can be made self-perpetuating (a so-called "chain reaction"). A few neutrons capable of starting the process are always present in natural environments, but the chain reaction is damped out because the neutrons produced by nuclear splitting are absorbed by other materials or are moving at the wrong speeds to split other nuclei. To make the reaction produce energy for useful purposes, the heavy atoms must be separated by material which will moderate the speeds of the neutrons without absorbing many. With appropriate arrangements of the heavy atoms and the moderating substance or substances, the reaction can be delicately controlled and hence can be used as a source of heat energy for driving machinery or generating electric power.

From a geologic standpoint the important question, obviously, is the availability of material containing the necessary heavy atoms. The element whose atoms have proved most suitable is uranium; another element, thorium, is a good possibility for future development. Uranium and thorium, we recall from an earlier discussion (Chapter 11), are elements that are naturally radioactive, meaning that the nuclei of their atoms decay spontaneously to give off tiny particles and thereby release energy; the rate of spontaneous change gives us a method of measuring geologic time, and the liberation of energy is one of the major sources of the earth's internal heat. But now we are dealing with a completely different property, the ability of these atoms to split or *fission* into big particles and thereby to give out energy in far larger amounts than the energy of radioactive decay. Perhaps there is a basic reason why fissionable atoms should also be spontaneously radioactive, but the two modes of behavior have nothing in common. The amounts of uranium needed commercially because of its radioactive properties are trivial, but the huge amounts needed for nuclear power plants have made uranium a prime target for geologic exploration.

Where would we expect to find ores of this strange metal? From its complex chemistry we might guess that its compounds will be concentrated by a variety of geologic processes, and correspondingly the search for uranium has brought to light many kinds of ores. In the United States the major deposits are in Mesozoic and Cenozoic sedimentary rocks in some of the western states, from Wyoming south to New Mexico (Plate 31). The original source of the metal is still in dispute, but certainly its compounds have moved long distances dissolved in groundwater percolating through porous sandstones. Far larger is a deposit in South Africa, a peculiar conglomerate of Precambrian age that also is a major source of gold; the argument has been loud and long as to whether the metals were concentrated mechanically in an ancient placer deposit, or chemically from hot solutions rising out of an hypothetical body of granitic magma somewhere beneath. Uranium minerals are also prominent in some dikes and veins cutting granites and metamorphic rocks near their contacts. Thus uranium has proved to be an exceedingly versatile element, and the geology of its occurrences is both difficult and fascinating.

Like all sources of energy derived from rocks, uranium ores are nonrenewable. What, then, of the future? If nuclear energy becomes a major source of power, will there be enough uranium to satisfy the world's needs? The answer is complicated by the fact that uranium has two fissionable isotopes, ^{235}U and ^{238}U; the former is scarce, making up less than one percent of natural uranium, but it is the one whose fission is most easily controlled and which is the source of power in most present reactors. If the world must depend entirely on ^{235}U, nuclear power will be a source of cheap energy for not much longer than the fossil fuels. But if the more difficult technology of deriving energy from ^{238}U can be mastered (in the so-called "breeder reactor"), resources of the element are adequate for many centuries.

The ability of nuclear fission to supply cheap energy on a large scale makes it a promising supplement or alternative to existing sources. Nuclear energy could hardly be used to power an automobile or airplane, but it could well take the place of fluid fuels in plants for generating electrical energy and so could extend the availability of petroleum and natural gas for a long future. Perhaps nuclear fission is mankind's great hope, the providential solution to our worries about sufficient energy to maintain a burgeoning population in a rapidly industrializing world.

Yet there is room for doubt, because nuclear power has handicaps that do not affect other methods of energy production. A reactor, as long as it is working well, seems an ideal energy source: it is clean, quiet, produces no smoke, and has no unsightly piles of fuel or ash nearby. But what happens if something goes wrong? Suppose the control mechanism fails—couldn't the reactor turn into a huge atomic bomb and devastate its surroundings? The possibility is exceedingly remote, but still a source of worry to many. Also, a reactor generates a good deal of highly radioactive material, whose radiation could be lethal to humans and other organisms if it should escape into the environment. This is most unlikely, because the radioactive waste is disposed of at a distance from the reactor by methods that will presumably keep it out of the environment permanently. But again, the inconceivable accident is remotely possible. Furthermore, the operation of a reactor requires a great deal of cooling water, and the water of course absorbs heat; if great volumes of hot water are circulated back into streams, lakes, or the ocean, how destructive to organisms will this be? These are the major uncertainties that cause apprehension about nuclear reactors. The dangers they represent, although enormously magnified in the popular press, are still real.

Worries about nuclear reactors can become highly emotional, and almost as much emotion is generated by environmental effects of the large-scale use of fossil fuels—smog, smoke belching from chimneys, landscapes mutilated by strip-mining, oil spills resulting from carelessness in offshore drilling or in handling oil shipments in harbors. The emotion poses a real dilemma. If strong enough, it can prevent or long delay the construction of new power plants. Yet civilized man requires the continual building of new plants to supply his growing appetite for energy. For the immediate future, a particular worry is the long time that must elapse between the decision to build a new plant and actually getting it into operation. If emotional response to the plans causes further delay, scheduling of energy supply for future needs becomes impossible. Perhaps the actual experience of widespread energy shortages will be needed before a reasonable balance can be struck between concern for environmental disturbance and risks of accident on the one hand, and our desire for abundant energy on the other.

The far-out possibilities

Reading the future from a crystal ball is always a pleasant pastime, but when the reading involves energy supplies it can become deadly serious. Much of our planning for the future depends on what estimates we can make about the exhaustibility of conventional supplies and about the possible development of entirely new sources. To this sort of crystal-gazing a geologist can bring his knowledge of materials and conditions within the earth, but in weighing the chances of new developments he must depend on the expertise of others. He can say with some confidence that the world has enough petroleum for a few decades, enough coal for a few centuries, enough uranium for several centuries if the breeder reactor proves practical but only enough for decades if we are limited to the fission of ^{235}U. More detailed geologic predictions are always subject to revision in the light of new discoveries, but the general order of times will not be changed. Beyond predictions of this sort a geologist must fall back on the expert knowledge of engineers, physicists, chemists, and economists. What additional possibilities can such experts see in the crystal ball?

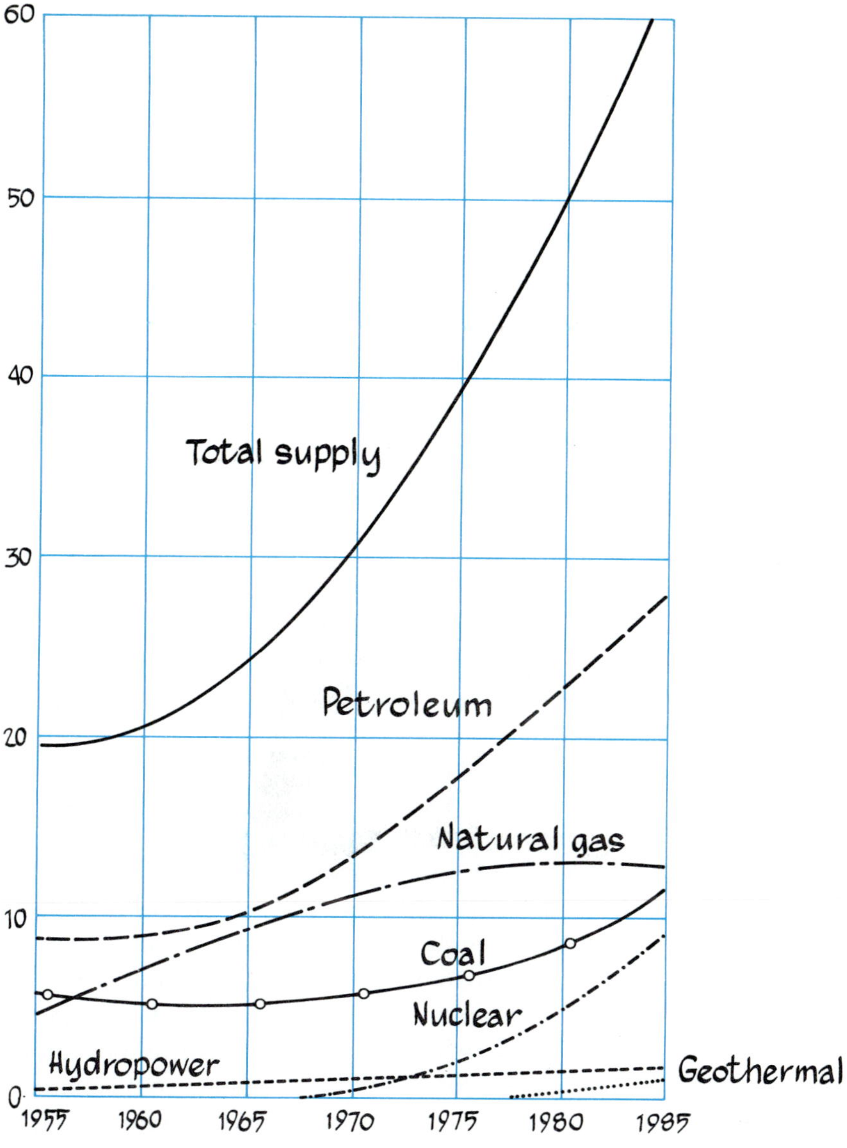

Fig. 17-11. Projected sources of energy in the United States, 1955–1985. The energy is expressed as millions of barrels of petroleum per day. (Shell Oil Company)

There are first two obvious sources of energy that we have mentioned before: the earth's internal heat and sunlight. Both of these are inexhaustible reservoirs of energy, and both have the great attraction that they can be developed and used with minimum disturbance of the environment. Sunlight is already being trapped on a minor scale for space heating, and geothermal power in a few places is making a minor contribution to supplies of electrical energy. Geologists are busy looking for other spots where the right combination of groundwater circulation and fairly recent volcanic activity will make possible additional geothermal plants with present technology. But the general problem of utilizing the earth's internal heat over wide areas, and the problem of converting sunlight to electricity on a large scale, are concerns of engineers rather than geologists. There is good reason to hope that one or both problems will eventually be solved, but at present this is only a dream for the future.

One other far-out possibility needs mention, to complete a canvass of imaginable energy sources. The energy that keeps the sun and other stars shining comes from a process in which matter is destroyed—or more properly, in which a small amount of matter is changed into energy. One of the reactions involved is the conversion of hydrogen to helium: four atoms of hydrogen have a slightly greater mass than one atom of helium, and during the reaction the excess mass is liberated as energy. The amount of energy so obtainable from a small amount of hydrogen is prodigious. If the reaction could be reproduced and controlled on the earth, our energy problems would vanish overnight. We could use the hydrogen of water as a "fuel," and the supply of water in the oceans is inexhaustible. This is another kind of "nuclear power," produced by the joining together of light atoms rather than the splitting of heavy atoms, hence called a *fusion* process in contrast to the *fission* process in a conventional reactor.

The catch in a fusion process is that the reaction takes place only at temperatures of a few million degrees, temperatures at which all ordinary matter is vaporized. So at present we can duplicate the process only in a device like an atomic bomb, where there is no problem of containment. The high temperatures are generated by uranium or plutonium in the bomb, and the hydrogen-helium conversion then makes the bomb far more destructive than it would be with uranium or plutonium alone (a "hydrogen bomb" in contrast to an "ordinary" atomic bomb). But if the energy is to be manageable for industrial purposes, the reaction must be made to take place in a confined region where its rate can be controlled—and how is this possible when any container would vaporize before the reaction even gets started? Conceivably the confinement might be managed by using powerful magnetic fields rather than a solid container, and this possibility has spurred much eager research. We are still far from the goal, but there is at least a chance that someday we will learn to construct small "suns" on the earth which can furnish cheap energy in hitherto unimaginable quantities.

Sunlight, geothermal power, hydrogen fusion: these are possibilities for the future, but at the moment our urgent concern with energy relates to supplies of coal, petroleum, natural gas, and uranium.

A summing up

The problem of supplying the world's energy needs over the next several decades is similar in many aspects to the problem of maintaining an adequate supply of metals. Both kinds of resources are limited in quantity, and neither kind, with trivial exceptions, is being renewed by present-day geologic processes. Both are being exploited at an ever-increasing rate, a rate that shows no sign of diminishing in the foreseeable future. The massive exploitation leads to increasing damage to the environment. As the richer concentrations of energy resources and metallic ore are exhausted, and as concern for the environment leads to stricter controls, cost must inevitably rise. No actual worldwide shortages have yet developed, but depletion of some resources in countries where exploitation has been most intense has become a source of political and fiscal worry. As shortages become more widespread and prices rise in consequence, standards of living and the balance of political power may well be altered. There seems little chance that the world will ever really exhaust its supplies of either metals or energy, but there is serious doubt that they will ever again be as cheap and readily available as they have been in the middle years of the 20th century.

Gloomy estimates of future scarcity can always be brightened by mention of two possibilities. One is the chance that increase in geologic knowledge will lead to discovery of hitherto unsuspected reserves of ore, petroleum, or natural gas. Such discoveries have confounded dire predictions many times in the past, and with the new understanding of earth processes provided by the hypothesis of plate tectonics our ability to locate ore deposits and oil fields may conceivably be greatly enhanced. The other possibility is a breakthrough in the technology of energy production, as suggested in the previous section. The first possibility is only a stopgap, a postponement of the inevitable day when rich concentrations of metallic ores and energy-producing substances will be exhausted. The second is a more revolutionary and more permanent solution, but would bring in its wake formidable environmental and economic problems. Whatever way we look, man's continued rapid exploitation of the world's nonrenewable resources will lead to problems whose solution will require profound changes in our accustomed ways of living.

We began this long inquiry into earth processes and materials back in the first chapter by standing on a cliff looking out over the sea. Part of our musing then concerned a question about our ability to derive materials necessary for civilized life from the nearby stretch of coastline. Only by the merest chance would we have found deposits of metals, or beds of coal, or

(David Muench)

structures that might trap petroleum, in that restricted area. Almost certainly we would have had to import most of life's necessities from elsewhere.

Now at the end of our inquiry we can picture all mankind in a similar position, looking out from the sun's third planet over the great ocean of space that separates us from the other planets and the stars. We have been musing in similar fashion about the adequacy of our planet to supply man's needs for the future. It is an ample storehouse, packed with the riches accumulated through its history of five billion years, but their quantity is finite and man has much to learn about using them wisely. The resources of the planet are enormous compared to a short stretch of seacoast, but ultimately they are subject to the same limitations. And there is this great difference: to the area of the seacoast we can bring in what we need from the outside, but for the third planet there is no outside. We may explore other planets, but to use them as sources of needed material is beyond the dreams even of science fiction. We are creatures of the earth, and to supply our wants the earth is all we have.

Books for Further Reading

The chapters you have read in this book are no more than a brief introduction to the study of the sun's third planet. If the study appeals to you and you want to pursue it further on your own, two avenues invite exploration. One is travel, travel with the deliberate intent of seeing rocks and landscapes in all their infinite variety. There is no better way to get acquainted with the planet you live on than to see for yourself its mountains and deserts and broad plains, and then to reflect on the rock structures that underlie them and the long history they record. The second avenue, perhaps more accessible, is provided by the many books on geology that you will find in any good library or bookstore. The following list is a sampling of the recent literature, not intended to be complete but only to serve as a guide to help you get started on further exploration of topics that interest you.

First a word about general books. Textbooks in elementary geology are very numerous, ranging in difficulty from those written for high school students to books studded with chemical formulas and mathematical equations. Two particularly good ones, at a level somewhat more advanced than this one, are

Principles of Geology, by J. Gilluly, A. C. Waters, and A. O. Woodford (4th ed.), published by W. H. Freeman and Company (1974)

Physical Geology, by R. F. Flint and B. J. Skinner, published by John Wiley and Sons, Inc. (1974)

Another good one, interesting because many of its examples are taken from the British Isles, is

Principles of Physical Geology, by Arthur Holmes (2nd ed.), published by Ronald Press (1965)

At a more advanced level, if you have a good background in physics and chemistry and would like to see how these subjects can be applied to geologic problems, you might look into

The Earth, by J. Verhoogen, F. J. Turner, L. E. Weiss, C. Wahrhaftig, and W. S. Fyfe, published by Holt, Rinehart and Winston, Inc. (1970)

These four books are devoted almost entirely to physical geology, the study of geologic processes going on at present and the record of such processes in the past as preserved in rocks and landscapes. If your interest inclines more toward historical geology, the sequence of events and the development of life during the planet's history, a good one to try is

Evolution of the Earth, by R. H. Dott and R. L. Batten, published by McGraw-Hill Book Company (1971)

Some books include both the physical and historical aspects of geology, for example

Geology, an Introduction, by R. L. Bates, W. C. Sweet, and R. O. Utgard (2nd ed.), published by D. C. Heath and Company (1973)

Perhaps your interest lies chiefly in some one part of the subject. If, for example, you find exciting the new developments in earth science embodied in the terms global tectonics, plate tectonics, sea-floor spreading, and continental drift, your reading list should include

Debate About the Earth, by H. Takeuchi, S. Uyeda, and H. Kanamori, published by Freeman, Cooper and Company (1970)

The authors of this book, all eminent geophysicists, have managed to express in popular and often humorous language much of the fun, frustration, and satisfaction that come from scientific work in a new and rapidly developing field. A somewhat more advanced treatment of the same ideas is presented in

Continental Drift, by U. Marvin, published by Smithsonian Institute Press (1973)

Are you concerned about the role of geology in environmental problems? Or about problems related to our dwindling supplies of metals and energy? Two books that emphasize such subjects are

Earthbound, by C. F. Park, Jr., published by Freeman, Cooper and Company (1974)

Geology, Resources, and Society, by H. W. Menard, published by W. H. Freeman and Company (1974)

Or are you an enthusiastic mineral collector? Most of the textbooks mentioned above contain tables that will help you identify the commoner minerals, but if you wish to go further into the study of less common minerals a good introduction is provided by

Mineral Recognition, by I. Vanders and P. F. Kerr, published by John Wiley and Sons (1967)

If fossils are your major interest, the following two books will provide expert guidance. The first is a general treatment of the record of past life, and the second contains more detailed descriptions of specific fossils:

The History of Life, by A. L. McAlester, published by Prentice-Hall, Inc. (1968)

Fossils, by W. H. Matthews, published by Barnes and Noble, Inc. (1962)

Perhaps you have spent time in the far north or in high mountains, and the behavior of glaciers has excited your curiosity. A popularly written and spectacularly illustrated book is

Glacier Ice, by A. S. Post and E. R. LaChapelle, published by the University of Washington Press (1973)

The radiometric methods of dating that have given us such assurance about dating geologic events through nearly 5 billion years of history are well described in

Ages of Rocks, Planets, and Stars, by Henry Faul, published by McGraw-Hill (1966)

Or if you prefer to concentrate on the oceans, a good introduction to a very complicated subject is provided by

Oceanography, by P. K. Weyl, published by John Wiley and Sons (1970)

To explore further the more violent aspects of geology, you might look into

Earthquake Country, by R. Iacopy (3rd ed.), published by Lane Books (1971)

for a good general treatment of earthquakes (although the emphasis is on earthquakes in California) or into

Volcanoes, by G. A. Macdonald, published by Prentice-Hall (1972)

for a survey of current knowledge about volcanoes.

In addition to these books, semipopular articles on topics of current geologic interest appear frequently in several magazines. *Scientific American* is the best known; a British equivalent is the *New Scientist*; two journals devoted more strictly to geological subjects are *Geotimes* and the *Journal of Geologic Education*. Both the books and the journal articles contain references to the original geologic literature, and if you become really wrapped up in a subject the next step would be to track down some of these references. If you go this far you will begin to get the flavor of professional geologic research.

Appendix

Large-Scale Data for the Earth

Mean radius	6,371 km (3960 mi)
Volume	1.09×10^{12} km^3 = 1.09×10^{27} cm^3
Mass	5.98×10^{27} g = 5.98×10^{21} tons
Density	
Mean for entire earth	5.52 g/cm^3
Mean for crust	2.8 g/cm^3
Continental crust	2.7 g/cm^3
Oceanic crust	3.0 g/cm^3
Surface area	5.10×10^8 km^2 = 5.10×10^{18} cm^2
Land area	1.49×10^8 km^2
Ocean area	3.61×10^8 km^2
Volume of oceans	1.37×10^9 km^3
Mass of oceans	1.41×10^{24} g
Density of sea water	1.028 g/cm^3 (normal seawater at 0°C)
Mass of ice in glaciers	22.83×10^{21} g
Mass of atmosphere	5.12×10^{21} g

Mean height of land above sea	875 meters
Mean depth of sea	3800 meters
Greatest elevation on land	8848 meters
Greatest known depth of sea	11,035 meters
Average thickness of continental crust	35 km
Average thickness of oceanic crust	5 km
Thickness of mantle	2881 km

Conversion Table for English and Metric Units

Length

1 meter (m)	= 100 centimeters (cm) = 1000 millimeters (mm)
	= 39.37 inches (in) = 3.28 feet (ft) = 1.09 yards (yd)
1 cm	= 10 mm = 0.39 in
1000 m	= 1 kilometer (km) = 0.62 mile (mi)
1 in	= 2.54 cm
1 ft	= 12 in = 30.5 cm = 0.305 m
1 mi	= 5280 ft = 1.61 km

Note that approximately a meter is 3 feet, a centimeter is half an inch, and a kilometer is a little over half a mile.

Area

$1\ m^2 = 10.76\ ft^2$
$1\ cm^2 = 0.155\ in^2$
$1\ km^2 = 0.39\ mi^2$
$1\ ft^2 = 0.09\ m^2$
$1\ mi^2 = 2.58\ km^2$

Volume

$1\ m^3 = 35.32\ ft^3$
$1\ cm^3 = 0.061\ in^3$
$1\ km^3 = 0.24\ mi^3$
$1\ ft^3 = 0.03\ m^3$
$1\ mi^3 = 4.17\ km^3$
1 liter (l) = 1000 milliliters (ml)
= approximately 1000 cm^3
= 1.06 quarts (qt)

Temperature

1 degree on the Celsius or Centigrade scale = 1.8 Fahrenheit degrees

0°C = 32°F
25°C = 77°F
100°C = 212°F
1000°C = 1832°F

Weight

1 kilogram (kg) = 1000 grams (g)
= 2.21 pounds (lb)
1000 kg = 1 metric ton
= 2.21 short tons

Numerical Data for the Planets

	Mean distance from sun, if earth's dist. is 1.00	*Mean diameter, if earth's diameter is 1.00*	*Mass, if earth's mass is 1.00*	*Mean density, (g/cm^3)*	*Period of revolution around sun*	*Period of rotation on axis (days)*	*Number of satellites*
Mercury	0.39	0.38	0.05	5.5	88 days	58	0
Venus	0.72	0.95	0.82	5.3	225 days	243	0
Earth	1.00	1.00	1.00	5.5	1 year	1.00	1
Mars	1.52	0.53	0.11	4.0	1.9 yrs.	1.03	2
Jupiter	5.20	11.2	318	1.3	12 years	0.41	12
Saturn	9.52	9.5	95	0.7	29 years	0.43	10
Uranus	19.16	3.7	15	1.7	84 years	0.45	5
Neptune	30.0	3.9	17	1.6	164 years	0.66	2
Pluto	40	0.5?	0.1?	?	247 years	6 ?	0

Earth: Mean distance from sun 93 million miles (150 million km)
Mean diameter 7920 miles (12,742 km)
Mass 5.98×10^{24} kg

Moon: Mean distance from earth 238,000 miles (382,000 km)
Mean diameter 2160 miles (3476 km)
Mass 7.35×10^{22} kg
Density 3.3 g/cm^3

Glossary

As an aid to reading the text, this glossary supplies definitions of most of the geologic terms that have been introduced. It includes also a few words that have not been used in this book, but which are common in geologic discussions and may be encountered in some of the suggested additional reading.

The definitions of many terms are taken, with permission, either verbatim or slightly modified from *Glossary of Geology*, copyright © 1972 American Geological Institute. This much more complete glossary is a good source to consult for meanings of words that go beyond the number that can be listed here.

Acid: a compound which in solution has a sour taste, the ability to change the colors of certain dyes, and the property of liberating hydrogen by reaction with active metals. An *acid gas* is a gas that will dissolve in water to form an acid. An *acid rock* (same as *felsic rock*) is a rock containing more than 65% silica in its chemical analysis and containing quartz as a major mineral constituent.

Alkali: a strong base; a substance that dissolves in water to give a solution with a bitter taste, a soapy feel, and an ability to cause changes in the color of dyes opposite to those produced by acids. The commonest alkali in nature is sodium carbonate, Na_2CO_3.

Alluvial fan: a gently sloping deposit, in the shape of an opened fan, of loose rock material deposited by a stream at the point where it issues from a narrow mountain valley onto a plain.

Amphibian: a cold-blooded vertebrate that in early life breathes by means of gills, then develops into a lung-breathing organism at maturity. Frogs, toads, and salamanders are examples of amphibians.

Anaerobic organisms: organisms, especially bacteria, that can live in the absence of free oxygen.

Andesite: a fine-grained, medium dark volcanic rock containing plagioclase and one or more of the mafic minerals (e.g. biotite, hornblende, pyroxene). It is the extrusive equivalent of *diorite.*

Anticline: a convex upward fold in bedded rocks.

Aragonite: a mineral which has the same composition as calcite ($CaCO_3$) but different properties such as hardness, density, and crystal form.

Aquifer: a rock body whose saturated permeable material conducts groundwater and yields water to supply wells and springs.

Archeozoic or **Archean:** the early part of Precambrian time. Defined as an era by some geologists, but its boundaries are not agreed upon.

Artesian well: a well that taps groundwater confined under an impermeable layer, and in which water rises above the water table. From some artesian wells water flows out on the surface.

Ash flow: a turbulent blend of unsorted, mostly fine-grained pyroclastics and high-temperature gas ejected from a fissure or crater. Also called a *pyroclastic flow*. The term *nuée ardente* refers to a large, swiftly moving ash flow ejected explosively and containing many incandescent fragments.

Asthenosphere: a layer in the upper mantle just below the lithosphere, in which seismic waves travel more slowly than in adjacent layers above and below. The asthenosphere is a zone of weakness where some magmas are generated.

Atmosphere of pressure: a unit of pressure equal to the normal pressure of the earth's atmosphere at sea level. It is equal to the pressure at the base of a column of mercury 76.0 cm (29.92 inches) high; equivalent to 14.66 pounds per square inch or 1.013 bars.

Atom: the ultimate particle of an element, consisting of a positively charged nucleus surrounded by negatively charged electrons.

Attitude: the position of a surface (bedding, fault, foliation, etc.) relative to the horizontal; expressed quantitatively by measurements of dip and strike.

Avalanche: a large mass of snow, ice, soil, or rock, or mixtures of these, falling or sliding very rapidly under the force of gravity.

Basalt: a dark-colored lava; more specifically, a mafic igneous rock composed chiefly of plagioclase and pyroxene in a fine-grained or glassy groundmass. It is commonly extrusive, but may occur in dikes as an intrusive.

Basin: a bowl-shaped depression in the land surface or in the sea bottom. A *structural basin* is a closed syncline or downfold in layered rocks. The *Great Basin* is the area of internal drainage in the western United States, including most of Nevada and parts of adjacent states.

Batholith: a mass of granitic rock with an exposed area of more than 40 square miles and with no visible floor.

Bedrock: a general term for the solid rock that underlies soil or other loose, unconsolidated material.

Biotite: black mica, a common silicate mineral containing potassium, aluminum, iron, and magnesium.

Blocks: in general, solid angular chunks of any size from pebbles to parts of continents. Specifically, solid chunks of material ejected from an erupting volcano.

Breccia: a coarse-grained rock consisting of angular and broken rock fragments cemented together in a fine-grained matrix; similar to a conglomerate except that most of the fragments have sharp edges and corners.

Breeder reactor: a nuclear reactor in which the "fuel," or fissionable substance, is plutonium that is generated by the reaction of fast neutrons with uranium-238. The reactor is said to "breed" its own fuel.

Calcite: a common mineral, usually colorless or light-colored, with prominent cleavage in three directions at oblique angles. Composition: $CaCO_3$. It is the principal constituent of limestone and marble and is a common mineral in veins.

Caldera: a large steep-walled basin or crater resulting from collapse of the top of a volcanic cone. The basin of Crater Lake is a good example.

Calcium carbonate: a chemical compound with the formula $CaCO_3$, the most abundant of the group of compounds called *carbonates.* In nature it occurs as the two minerals *calcite* and *aragonite.*

Caliche: a material composed of soluble calcium salts and impurities such as gravel, sand, clay, or silt, found as a layer on or near the surface in arid and semi-arid regions. The most abundant constituent is generally calcium carbonate.

Carbon dating: determination of rock ages in years by measuring the concentration of carbon-14 remaining in organic material. The method is useful only for material younger than 50,000 years.

Carbonate minerals: minerals consisting of the chemical compounds called *carbonates,* which contain a metal united with the atom-group CO_3. The two commonest carbonate minerals are *calcite,* $CaCO_3$, and *dolomite,* $CaMg(CO_3)_2$.

Catastrophism: a theory that the earth's crust has been greatly modified by short-lived, violent, usually worldwide events different in kind from events in ordinary experience. Also, a theory that changes in organisms during geologic time are explained by recurring catastrophes, followed by creation of different organisms.

Chert: a hard, very fine-grained sedimentary rock, consisting principally of microcrystalline quartz. It occurs commonly as nodules or concretions in limestones and dolomites, but may occur as beds. Agate, jasper, and flint are colored varieties.

Cinders: fragmental material ejected in a solid or nearly solid state, consisting of fresh lava erupted by a volcano and rapidly cooled, generally porous because of holes left by expanding gas. A *cinder cone* is a steep-sided hill or mountain built up from fragmental material ejected during a volcanic eruption.

Cirque: a steep-walled, flat-floored, half-bowl-shaped hollow, situated high on a mountainside and commonly at the head of a glacial valley, produced by the erosive activity of a mountain glacier.

Clay: used in two senses, to mean (1) a sedimentary material of any composition and with particles less than 0.004 mm in diameter, or (2) a fine-grained sedimentary material consisting chiefly of one or more of the clay minerals.

Clay minerals: light, soft silicates of aluminum and hydrogen. Some clay minerals contain also magnesium, iron, or potassium. *Kaolinite* and *montmorillonite* are common clay minerals.

Cleavage: the property of minerals to split along planes which are determined by particle arrangement in their crystal lattices. Mica flakes, for example, have perfect cleavage in one direction.

Coal: a combustible rock containing a high percentage of carbonaceous material originating as partly decayed plant remains compacted and hardened over a long period of time.

Columnar jointing: hexagonal or pentagonal parallel columns in basalt flows and sometimes in other extrusive and intrusive rocks, formed by contraction during cooling.

Conglomerate: a coarse-grained sedimentary rock whose coarser particles, of varying sizes and shapes, are set in a fine-grained matrix of silt or sand.

Continental rise: the area of ocean floor between the *continental slope* and the ocean deeps. In general it has smooth topography and a very gentle incline.

Continental shelf: the area of flat or gently sloping ocean floor along a continental margin, between the shoreline and the *continental slope.*

Continental slope: the area of ocean floor between the *continental shelf* and the *continental rise.* It has a steeper slope (from 3° to 6°) than either the shelf or rise.

Convection: movement in a fluid produced by unequal heating, in which material rises over a heated area and descends in cooler regions. Specifically, a supposed mass movement of mantle material, rising under mid-ocean ridges and descending near some continental margins.

Coral reef: a limestone reef or ridge chiefly composed of fragments of coral and other organisms, coral sands, and solid limestone, often built up around a framework of other organic remains, especially those of calcareous algae. Especially abundant along continents and islands in warm parts of the Pacific Ocean.

Core: the central part of the earth's interior, divided into an *inner,* probably solid core and an *outer* liquid core. The core has a radius of nearly 3,500 km, just over half the earth's radius.

Crater (volcanic): a bowl-shaped depression at the mouth of a volcano.

Cross-bedding: a pattern in a layer of stratified rock in which minor thin beds slope at various angles to the original depositional surface or to the contact of the layer. Cross-bedding is caused by changing currents of air and water.

Crust: the outermost layer or shell of the earth, usually defined as the part above the Mohorovičić discontinuity.

Crystal: a homogeneous, solid body having a regularly repeating arrangement of its particles which may be outwardly expressed by plane faces.

Crystalline solids: solids whose particles are arranged in regular geometric patterns, and which show flat crystal faces if they have grown without interference from neighboring grains. Nearly all minerals are crystalline solids.

Dacite: a fine-grained volcanic rock, similar to andesite but containing quartz. Dacite is approximately the extrusive equivalent of granodiorite.

Deep: a large depression of the ocean floor.

Delta: a flat, alluvial land area deposited at or near the mouth of a river, resulting from accumulation of sediment supplied by the river in such quantities that it is not removed by tides, waves, and currents.

Density: the mass per unit volume of a substance, usually expressed as grams per cubic centimeter.

Detrital material: fragmental material, such as gravel, sand, and clay, which has been worn from older rocks and removed from its place of origin, usually by water.

Diabase: an intrusive igneous rock, principally composed of plagioclase and pyroxene, with a grain size intermediate between basalt and gabbro.

Diatoms: single-celled, microscopic plants which grow in fresh or marine waters. Their skeletons are composed of silica and may accumulate in enormous numbers to form siliceous sediment. The sediment and the rock formed from it are called *diatomite.*

Differential erosion: the wearing down of land features at differing rates, softer rocks being worn down more rapidly than harder, more resistant rocks. The more resistant rocks are left as ridges or cliffs.

Dikes: igneous intrusions of sheet-like form that cut across existing rock layers.

Dinosaurs: a group of extinct reptiles that flourished during the Mesozoic Era. Some of the dinosaurs were the largest land animals in the earth's history.

Diorite: a plutonic rock composed of plagioclase and one or more ferromagnesian minerals. It is approximately the intrusive equivalent of *andesite.*

Dip: the angle that a structural surface, e.g. a bedding plane, makes with the horizontal, measured perpendicular to the *strike* of the surface.

Discontinuity: a surface in the earth's interior at which seismic-wave velocities change abruptly.

Dolomite: a mineral with the composition $CaMg(CO_3)_2$, similar to calcite in its properties but less reactive with acid. Also a rock consisting chiefly of this mineral.

Dome: a dome-shaped land form or rock mass which roughly resembles the dome of a building, e.g. the rounded granite peaks found in Yosemite, California. Also, an anticline-like structure in which beds dip away in all directions. Also, a dome-shaped mass of solidified lava, formed by extrusion of highly viscous lava. A *dome volcano* is a volcano consisting of a lava dome surrounded by pyroclastics.

Dune: a mound of loose granular material (generally sand but sometimes volcanic ash), which is moved by winds from place to place, always retaining its characteristic shape.

Earthquakes: sudden movements in the earth caused by release of strain along faults or by the movement of gases and molten rock underground during volcanic eruptions.

Electrons: tiny negative particles which, together with a positively charged nucleus, make up an atom.

Epicenter: the point on the earth's surface directly above the focus of an earthquake.

Epoch (geologic): a time unit which is a subdivision of a *period.*

Era (geologic): one of the major time units, comprising two or more *periods.*

Erosion: processes (e.g. erosion by streams, wind, waves, glaciers) by which rocks are worn away and their debris removed. The word may be used to include the slow decay or weathering of rocks, or it may be restricted to the active processes of wearing away and removal.

Eruption cloud: the cloud issuing from an erupting volcano, which contains volcanic gases, water droplets, volcanic ash, and generally larger solid fragments.

Estuary: a partially enclosed body of water where tidal water from the sea mingles with fresh water flowing down from a river valley.

Evolution: gradual developmental changes in life forms from simple organisms of the distant past to the complex ones of the present.

Extrusive rocks: igneous rocks that have been ejected onto the earth's surface during a volcanic eruption.

Fault: a surface of rock fracture, along which displacement has taken place. In a *dip-slip fault* the important movement is parallel to the dip of the fault; in a *strike-slip fault* the movement is parallel to the fault's strike. A *thrust fault* is an oblique fault in which the side overlying the fault plane has apparently moved up with respect to the other side. A *normal fault* is an oblique fault in which the side above the fault plane has apparently moved down with respect to the other side.

Fault-block mountains: mountains formed by large-scale normal faulting which divides an area into structural blocks of varying heights and directions.

Fault breccia: a coarse-grained rock composed of angular fragments resulting from crushing or shattering by fault movements.

Fault gouge: soft, clay-like uncemented pulverized material found in a fault zone, resulting from crushing during fault movement.

Fault scarp: see *scarp.*

Feldspar: the commonest of all minerals, making up 60% of the earth's crust. Actually feldspar is a group of minerals, all light-colored silicate minerals containing aluminum and one or more other metals. Two common feldspars are *orthoclase* (potassium feldspar) and *plagioclase* (sodium-calcium feldspar). Feldspar is an important mineral in most rocks.

Felsic rocks: igneous rocks composed chiefly of light-colored minerals, especially quartz and feldspar, hence with chemical compositions high in silica. Same as *acid rocks* and *silicic rocks.*

Ferromagnesian minerals: a large group of silicates of iron and magnesium, usually dark green to black in color, occurring as the most abundant dark-colored constituents of common rocks. Olivine, pyroxene, amphibole, and black mica are examples. *Mafic minerals* is an approximate synonym.

Field: as used by geologists, "the field" refers to a geological area currently under study, as opposed to related work carried on in the laboratory. In more special senses, an *electric field* is the region around an electric charge in which forces act on other charges; a *magnetic field* is the region around a magnet; and a *gravitational field* is the region around a large material object in which gravitational forces are perceptible.

Fission: splitting, by particle collision, of a heavy atomic nucleus into a pair of approximately equal-sized fragments plus some neutrons. The reaction releases an enormous amount of energy.

Floodplain: an area of relatively smooth land that borders a river and that may be covered by its waters during times of flood. The plain is built up of alluvium deposited by the river during flood stages.

Focus (of an earthquake): the point within the earth which is the initial point of rupture of an earthquake and the origin of its elastic waves.

Foliation: a planar arrangement of flat or elongated mineral grains in parallel layers to give a banded or layered appearance. A highly foliated rock always contains mineral grains that occur as flakes (e.g. mica) or as long needles (e.g. some ferromagnesian minerals).

Fool's gold: any mineral that superficially resembles gold. The commonest examples are pyrite and partly weathered biotite.

Foraminifera: one-celled animals or protozoa commonly found in marine to brackish environments. Some varieties of foraminifera which have shells of calcium carbonate make up one kind of deep-ocean ooze, and other varieties are important as fossils.

Fossil fuel: any deposit of organic material that can be used for fuel, e.g. coal, petroleum, natural gas.

Fossils: remaining parts, or traces of parts, of plants and animals preserved in rocks or sediments. A *fossil assemblage* is a group of fossils found together in a given stratum or locality.

Fumaroles: volcanic vents which emit hot gases of varying composition. They are characteristic of the late stage of a volcano's activity.

Fusion: the combination of two light atomic nuclei to form a heavier nucleus. Like *fission,* fusion releases great amounts of energy.

Gabbro: a dark-colored, intrusive igneous rock, composed principally of plagioclase and pyroxene, approximately the intrusive equivalent of *basalt.*

Galena: a heavy, fairly soft gray mineral (formula PbS) with metallic luster and excellent cleavage. It is the most important ore mineral of lead and also one of the important sources of silver.

Garnet: a brittle, hard, relatively transparent silicate mineral, with a vitreous luster and no cleavage, usually of a red or brown color. It is most commonly found in metamorphic rocks, such as gneisses and schists.

Geochronology: study of time in relation to the history of the earth.

Geologic time: the elapsed period, or any major part of this period, between the creation of the earth and the present.

Geothermal energy or **Geothermal power:** useful energy that can be obtained from naturally occurring hot water and steam found in volcanic zones, whose surface manifestations include hot springs and fumaroles.

Geothermal gradient: the rate at which the temperature of the earth's interior increases with depth.

Geysers: hot springs that intermittently erupt jets of steam and hot water. They are caused by groundwater coming into contact with underground heat sufficient to produce steam under conditions that inhibit free circulation.

Glaciers: large ice masses which survive over a number of years and which move slowly outward and downward as a result of their own enormous weight. The term is used for masses of all sizes from small mountain glaciers to continental ice sheets.

Glass: a noncrystalline material formed by the rapid cooling of a viscous liquid. Natural glasses of interest in geology are formed from silicate liquids with compositions like those of igneous rocks.

Global tectonics: tectonic processes related to very large-scale movement of parts of the crust or material within the earth. Often used as a synonym of *plate tectonics,* but properly a more general term.

Gneiss: a coarse-grained foliated and banded metamorphic rock, produced from many other rocks by conditions of high temperature and pressure, and commonly containing large amounts of feldspar, quartz, and mica.

Graded bedding: a type of bedding in which each layer displays progressive change in particle size. Most commonly the coarser grains are at the base of a layer and the fine grains at the top.

Granite: a quartz-bearing plutonic rock whose origin is still in dispute. Most granite is probably an igneous rock, but some may be the product of metamorphism and partial melting of older rocks.

Gravitational field: the region around a material object in which other objects experience an attractive force. Theoretically any piece of matter has a gravitational field, but practically a field is perceptible only near a large object like a planet or star.

Gravity: the attractive force which two or more bodies exert upon one another, and which is directly proportional to the product of their masses and inversely proportional to the square of the distance between their centers. Specifically, the attraction of the earth for a unit of mass at any point.

Gravity anomaly: the difference between the gravity actually measured at a given point and that which has been calculated theoretically for that point.

Great Basin: a 120,000 square mile area of interior drainage in the western part of the United States. It includes most of Nevada and portions of California, Idaho, Oregon, Utah, and Wyoming.

Half life of a radioactive element is the time required for the decay of one-half of the original quantity of the element.

Hard water: water that does not lather readily when used with ordinary soap and that forms a scale when it evaporates. Hardness is generally due to the presence of dissolved calcium and magnesium salts.

Hematite: a mineral consisting of ferric oxide, Fe_2O_3. It is commonly reddish or brownish and is always red when powdered. It is the principal ore of iron.

Hornfels: a hard, fine-grained, unfoliated metamorphic rock produced by contact metamorphism at the border of an igneous intrusion. The most common hornfelses are produced by metamorphism of shale.

Hydrocarbons: gaseous, liquid, or solid organic compounds, consisting solely of carbon and hydrogen. Crude oil is a complex mixture of hydrocarbons.

Ice Age: the most recent time in geologic history when glaciers were widespread; same as the *Pleistocene Epoch.*

Igneous rocks: rocks formed by the cooling of thick, viscous siliceous liquids resembling molten glass.

Inorganic: relating to material that contains no complex carbon compounds; relating to material that is not now, and has not been, part of living substance.

Intrusive or **Intrusion:** a mass of igneous rock formed from magma that was intruded into other rocks underground but that did not reach the surface as a liquid. Dikes and batholiths are examples of intrusives.

Ionic bond: a linkage between parts of a compound that depends on attraction between positive and negative ions. For example, the linkage between sodium and chlorine in NaCl is largely an ionic bond between Na^+ and Cl^- ions.

Ionic compound: a compound whose particles are ions rather than atoms or molecules. The ions are formed by transfer of electrons from one kind of atom to another. Most ionic compounds are crystalline, and most minerals are at least in part ionic compounds.

Ions: atoms or molecules which possess an electric charge. An atom or molecule gains a positive charge by losing one or more electrons; it gains a negative charge by attaching to itself electrons in excess of its normal complement.

Iron formation: low-grade sedimentary iron ore, typically thin-bedded or finely laminated, and commonly containing layers of chert. Most iron formation is Precambrian.

Island arcs: curving chains of islands which rise from the deep sea floor near continental margins. The Aleutians are a good example of an island arc.

Isostasy or **Isostatic balance:** the condition of equilibrium, comparable to floating, of large parts of the brittle crust resting on the plastic mantle. As a result of isostasy the weight per unit area of large parts of the crust is the same, which means that high-standing parts of the earth are underlain by more lightweight material than lower parts. When material is moved from one block to another, say by erosion or volcanic activity, isostatic balance is restored by movement in the plastic mantle beneath.

Isotopes: two or more species of the same chemical element. Atoms of isotopes have the same number of protons in their nuclei, but different numbers of neutrons, which means that their atomic weights are different.

Joints: cracks or fractures in rocks.

Kaolinite: the simplest of the clay minerals and the chief constituent of kaolin. Composition $H_4Al_2Si_2O_9$.

Karst: topography that is characterized by caves, sinkholes, and underground drainage. It is formed over limestone, dolomite, or gypsum by extensive dissolving of the rock.

L waves: surface waves; seismic waves that travel along the earth's surface, and reach distant points later than P waves and S waves.

Landslides: a general term for rapid or fairly rapid (at least a foot per year) downslope movement of masses of soil and rock; also, the land forms produced by such movement. Many kinds of landslide are recognized, depending on speed and the nature of the material, e.g. talus, debris slide, avalanche, mudflow.

Lava: molten rock after it has been expelled from a volcano (see *magma*). The term is also used for the rock after it has cooled and solidified.

Lava flow: the outpouring of molten material from the crater of a volcano or from fissures. The term also refers to the body of rock so produced after it has cooled and solidified.

Leaching: the selective removal or dissolving out of soluble constituents from a soil, ore body, or rock by the action of percolating water.

Levee (natural): a low ridge or embankment of sand and silt, built by a stream on its floodplain and along both banks of its channel. An *artificial levee* is an embankment built along a stream channel, often on top of the natural levee, to protect adjacent lands from inundation.

Lignite: a brownish-black coal, intermediate between peat and bituminous coal, with a low heat value per pound.

Limestone: a sedimentary carbonate rock consisting chiefly (more than 95%) of calcite or aragonite.

Limonite: a general field term for amorphous, dark- or yellowish-brown ferric oxides of uncertain identity. Generally a major constituent is goethite, FeOOH. Limonite occurs commonly as a secondary material formed when iron oxidizes, and is a minor ore of iron.

Lithification: formation of solid rock from newly deposited unconsolidated sediment by a variety of processes such as cementation, compaction, recrystallization, etc.

Lithosphere: the outer part of the solid earth, above the asthenosphere, including the crust and uppermost part of the mantle. Also used in a general sense to mean the solid part of the earth, as opposed to the atmosphere and hydrosphere.

Longitudinal waves: (see P waves.)

Longshore currents: movement of water roughly parallel to the shore of a lake or the ocean, produced when waves approach the shore at an angle.

Mafic rocks: igneous rocks composed chiefly of dark-colored ferromagnesian minerals, hence with chemical compositions rich in iron and magnesium. Same as *basic rocks.*

Magma: molten rock underground. When it comes to the surface of the earth, magma is called *lava.*

Magnetic field: a region in which magnetized bodies or electric currents are affected by magnetic forces.

Magnetite: a heavy, black, strongly magnetic mineral of iron (Fe_3O_4). An important ore of iron.

Magnetometer: an instrument that measures the strength and direction of magnetic fields of rocks, or of the earth.

Magnitude of earthquakes: a measure of the strength of earthquakes, or the strain energy released by earthquakes, as determined from seismographic records.

Mammals: warm-blooded vertebrates who suckle their young.

Mantle: the part of the earth above the core and below the crust, with a thickness of about 2,900 km.

Marble: a metamorphic rock consisting principally of recrystallized calcite and/or dolomite.

Mass spectrometry: the process of measuring the masses of atoms or molecules by giving them electric charges and causing them to move through electric and magnetic fields, thus separating them into a mass spectrum.

Matrix: finer-grained material which fills the interstices between the larger grains and fragments of igneous and sedimentary rocks.

Meanders: a series of fairly regular, sinuous loops and winding curves characteristic of the channel of a stream on a floodplain.

Metamorphic rocks: rocks produced from older rocks by the action of heat and pressure below the surface. During metamorphic changes the rock remains entirely or in large part solid.

Meteor: a solid object from space which, entering the earth's atmosphere, becomes incandescent from friction and burns up before any part of it can reach the earth's surface. Objects that do reach the earth before being vaporized are called *meteorites.*

Methane: the simplest hydrocarbon, CH_4, a colorless, odorless, inflammable gas. It is the principal constituent of natural gas.

Mica: any of a group of silicate minerals characterized by perfect cleavage in one direction, which allows thin plates of the mineral to be split off. The commonest micas are *muscovite* (white mica) and *biotite* (black mica).

Mineral: a naturally formed, homogeneous material with a definite or fairly definite chemical composition and definite or fairly definite physical properties.

Moho or **Mohorovičić discontinuity:** the boundary surface or sharp seismic-velocity discontinuity that separates the earth's crust from the mantle. Its depth ranges from 5–10 km beneath the ocean floor to about 35 km below the continents and as much as 70 km under some mountain ranges.

Molecules: the smallest particles of elements or compounds, consisting of one or more atoms. In ionic solids, however, no separate molecules exist, because ions of opposite charge are units in a structure extending indefinitely in three dimensions. Thus in sodium chloride there are no NaCl molecules, but Na^+ and Cl^- ions in a 3-dimensional array.

Moraine: piles of unsorted, unstratified debris around the end and along the sides of a glacier. As the glacier melts back, these are left as low hummocky ridges.

Natural gas: any of the gaseous hydrocarbons, or mixtures of hydrocarbons, generated below the earth's surface. Methane is generally the major constituent.

Neutrons: tiny uncharged particles which have approximately the same mass as *protons* and which are constituents of all atomic nuclei except those of hydrogen. A free neutron is unstable and decays radioactively into a proton and an electron.

Nuclear reactors: devices for producing controlled power from nuclear fission.

Nucleus: a tiny, positively charged particle, around which negatively charged electrons are patterned to form an atom. The hydrogen nucleus consists of a single proton; all other nuclei contain protons and neutrons.

Nuée ardente: a turbulent, swiftly moving, gaseous cloud emitted from a volcano and containing ash and other pyroclastics in its lower part. It is sometimes incandescent, hence the name "glowing cloud."

Obsidian: a glassy volcanic rock, usually dark-colored or black, characterized by smoothly rounded surfaces when fractured (conchoidal fracture). Most obsidian has the chemical composition of rhyolite.

Offshore bar: a low, elongate sand ridge, built chiefly by wave action, located at some distance from the shore and generally parallel to it.

Oil shale: brown or black shale from which liquid or gaseous hydrocarbons can be distilled.

Olivine: a common greenish or brownish rock-forming mineral, usually found in gabbros, basalts, and peridotites. Composition: $(Mg,Fe)_2SiO_4$.

Oozes: sediments consisting of at least 30% skeletal remains of open-ocean marine organisms, the rest being clay minerals.

Ores: materials in which metals occur naturally and from which such metals can be profitably extracted.

Organic: relating to material containing carbon compounds, usually with hydrogen bonded to the carbon atom. Most such material is now or has been part of living organisms.

Orogeny: the process by which structures in mountainous areas were formed, including folding and faulting in the higher layers and metamorphism and plutonic intrusions in the deeper layers. *Orogenic belts* or *orogenic zones* are elongate regions in which orogeny has taken place or is taking place.

Orthoclase: one of the common varieties of feldspar. A hard, light-colored silicate mineral with two good cleavages at right angles. Composition: $KAlSi_3O_8$.

Outcrop: an exposure of bedrock at the earth's surface.

Oxide: a compound containing only oxygen and another element. Hematite, Fe_2O_3, and cassiterite, SnO_2, are examples of oxides that occur as minerals.

P waves: longitudinal waves produced by an earthquake; waves in which particle motion is back and forth in the direction of propagation. P waves are the fastest earthquake waves, the ones recorded first on a seismogram. Also called *primary waves, pressure waves, push-pull waves.*

Parabola: the curve followed by a stone thrown upward at an angle. Mathematically it is the curve described by the equation $y = ax^2$.

Pedology: soil science.

Pegmatite: a very coarse-grained (most grains more than 1 cm in diameter) igneous rock, generally occurring as irregular dikes or lenses, especially near the margins of batholiths. Most pegmatites are mixtures of quartz and feldspar, similar in composition to granite.

Perched water table: the surface of saturation of a body of groundwater separated from an underlying main body of groundwater by an unsaturated zone.

Peridotite: a coarse-grained plutonic rock, composed chiefly of olivine with or without other mafic minerals and containing little or no feldspar.

Period (geologic): the fundamental unit of the geological time scale, a subdivision of an *era.*

Permeability: capacity of soil, sediment, or porous rock to transmit a fluid.

Phenocrysts: in an igneous rock with grains of two distinctly different sizes (a *porphyritic* rock), the larger grains are called phenocrysts. Many volcanic rocks and some granites have conspicuous phenocrysts.

Photosynthesis: the reaction by means of which plants manufacture carbohydrates out of water and carbon dioxide by absorbing the sun's energy through the green parts (chlorophyll) of their leaves. Another product of the reaction is free oxygen.

Piedmont: land which lies or has been formed at the base of a mountain. The "Piedmont" of the eastern United States is the area of gentle slopes between the Appalachian Mountains and the coastal plain.

Pillow lava: lava characterized by discontinuous pillow-shaped masses ranging from a few centimeters to a meter or more in diameter. The lava is generally basalt or andesite, and was extruded under water or flowed into water.

Placer deposit or **Placer:** a surface mineral deposit formed from weathered debris by mechanical concentration of mineral particles. The mechanical agent is usually a stream, but some placers are formed by the action of wind or waves. The most common placers are those of the heavy metal gold.

Plagioclase: one of the common varieties of feldspar. A hard, light-colored silicate mineral with two good cleavages approximately at right angles. Composition: a mixture of $NaAlSi_3O_8$ and $CaAl_2Si_2O_8$.

Planets: non-luminous celestial bodies traveling in elliptical orbits around a central sun. The known planets of the solar system are Mercury, Venus, Earth, Mars, Jupiter, Saturn, Uranus, Neptune, and Pluto.

Plasticity: the property of a solid to change its shape permanently without rupture in response to a force, the change once begun being continuous without increase in force. Thus clay yields permanently to a force, and changes its shape continuously as long as the force acts.

Plate: one of the large, solid, nearly rigid blocks of material, comprising both the earth's crust and a part of the upper mantle, which in the theory of plate tectonics move about independently. The continents form parts of the plates and move with them.

Plate tectonics: a geologic theory in which the outer part of the earth is regarded as composed of a small number of large, thick, rigid plates which float on a viscous underlayer in the mantle and move about independently of one another.

Playa lakes: shallow, intermittent lakes of arid or semi-arid regions. Drying up in summer, they leave a flat, vegetation-free area known as a *playa.*

Pleistocene: the glacial epoch or Ice Age, occurring after the Pliocene and before the Holocene (or Recent).

Plug (volcanic): a vertical, pipelike body of igneous rock that represents the conduit to a former volcano.

Plutonic: refers to igneous rocks which are formed under the earth's surface by crystallization of magma. Same as *intrusive.*

Porphyritic: said of the texture of an igneous rock in which larger crystals (phenocrysts) are set in a finer groundmass which may be crystalline or glassy or both. A rock with this characteristic is often called a *porphyry.*

Potash: potassium carbonate, K_2CO_3. The term is also used for other compounds containing potassium, e.g. potassium chloride, potassium oxide, etc. Potash is used chiefly as fertilizer.

Precipitate: a solid that forms as the result of a chemical reaction taking place in a solution.

Pressure wave: (see *P wave.*)

Proterozoic: a name sometimes used for the more recent part of Precambrian time (see Archeozoic).

Pumice: a light-colored volcanic rock consisting chiefly of glass blown full of small holes by gas escaping as the lava cooled. It is sometimes so light that it will float on water.

Pyrite: a common mineral, FeS_2. Its metallic luster, brassy-yellow color, and lack of cleavage have led to its being mistaken for gold, hence the name "fool's gold" commonly used.

Pyroclastic material: solid fragments formed by volcanic explosions, including material of all sizes from fine-grained ash to large blocks and bombs.

Pyroxene: any of a group of dark silicate minerals which are common constituents of igneous rocks.

Quartz: a crystalline form of silicon dioxide (SiO_2), the commonest rock-forming mineral next to feldspar. Very hard, no cleavage. Abundant as a constituent of many rocks; when well crystallized, forms colorless or light-colored hexagonal prisms and pyramids ("rock crystal").

Quartzite: a metamorphic rock consisting mainly of quartz and formed by recrystallization of sandstone or chert during either regional or contact metamorphism. Also used for an unmetamorphosed sandstone consisting chiefly of quartz grains so completely cemented by silica that the rock breaks across or through the grains rather than around them.

Radioactivity: spontaneous decay of the atoms of certain isotopes to form new isotopes. Radioactive decay is accompanied by the emission of helium ions (alpha particles) or electrons (beta particles) and gamma rays, and by the generation of heat.

Radiolaria: one-celled animals (protozoa) characterized by a siliceous skeleton and by their habitat in an open-ocean environment.

Radiometric age: an age expressed in years and calculated from amounts of radioactive elements and their decay products. Also called *absolute age.*

Recrystallization: formation of new mineral crystals in an essentially already-formed rock. Recrystallization takes place under physical and chemical conditions imposed at depth and which differ from the conditions under which the rock was originally formed.

Remote sensing: measurement of properties by a recording device that is not in contact with the object or occurrence under study. Various devices are so employed, including cameras, lasers, magnetometers, radar systems, and the like.

Rhyolite: a light-colored volcanic rock containing quartz and potassium feldspar, generally with a partly glassy groundmass. Rhyolite is the extrusive equivalent of granite.

Rift valley and **Rift zone:** the deep, central cleft in the crest of a mid-oceanic ridge. More generally, any system of fractures in the crust; often used for an elongate series of clefts in a volcanic area, and for major zones of recurrent faulting. For the latter, the San Andreas rift valley and the East African rift valleys are good examples.

Runoff (water): precipitation which is not absorbed by the soil, usually appearing as temporary surface streams or sheetwash.

S waves: transverse waves produced by an earthquake; waves in which particle motion is at right angles to the direction of propagation. S waves travel more slowly than P waves, hence are recorded later on a seismogram. Unlike P waves, they cannot travel through liquids or gases. Also called *secondary waves, shear waves,* and *shake waves.*

Sandbars: low ridges built up by currents or wave action in river channels, lakes, or shallow sea water.

Sandstone: a sedimentary rock composed chiefly of grains ranging from 0.1 to 2 mm diameter, generally with a silt or clay matrix and cemented by silica, iron oxide, or calcium carbonate.

Satellites: relatively small secondary bodies which move in orbits around larger primary bodies. For example, the moon is a satellite of Earth, Ganymede is a satellite of Jupiter, etc. The term is also used for man-made objects propelled into orbit around the earth.

Saturation: a word used in many senses. With reference to groundwater, saturation refers to a condition in which the interstices (open spaces between the grains) of a rock or soil are completely filled with water.

Savanna: an open, grassy, level plain, devoid of trees, often found in sub-tropical and tropical regions. The term is also used for the marshy alluvial flats along the southeast Atlantic coast of the United States.

Scarp: a relatively straight, steep cliffline produced by faulting or erosion.

Schist: a spangled-appearing foliated metamorphic rock which readily splits into thin flakes or slabs because of the well developed parallelism of at least half of its minerals.

Scoria: cindery, often reddish crust on the surface of basaltic or andesitic lava, with many cavities formed by the escape of gases as the lava solidified. Heavier and darker than pumice.

Seamounts: flat-topped or peaked elevations of the sea floor, 1000 meters or higher.

Sea stacks: small, pillar-like, isolated rocky masses occurring near a marine cliff and detached from a headland principally by wave erosion.

Sedimentary rocks are formed by the consolidation of loose, layered sediments, formed as detrital material by erosion or as chemical or organic precipitates.

Seismic waves: used generally to refer to all waves produced by earthquakes or generated artificially by explosions.

Seismogram: a record obtained on a seismograph of the vibrations of the earth, especially those produced by earthquakes.

Seismograph: an instrument that records the vibrations produced by earthquakes and other vibrations of the earth.

Shale: a fine-grained sedimentary rock resulting from the consolidation of silt, mud, or clay. It is commonly laid down in thin parallel layers, which readily separate into slabs.

Shear wave: (see S wave.)

Sheetwash: water from heavy rain that flows down a slope as a more or less continuous sheet rather than concentrated into rivulets.

Shield volcano: a volcano with the shape of a broad, flattened dome, built by highly fluid basaltic lava flows. Shield volcanoes are typical of Hawaii.

Silica (silicon dioxide, SiO_2): a hard, insoluble solid with a high melting point, occurring naturally as the mineral quartz and several less common minerals. It is the chief constituent of opal, sand, and chert.

Silicates: compounds containing metals, silicon, and oxygen. Sodium silicate, Na_2SiO_3, is a simple example. Silicate minerals are the most abundant of all minerals.

Sill: a tabular igneous intrusion that parallels the bedding or other prominent structure of the surrounding rock.

Silt: sediment with particle sizes between fine sand and clay; diameters in the approximate range 0.004–0.1 mm. *Siltstone* is a rock made up of particles dominantly in this size range; like shale but gnerally less finely laminated.

Slate: a fine-grained, compact metamorphic rock formed from shale or volcanic ash, finely foliated so that it splits readily into thin plates.

Slickensides: smoothly striated, polished surfaces resulting from friction along a fault plane.

Slopewash: soil or rock fragments which have been moved down a slope by unchanneled running water. The term may also refer to the process of moving slopewash material.

Soft water: water that lathers readily with ordinary soap, because it contains only a small amount of soluble calcium and magnesium salts.

Soil: material consisting of weathered rock and organic matter, capable of supporting the growth of land plants. In engineering practice the term is used for any unconsolidated material overlying bedrock.

Soil science: study of the properties, formation, and classification of soils. Same as *pedology*.

Sole marks: small convex irregularities found on the original underside of a sandstone or siltstone bed, formed by filling of a primary sedimentary structure (crack, groove, or other depression) on the upper surface of the underlying bed.

Sorting: natural selection and separation of sedimentary particles of some particular characteristic, such as size, shape, or density. It is commonly brought about by the action of running water, waves, or wind.

Specific gravity: the density of a substance relative to the density of water. An acid with a specific gravity of 1.26, therefore, has a slightly greater density than water. In the metric system density and specific gravity are numerically almost identical.

Speleology: exploration of caves and the scientific study of their genesis, morphology, and mineral deposits.

Springs: places where groundwater flows naturally from soil or rock onto the land surface or into a body of surface water.

Stalactites: conical or icicle-like formations which grow downward from the ceilings of caves. They are deposited by dripping water, and are usually composed of calcium carbonate.

Stalagmites: formations similar to stalactites but which grow upwards from cave floors.

Stratification: the formation or deposition of sedimentary material in layers, or the layered structure so produced. The *strata* may be distinguished by differences in internal structure, mineral composition, texture, color, etc.

Stream erosion: the wearing away and removal of soil, bedrock, or other exposed material from its channel by a stream.

Stream gradient: the slope of a stream, measured in the direction of the flow. More precisely, it is the angle between the water surface and the horizontal.

Striae or **Striations:** scratches or small furrows on rock surfaces, usually roughly parallel to one another. They are commonly caused by glaciers, but may also result from faulting or landslides.

Strike: the direction that a structural surface, e.g. a bedding plane or fault plane, shows as it intersects the horizontal. The strike is the compass direction of the line of intersection, e.g. N30°W or N85°E. To describe the orientation of the surface completely requires also the *dip,* which is the angle it makes with the horizontal as measured perpendicular to the strike.

Strip-mining: surface mining in which the ore is exposed by removal of the overlying material. Also known as *opencut mining* or *openpit mining.*

Structural dome: an uplift or anticlinal structure in sedimentary rocks, circular or elliptical in outline, in which beds dip gently away in all directions. A *structural basin* is a similar synclinal structure.

Subduction zone: an elongate region along which a crustal block descends relative to another crustal block.

Syncline: a concave-upward fold whose core contains the stratigraphically younger rocks.

Talus: angular rock fragments of varying sizes which have broken off from and are lying near the base of a steep slope or cliff. A *talus slope* is the steep, concave-upward slope formed by the accumulation of talus.

Tectonics: study of the structure and deformation of the upper part of the earth's crust. Tectonics usually deals with large-scale geologic features, while *structural geology* deals with smaller-scale features, but the distinction is not sharp.

Terrace: a long, narrow, level or gently inclined surface bounded on one side by a steeper ascending slope and on the other by a steeper descending slope; a large bench or step-like ledge breaking the continuity of a slope. Generally along the margin and above the level of a body of water, marking a former water level; for example, a *stream terrace* or a *marine terrace.*

Tidal wave: an inaccurate term, often used popularly for a *tsunami* or for a large storm wave.

Till: a heterogeneous mixture of clay, sand, gravel, and boulders of various sizes and shapes, unsorted and unstratified, deposited directly by and underneath a glacier without subsequent reworking by water.

Topography: the physical features of any part of the earth's surface, especially land, and including man-made features as well as its natural relief.

Transform fault: a strike-slip fault which characteristically occurs along mid-ocean ridges and along which the ridges are offset.

Transverse waves: (see S waves.)

Traveltime curve: a graph showing the time of travel of seismic waves from the focus plotted against distance along the earth's surface. Also called *time-distance curve.*

Trenches: long, narrow depressions in the sea floor, parallel to the trend of a nearby continent or island arc, and situated between the continental margin and the deep part of the ocean. They are about 2 km deeper than the surrounding ocean floor and may be thousands of kilometers in length.

Tsunami: a large sea wave produced by a disturbance of the ocean floor, such as a shallow submarine earthquake. Volcanic eruption, subsidence, or other submarine activity may also cause tsunamis. Also called *seismic sea waves;* popularly but wrongly called *tidal waves.*

Tuff: volcanic ash and dust which have been consolidated. Besides pyroclastic material, tuff may contain sediments such as clay or sand.

Turbidity current: a bottom-flowing current laden with sediment kept in suspension by extreme turbulence, moving swiftly down a subaqueous slope under the influence of gravity.

Ultramafic rocks: igneous rocks composed chiefly of ferromagnesian minerals, with little or no feldspar.

Unconformity: a substantial break or gap in the geologic record where one rock unit is overlain by another which is not stratigraphically related to the former. An unconformity results from changes that cause deposition to cease for a considerable time; it usually implies uplift and erosion, with loss of part of the geologic record.

Vein: a mineral filling of a fracture in rock, in tabular or sheet-like form. The commonest vein minerals are quartz and calcite.

Vertebrates: animals which have a backbone or spinal column.

Vesicle: a cavity of variable shape in a lava, formed by entrapment of a gas bubble during solidification of the lava. A lava with abundant vesicles is said to be *vesicular*. *Pumice* is a light-colored felsic lava with many small vesicles. *Scoria* is a vesicular, cindery, often reddish crust on the surface of andesitic or basaltic lava.

Viscosity: the degree to which a substance offers internal resistance to flow. For example, molten glass has a much higher viscosity than water.

Volcanic ash: fine unconsolidated particles of pyroclastic material. The term is sometimes used as a synonym for tuff, which is a consolidated ash.

Volcanic bombs: pieces of hot, viscous material of varying sizes forcibly projected upward and outward by an erupting volcano. Bombs receive their rounded, often elongated shapes while in flight.

Volcano: an opening in the earth's crust through which molten rock (magma), associated gases, and ash are expelled. The term may also refer to the cone-shaped hill or mountain that marks previous volcanic activity.

Water table: after a heavy rain, water drains out of the upper part of a soil layer, but the lower part usually remains saturated for a considerable time. The fluctuating surface of the saturated part of the ground is called the water table.

Water witch: a forked stick allegedly capable of detecting the presence of water underground; also, a person who uses such a stick. Such a person is often called a *dowser*. Water witch is also used to describe instruments, especially electrical, which respond to the presence of water.

Weathering: disintegration of rocks and accompanying changes in their color, texture, composition, firmness, or form by exposure to atmospheric agents, with little or no transport of the loosened or altered material.

Welded tuff: a pyroclastic rock that has been hardened by the action of its own heat and that of accompanying hot gases and by compression by overlying material.

Questions for Chapters

Chapter One

1. Near the base of a sea cliff, caves are often found extending a few meters or tens of meters into the cliff. Other caves are sometimes seen extending completely through a rocky nose or through a sea stack, forming "natural bridges." How are such caves formed?
2. Tourists, gawking at a spectacular sea cliff, often exclaim something like: "Wow—there must have been a big earthquake here to make a cliff like this!" What evidence could you suggest to show that this hypothesis is wrong (or at least unnecessary)?
3. Suppose that you could make a vertical cut through a sea cliff perpendicular to the shoreline, and that you then look at the face of the cut. Make a sketch to show what you would see, especially the slope of the land surface both above and below sea level. (Geologists call this a "cross-section" or "profile.") Assume that the cliff face is about 100 feet high, and that the land surface 1000 feet back from the cliff has an altitude of 200 feet above sea level; and assume that the depth of sea water 1000 feet offshore is 100 feet. Show positions of the beach and a sea stack or two, and indicate where sediment is being deposited.
4. If erosion is causing the cliff to retreat landward at an average rate of an inch per year, show on your sketch what the profile of the land surface looked like 1000 years ago, and what it will look like 1000 years in the future.
5. Suppose that a succession of earthquakes raises the land surface by about 100 feet in a fairly short time, so that land now beneath the sea will be exposed to the air. Predict what will happen, and what the landscape will look like after erosion has been at work on it for a few hundred years. Can you see any evidence that this sequence of events might have occurred in the area shown in Fig. 1-6?

Chapter Two

1. If you were seeking answers to the following questions, what sort of scientist would you approach for expert opinion—geologist, chemist, physicist, biologist, geochemist, geophysicist, paleontologist, hydrologist, astronomer, oceanographer, soil scientist? (For some of the questions the answer will not be clear-cut, because specialists of more than one kind could help you.)
   a) Why does the earth have a magnetic field?

b) You find a rock containing a fossil bone; from what sort of animal did the bone come?
c) What substances are present in a sample of volcanic gas?
d) Are there volcanoes on the moon?
e) You have bought some land and want to drill a well; how can you decide where to drill?
f) Are the Rocky Mountains older than the Appalachian Mountains?
g) Why are soils in the tropics often red?
h) Why are there more earthquakes in California than in Mississippi?
i) Why do some volcanoes have the shape of an almost perfect cone?

2. Scientists are often accused of having such narrow, specialized interests that they cannot describe their work intelligibly to a layman. Do you think that scientists should try to change this image? What advantage is there to a scientist in narrow specialization? What disadvantage?
3. Geologists working around an erupting volcano are often caught in sudden heavy downpours, and the rain is often noticeably acidic. How would you explain these observations?
4. What do you imagine the inside of an erupting volcano is like? Sketch a cross-section through the middle of a volcano to illustrate your ideas.
5. Craters at the tops of volcanoes have many different shapes—some broad and shallow, some deep and steep-sided. Suggest possible reasons for the various shapes.
6. Suppose a volcanic eruption occurs on the sea bottom instead of on land. Would you expect there to be much difference in the nature of the eruption? In the kind of hill or mountain formed?

Chapter Three

1. Fig. 3-25 is a sketchmap of a river flowing across a floodplain. Draw a cross-section along the line ABCD, showing the river channel and the sediments beneath. Indicate on the section the position of the natural levees. Would you expect the deepest part of the channel to be closer to C or to B? Why? In what part of your cross-section would you expect the floodplain sediments to be most coarse-grained? Why?
2. Over a long period of time, how would you expect the path of the river in Fig. 3-25 to change? To answer this question, consider the effect of the river on its banks at points B and C. In time of high water, when the river is moving rapidly, at which of these points would the bank be eroded most rapidly? Look also at E and F: would one be eroded faster than the other? If you project these different rates of erosion into the future, how will the river-course be altered?
3. A river on a floodplain commonly has a very sinuous course, made up of sweeping curves first in one direction and then in the other (Fig. 3-12). The curves are called *meanders*, and the river is said to be a *meandering stream.* From your answer to Question 2, can you explain why meanders develop?

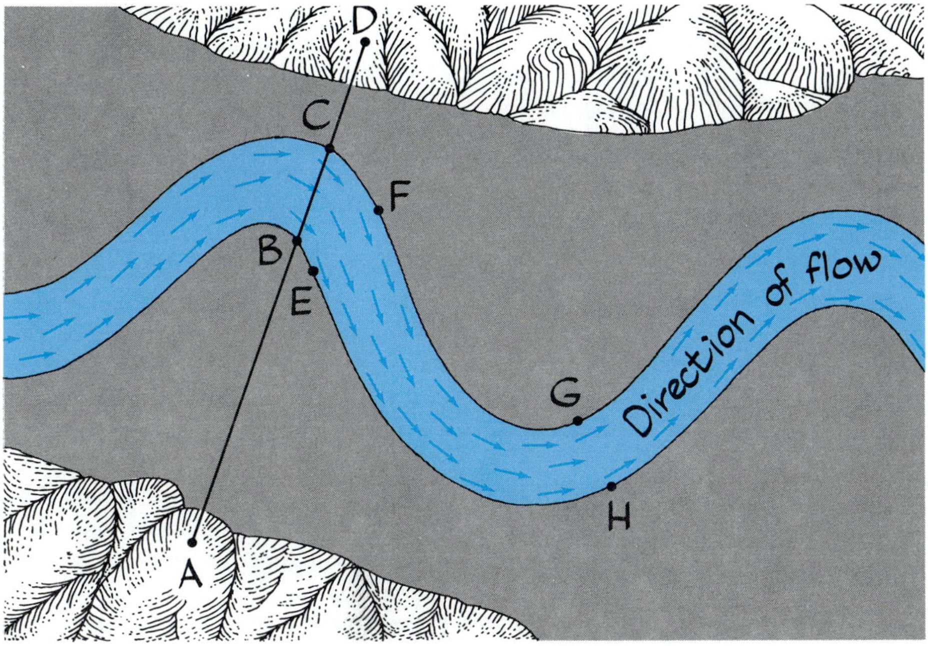

Fig. 3-25.

4. On some floodplains, especially where a river is carrying a great deal of sediment, a different pattern develops in which the river breaks up into a network of intertwined channels (Fig. 3-8). Such a river is called a *braided stream.* Why should this pattern develop rather than meanders?
5. The valley of the Nile in Egypt is a narrow floodplain flanked by desert hills on either side. The river, flowing from sources in the high mountains of equatorial Africa, used to rise out of its channel each year (before the Aswan Dam stopped the flooding) and cover most of the plain with shallow water. Does this suggest a reason why the plain should still be extraordinarily fertile, despite intensive cultivation for several thousand years?
6. The Yellow River (Huang Ho) of northern China is another river with a source in distant mountains that flows across a thickly populated plain (or series of plains, separated by hilly areas). The rainfall in the source area is less uniform than in Africa, so that occasionally the plains are inundated by disastrous floods that have earned the river the name of "China's sorrow." In an effort to prevent flooding, the inhabitants have built artificial levees or dikes on top of the natural levees. The river has deposited sediment in its channel, raising its bed and increasing the flood danger, so that the inhabitants have had to make the levees still higher. Over thousands of years this sequence of events has raised the river to levels that in places are several tens of meters above the surrounding plain. Most of the time the river is contained in its

channel, but when floods do occur they are more disastrous than they would be in a natural state. Explain the behavior of the river, especially its tendency to raise its bed above the surrounding level.

7. Experience with the Huang Ho suggests that building artificial levees is not an effective long-time protection against flooding. Can you suggest other possible ways to reduce the danger of flooding from large rivers meandering across a floodplain—for example, in the plains along the lower course of the Mississippi?
8. The Black Hills of South Dakota are an isolated group of hills rising above the high plains of the western Dakotas and eastern Wyoming. Suppose someone asks you why the hills are here. You cannot give a complete answer at this point, because we have not looked into the geology of the area. But perhaps you can analyze the question: What does it really mean? What sort of answers are possible? What kind of information would you need to work out the answers?
9. Volcanic cinder cones consist of loose, light fragmental material that would be easily carried by running water, yet many fresh-appearing cinder cones are more than a million years old. Why should these structures be so apparently resistant to erosion?
10. In some places in the western states flat-topped ridges (often called "table mountains") have abundant lava at their tops. Can you suggest an explanation for the flat tops of these ridges?
11. Alluvial fans are more conspicuous in the western part of the United States than in the east. Suggest an cxplanation.
12. Alluvial fans are deposits made by streams carrying abundant debris, at points where the streams are slowed as they emerge from mountains onto flat plains. Similar deposits called *deltas* form where a stream has its motion suddenly slowed by entering a lake or a quiet part of the ocean. The two kinds of deposit consist of similar materials and form for the same basic reason—the loss of carrying capacity of running water because its speed is reduced. In what ways would you expect deltas and alluvial fans to be similar, and in what ways different?
13. Draw a cross-section through two or more Appalachian ridges, at right angles to the ridge crests, to show the relation of the topography to the underlying rock layers.
14. Both in the Appalachian Mountains and in eastern Nevada many ridges have straight, steep fronts. In Nevada a common explanation is upward movement of a ridge along a fault, in a succession of earthquakes over a long period of time. An entirely different explanation is used for the Appalachian ridges. Why?
15. Chesapeake Bay is a large estuary extending more than a hundred miles inland from the Atlantic Ocean, with a large river (the Susquehanna) flowing into its upper end. Is it reasonable to suppose that the estuary has been formed by erosion by the river? Why or why not? Could the estuary have been formed

by erosion by ocean waves? Are there other possible explanations for the estuary?

Chapter Four

1. Suppose that you find an outcrop of dark-colored, fine-grained rock. It could be any one of several rock types: basalt, black shale, black limestone, or dark-colored chert (flint). What tests would you use to decide which of these rocks it is?
2. A friend asks you to identify a specimen of light-colored rock in which you can see with your handlens abundant grains of quartz and feldspar. On what basis can you decide whether the rock is granite or sandstone? Sometimes this distinction is troublesome, especially if the granite has been weathered. If you are uncertain, you could ask your friend to take you to the outcrop where he found the specimen. What would you look for in the outcrop to help in the identification?
3. Lava often has easily visible crystals scattered through a fine-grained groundmass (Fig. 4-25). You can see nothing in the groundmass with a handlens, but in a thin slice of the rock under a high-powered microscope you can observe many tiny intergrowing crystals. A rock of this sort, with crystals of two conspicuously different sizes, is called a *porphyry*, or is said to have *porphyritic texture;* the bigger crystals are *phenocrysts*, and the smaller ones make up the

Fig. 4-25. A porphyritic lava in hand specimen and in thin section. The large white crystals (phenocrysts) in each are plagioclase feldspar. (Joachim Hampel and R.S. Fiske, U.S. Geological Survey)

groundmass. Remembering that big crystals take a longer time to grow than small ones, can you suggest an explanation for the porphyritic texture?

4. In a specimen of the light-colored lava called rhyolite you can generally distinguish three kinds of phenocrysts: clear glassy ones with broken surfaces and rounded or irregular shapes, glassy or chalky ones with straight edges and flat surfaces, and scarcer very shiny black ones. What are these minerals?
5. Specimens of quartz often have flat surfaces. How could you demonstrate that these are not cleavage surfaces? What is the explanation for the flat surfaces?
6. Table salt and table sugar are properly called "crystalline solids," yet the individual grains are not separate, complete crystals. In what sense are these substances crystalline solids? How could you demonstrate that they are in fact crystalline? Can you suggest a familiar transparent solid substance that is not crystalline?
7. In outcrops you often see white streaks cutting the rock, the streaks consisting of material evidently filling cracks and probably deposited by water that has percolated through the rock. Such material is called a *vein* (or *veinlet* if it is very small). The most common materials found in veins are quartz and calcite. Suppose you are studying an outcrop and you have no acid with you. How could you tell whether veins in the outcrop consist of one of these minerals or the other?
8. What is the meaning of the following statements?
   a) Conglomerate is a detrital sedimentary rock.
   b) Chert is a sedimentary rock formed from a chemical precipitate.
   c) Andesite is an extrusive igneous rock.
   d) Diorite is an intrusive igneous rock.
   e) Limestone is formed by recrystallization of a limy sediment.
   f) Pyroxene has a higher density than kaolinite.
   g) Biotite is an example of a silicate mineral.
   h) Most minerals are ionic compounds.
9. Is it correct to say that chert is always harder than limestone? That granite is always harder than sandstone? Explain.
10. Distinguish between (a) atoms and ions, (b) electrons and ions, (c) crystal and cleavage fragment, (d) oxide minerals and silicate minerals, (e) plutonic rocks and volcanic rocks.

Chapter Five

1. Suppose that a turbulent stream is carrying fragments of different sizes, ranging from boulders through pebbles and sand grains to particles of silt and mud. Would you expect the process of rounding of the fragments (by collisions with each other and with bedrock in the stream channel) to be more effective for large fragments or for small? Would you expect to find some sizes of fragment that have not been affected by rounding at all? If you have some specimens of conglomerate at hand, you can probably check your conclusion by looking at the shapes of pebbles and sand grains.

2. Would you expect rounding of sand grains to be more effective or less effective in wind currents than in water currents of the same average speed? Would you expect sorting of clay particles from sand grains to be more effectively accomplished by wind or by moving water? If you are in doubt about your answers, comparison of the Coconino Sandstone (wind-deposited sand) and Tapeats Sandstone (marine sandstone) from the Grand Canyon would help (using either actual specimens or published descriptions).
3. A valley glacier is a tongue of ice moving slowly downhill and melting at its lower end. As it moves, it pushes forward a pile of debris consisting of material that the ice has scraped or dug from its bed plus rock fragments that have fallen on the ice from the valley walls. Would you expect that the debris pile (moraine) would show any sorting or rounding of fragments? The material of the moraine is in part worked over by streams of melt-water coming from the glacier; if the streams deposit some of the material within a short distance of the moraine, how would this material differ from that in the moraine itself?
4. Consider a bed of volcanic ash formed by material settling out of an eruption cloud. Would you expect the ash to show cross-bedding? Graded bedding? Explain.
5. You are told that material exposed in a roadcut is part of an old landslide. What characteristics (sorting, rounding, composition, and so on) would you expect the material to show?
6. A turbidity current in a lake or the ocean is a rapidly moving tongue of water and sediment in a state of violent commotion, the sediment remaining in suspension because of the extreme turbulence. When the current stops, the sediment must settle to the bottom. A common characteristic of such deposits is graded bedding. Explain how the grading is produced.
7. On the continental shelf along a straight coastline, what (ideal) distribution of gravel, sand, and mud would you expect to find? What kind of sediment would you expect to find on the continental slope? Why is this ideal distribution so often not found?
8. The amount of sand on a beach is determined by a balance between the rate of supply (from streams, wave erosion, and longshore currents) and the rate of removal by wave action. Any interference with supply, for example by construction of a breakwater, may change the amount of sand at a particular place drastically and unexpectedly. Suppose that a breakwater perpendicular to the shore is constructed on a north-south coast where the principal supply of sand is a longshore drift from the north. Where would the beach accumulate more sand than normal? Where less? Where, then, would there be increased danger of wave erosion on the shore?
9. Draw cross-sections to show (a) an unconformity representing erosion of granite followed by sedimentation, and (b) an unconformity representing erosion of tilted sedimentary strata followed by sedimentation.
10. The sedimentary bed immediately above an unconformity is often a conglomerate ("basal conglomerate"). Can you suggest a reason?

Chapter Six

1. Suppose that the sediment accumulating at present in one of the lakes of northern Minnesota is buried by later sediments, and then a million years from now is partly exposed by erosion. What kind of fossils might geologists of the future find in the old lake sediments? How complete a record would there be of the present life in and around the lake?
2. You are studying a Jurassic limestone in Nevada, and in some outcrops you find abundant fossil remains of corals clustered in large colonies. What conclusions can you draw about conditions in this part of Nevada during the Jurassic Period?
3. Suppose that much of the coral material in Question 2 consists of silica, so that the fossils stand out on weathered surfaces because the silica is less soluble than the surrounding limestone. You conclude that the original calcium carbonate of the coral has been replaced by silica deposited out of solution in water that percolates through the rock. Why do you think this conclusion is more reasonable than to suppose that the living corals made their shells out of silica?
4. The evolution of the human race has been particularly difficult for geologists to trace, because fossil remains of man's prehistoric ancestors are very scarce. Can you suggest possible reasons for the scarcity?
5. One of the great mysteries of geologic history is the sudden extinction of the dinosaurs at the end of the Mesozoic Era. From their fossil remains we know that these creatures were abundant up to the very end of the era, but not one species of dinosaur survived into the Cenozoic. If you were speculating about possible reasons for this disappearance, what suggestions could you make?
6. You find a sandstone layer containing abundant shell fragments. Some of them you can recognize as broken pieces of clam shells, snail shells, and barnacles, but despite long search you find no complete, unbroken shells. What can you conclude about conditions under which the original layer of sand was deposited?
7. At the petrified forest in Yellowstone Park, flat-lying layers of tuff (partly consolidated volcanic ash) are exposed on a steep slope. Some of the layers have forms of logs and tree stumps well preserved, but other layers lack them. Under what conditions was this deposit formed?
8. The evolutionary development of insects and of birds is little known. Suggest a reason.
9. In the eastern part of the United States an unconformity separates beds of Permian age from beds of Triassic age, but in some parts of the western states there is a continuous sequence of marine sedimentary beds from Permian into Triassic. What does this difference mean with regard to the geologic history of this time interval? In some places where the sequence is continuous, geologists have difficulty deciding just where the division between the two periods should be placed. On what basis can such a decision be made?
10. The fossil record permits us to trace the development of living forms from the Cambrian Period to the present, roughly 600 million years. The fact that life

has been able to flourish on the earth's surface for this long period should give an indication of how constant the temperature and compositions of the ocean and atmosphere have remained. Roughly, what are the limits of temperature and composition that life can tolerate? Do you know of any object in the universe besides the earth on which conditions lie within these limits?

11. The Precambrian rocks of the Grand Canyon can easily be divided into two groups separated by the lower of the two unconformities (Fig. 5-23): unmetamorphosed sedimentary rocks above the unconformity, highly metamorphosed schists below. The former are sometimes assigned to a so-called "Proterozoic Era," and the latter to an "Archeozoic (or Archean) Era." Why is this division of the Precambrian into eras not suitable for a worldwide classification?

Chapter Seven

1. The crust of the earth will be defined later (Chapter 13) as an outer layer whose lower boundary is a level where the speeds of earthquake waves increase abruptly. This level is somewhat variable, but averages about 35 km under the continents and about 5 km under the floor of the ocean. What is the average pressure at the base of the crust under the continents and under the oceans? (Use 4000 meters as the average depth of the oceans.)
2. In a famous science-fiction story of the last century ("Journey to the Center of the Earth"), Jules Verne described a cave in which an exploring party found a route that took them close to the earth's center. Why would such a journey be impossible? Roughly, how far might you expect to go beneath the surface in natural caves?
3. Why is it inaccurate to speak of the "melting point" of a piece of basalt or a piece of granite?
4. When a flow of molten basalt moves over a gravel deposit, it sometimes picks up pebbles and boulders from the gravel and carries them long distances. What changes would you expect to see in a pebble of (a) shale (b) limestone (c) granite that had been carried in such a flow?
5. By what characteristics could you distinguish (a) slate from hornfels? (b) hornfels from limestone? (c) slate from schist? (d) marble from quartzite? (e) gneiss from diorite?
6. Some varieties of sandstone and schist consist essentially of the two minerals quartz and mica. The sandstone is said to be *layered* or *bedded* while the schist is *foliated.* Explain the meaning of these terms, and tell how you could distinguish the two rocks in the field.
7. If you take a trip by rubber boat down the Grand Canyon, you will see many outcrops of schist intruded by granite dikes, and in some places you will find both of these rocks overlain unconformably by layers of sandstone. Draw a cross-section to show these relations. Which of the three rocks is the oldest? Which the youngest?

8. Fossils are not common in metamorphic rocks, but are sometimes found. Would you expect the chance of finding them to be greater in foliated rocks or unfoliated rocks? Why?
9. Which would you expect to be more resistant to erosion, shale or slate? Quartzite or hornfels? Peridotite or granite?
10. Of the following rocks, which ones have been molten at some time in their history? andesite, hornfels, gneiss, limestone, chert, peridotite, gabbro, schist, marble, conglomerate.
11. If you were to obtain chemical analyses of the following rocks, which ones would you expect to have abundant silicon? Which would have abundant iron and magnesium? rhyolite, sandstone, limestone, basalt, chert, peridotite, obsidian.
12. Suppose that you find an outcrop showing a layer of basalt between two layers of sandstone. Two interpretations are possible: (1) the basalt might be a lava flow that moved over one of the sandstone beds, and then the other sandstone was deposited on top of it, or (2) the basalt might be a sill that was intruded between the two sandstone beds. What evidence would you look for to decide which interpretation is more probable?

Chapter Eight

1. Gabbro is an intrusive igneous rock consisting chiefly of two kinds of minerals: plagioclase feldspar (sodium-calcium aluminum silicate) and a ferromagnesian mineral (most commonly the variety called *pyroxene,* a silicate of calcium, magnesium, and iron). How would the appearance of the rock change as it weathers? What are the principal products of complete weathering in a humid temperate climate?
2. Weathering along joints in a sandstone often produces a change in color, and if the joints are closely spaced and approximately parallel the color bands can easily be confused with bedding. If you are examining such a rock in the field, how can you decide whether the banding represents bedding or is only discoloration due to weathering?
3. Weathering and erosion can reduce any kind of rock to fragments the size of pebbles, but the pebbles of some kinds of rock are much more durable in sedimentary environments than those of other kinds. Of the following kinds of rock, which one or ones would you expect to find as abundant pebbles in a conglomerate? Which would be very rare or absent in most conglomerates? limestone, chert, tuff, quartzite, basalt, diorite, slate, gypsum, shale.
4. Soil scientists speak of a vertical section through a soil as a *soil profile,* and designate the different layers as *horizons:* the A horizon at the top, generally containing much organic matter and at least partly leached of soluble constituents by rainwater percolating downward; the B horizon, in which some of the material from the A horizon is deposited, especially clay and iron oxides; and the C horizon at the bottom, consisting of rock only slightly weathered and which grades downward into fresh rock. In what sort of topographic and

climatologic environments would you expect to find soil profiles well developed? In soils developed by the weathering of granite, how would you expect the A and B horizons to change as you investigate soils in a warm, humid climate, then those in a cool, moderately humid climate, and then those in a desert climate?

5. In a sequence of sedimentary beds an unconformity represents an old surface of erosion, and on such a surface the weathered material must at one time have formed a soil. Would you expect to find remnants of such old soils commonly preserved along unconformities? If you find a layer that you think might represent an old soil, how could you decide whether it is actually a buried soil or just another layer in the sedimentary sequence?
6. Suppose that you find exposed in a roadcut material consisting of rock fragments ranging in size up to big boulders embedded in a fine-grained matrix. Such material might have formed in several ways: as a mudflow, as an alluvial-fan deposit, as till in a glacial moraine, as pyroclastic debris from a volcanic eruption. What would you look for in the deposit to help you decide which of these origins is the most probable?
7. Lakes are common in mountain valleys that were occupied by glaciers during the Pleistocene, but very uncommon in valleys that have been eroded only by streams. Explain.
8. Very rugged mountain topography, with vertical cliffs, steep narrow ridges, and sharp pinnacles, indicates that valley glaciers have been active in the fairly recent geologic past. Explain how glacial erosion can produce such topography.
9. Parts of the north central states and much of central and eastern Canada are characterized by numerous lakes and swamps and by river systems that do not show the usual well organized dendritic patterns. In very general terms, how long would you expect such features to persist (thousands of years, tens of thousands, millions . . .)? What sort of changes would you expect to take place?
10. In parts of the southern continents there is good evidence for the widespread existence of glaciers long before the Pleistocene, roughly about 250 million years ago (Permian Period). Of the usual kinds of evidence for past glacial activity, which would you expect to be still recognizable in a region where glaciation is as old as this?
11. A real estate developer is planning a new subdivision in a hilly area where the underlying rock consists of alternating beds of sandstone and shale with a general dip of about 40° to the east. What advice could you give him about the location and construction of roads and houses so as to minimize the danger of landslides? What landscape features would you look for in order to spot places where sliding is most probable?
12. Meteor impact is an important process of erosion on the moon but not on the earth. Why?
13. A common rock called *dolomite,* which resembles limestone in appearance, has the composition $CaMg(CO_3)_2$ (calcium magnesium carbonate). By analogy

with the behavior of limestone, what guess could you make about the reactions involved in the weathering of dolomite? What substances would be found in solutions resulting from this weathering process?

14. *Caliche* is a deposit of white material, usually consisting chiefly of calcium carbonate, that forms at the surface of the soil or as a layer a short distance below the surface. Why should caliche be common in arid regions but rare or absent in soils of humid regions?

Chapter Nine

1. In hand specimens both basalt and limestone appear to be very impermeable, yet springs are not uncommon in areas underlain by these kinds of rock. Explain.
2. Wells drilled in the lower part of an alluvial fan are often found to be artesian wells. Why should this be true?
3. Draw a diagram to show the relation of the water table to lakes and streams in a region of fairly humid climate. In arid parts of the western states this relation commonly does not hold. Explain.
4. Would you expect to find perched water tables in a region underlain largely by limestone? By massive sandstone? By granite? By the floodplain deposits of a large river? Explain.
5. Fresh-water springs are sometimes found emerging from the sea bottom at short distances offshore. Draw a diagram to show geological conditions that would make this possible.
6. *Travertine* is a variety of calcium carbonate that accumulates near springs, and that often has a layered structure which makes it suitable for polishing as a decorative stone. If a spring emerges from a layer of limestone, describe the chemical reactions that lead to the deposition of travertine around the mouth of the spring. Why is travertine generally more abundant near hot springs than near cold springs?
7. Why are springs commonly found along faults?
8. Detailed maps of western Kentucky have the peculiarity that many streams appear to flow into depressions and simply stop. What does this behavior suggest about the underlying rock?
9. Disposal of highly radioactive waste materials is a serious problem in the development and maintenance of nuclear power reactors, because such materials must be kept out of any possible contact with living things for many hundreds of years. One suggestion is to bury the radioactive material in caverns excavated in granite. Would this provide complete isolation, or would some of the radioactive elements eventually get into groundwater? (There isn't a yes-or-no answer to this question. In discussing it, remember that the times involved are very long.)
10. Large springs are more common in areas underlain by basalt flows than in areas underlain by granite. Why?

Chapter Ten

1. An anticline and syncline involving a series of beds of various kinds of sedimentary rocks have been eroded to a nearly level surface. Sketch a geologic map of the surface, showing the beds by different patterns and indicating the structure by many dip-and-strike symbols.
2. Sketch a geologic map of an eroded anticline whose axis is tilted toward the north—a so-called *plunging anticline.* The rock layers, in other words, have been bent into an arch, the arch has been tilted, and then the region has been eroded to a nearly level surface. Use enough dip-and-strike symbols to make the structure clear.
3. Now suppose that the plunging fold of Question 2 is cut by a vertical fault in an east-west direction, so that the part of the fold north of the fault has moved down a few feet with respect to the part south of the fault. Again make a geologic map of the eroded surface.
4. Sketch a geologic cross-section through a series of beds cut by a normal fault and by an andesite dike. Suppose first that the dike is younger than the fault, and then make another sketch in which the dike is older than the fault.
5. Suppose you are studying the geology of one of the long Appalachian ridges. How could you prove that the ridge is part of a large eroded fold rather than a fault-block range like one of the ranges in Nevada?
6. In a roadcut you see sedimentary beds toward the top of the cut separated from beds at the bottom by a surface with a gentle dip, say about 10°. Two possibilities come to mind: the surface may be an unconformity representing erosion of the lower beds before the upper ones were deposited, or it may be a thrust fault along which the upper beds have moved over the lower ones. What evidence would you look for to decide between these alternatives?
7. Columnar jointing is a common phenomenon in basalt, but is never observed in granite. Explain.
8. A large strike-slip fault like the San Andreas is often marked in the topography by a conspicuously long, straight valley. Why should such valleys develop along this kind of fault?

Chapter Eleven

1. Is the fault in Fig. 10-2 older or younger than the fold?
2. Arrange in order from oldest to youngest the geologic events shown by the cross-section in Fig. 11-10.
3. Along the Rocky Mountain front in Glacier National Park at several points Precambrian sedimentary rocks can be seen on top of Cretaceous shales, with a thrust fault between them. From this information alone, how closely can you date the faulting movement?
4. What radiometric method or methods could you use to date (a) a Precambrian diorite? (b) a tree stump preserved in a late Pleistocene till? (c) a Pliocene rhyolite? (d) a sample of rock brought back from the moon?

5. Why is it generally impossible to date directly by radiometric methods a limestone? a shale? a quartzite?
6. The radioactive decay of uranium gives both helium and lead, but for age measurements the uranium/lead ratio is used in preference to uranium/helium. Why? The decay of potassium-40 gives both argon and calcium, but potassium/argon ratios are used in preference to potassium/calcium. Why?
7. What assumptions are involved in the use of uranium/lead ratios for finding geologic ages?
8. Uranium/lead measurements made on weathered uranium minerals generally give different results from those made on fresh material. Why should this be?
9. Why can't the carbon-14 method be used for dating materials older than about 40,000 years?

Chapter Twelve

1. Contrast the shapes of a typical shield volcano (e.g. Mauna Loa in Hawaii), a typical composite volcano or *stratovolcano* (e.g. Fuji-san in Japan or Mt. Rainier in Washington), and a typical volcanic plug or dome (e.g. Lassen Peak). Explain why the shapes are different.
2. Welded tuffs cover large areas in some of the western states, and in earlier geological surveys many of them were mapped as "rhyolite" or "dacite." Why was this mistake so common? Can you suggest criteria by which you might be able to distinguish a welded tuff from a lava?
3. Many large volcanoes have at their tops a huge basin or "crater," much larger than the actual crater from which eruptions take place. Such a basin, called a *caldera,* is outlined by curving faults, and evidently originates by the collapse or downward dropping of the inner part of the volcanic structure. Crater Lake in Oregon is a good example of a caldera, except that the filling with water largely conceals the bounding faults. From the general mechanism of volcanic eruptions, can you suggest a reason why calderas might form?
4. Volcanoes and major earthquakes are popularly supposed to be closely related. Cite evidence indicating that the relationship in general cannot be very close.
5. Many of the mountain ranges in the Great Basin are described as fault-block mountains, despite the fact that no earthquakes have occurred along their bases in historic time. What is the evidence that these ranges have formed in large part by slow uplift of the mountains with respect to adjacent valleys along normal faults?
6. Why is magnitude, for scientific purposes, a more satisfactory measure than intensity by which to express the size of an earthquake?
7. If the seismogram of an earthquake recorded at a given station shows a time difference of 5 minutes between the first arrivals of P and S waves, how far was the earthquake from the station? What additional information is needed to determine the location of the earthquake epicenter?
8. In the course of geologic mapping you find that rocks are offset along a fault. What would you look for in order to decide whether the fault is active—in

other words, whether earthquakes may be expected to occur every few years or decades, or whether movement on the fault ceased millions of years ago?

9. Volcanoes in which the lava is principally basalt usually differ in both shape and kind of eruption from volcanoes whose lava is chiefly rhyolite. What are the principal differences, and to what characteristics of the lavas are they related?

Chapter Thirteen

1. Travel-time lines for earthquake waves (plots of first-arrival times against distance from the epicenter, as in Fig. 13-3) are curves rather than straight lines. Why should this be? What information can be obtained from the amount of curvature?
2. Gabbro and peridotite are both dark-colored intrusive igneous rocks. How do they differ in (a) appearance, (b) mineral composition, (c) chemical composition?
3. When a major earthquake sets up waves that travel completely through the earth, seismographs in a zone between about 102° and 143° from the epicenter record neither P waves nor S waves (except feeble ones that have followed complicated paths involving reflection and refraction from various discontinuities). This zone is called the *shadow zone* (Fig. 13-7). Why are S waves absent from this zone? Why are P waves absent? (In answering this last question, note that the speeds of P waves are considerably slower in molten iron than in solid silicate rocks.)
4. What radioactive elements are chiefly responsible for heat production in the earth's crust? For one of these elements, name a mineral in which it is a major constituent, and name an important daughter-product of its radioactive decay.
5. We have no direct information about the composition of the interior of the planet Mars, but a good guess would be metallic iron and ferromagnesian silicates as major constituents. What is the basis for such a guess?
6. In determining past orientations of the earth's magnetic field, why is basalt a better rock to use than rhyolite? Than shale? Than mica schist?
7. A rock like chert or limestone, which in the laboratory is apparently brittle and breaks under stress, is often found in nature in thin beds which are intricately folded, so that the rock must have been capable of plastic flow. How can you account for this apparent discrepancy?
8. A high mountain range adjacent to a plain (like the Front Range of the Rockies in central Colorado) should seemingly be a greater weight on the asthenosphere than a plains area of comparable size. Yet if this is true, the material of the asthenosphere should slowly flow from under the mountains to under the plain until the mountain area and plains area stand at the same level and their weights are therefore equalized. What can you suggest about the nature of the materials making up the mountains and the plain that might prevent this from happening?

9. The persistence of mountains and plains, as noted in Question 8, requires that there be a difference of some sort in the materials, or in the amount of material, below the surface of the two kinds of topography. Now apply the same sort of reasoning to still larger features, the continents and the ocean basins. Can you suggest a correlation between the average thicknesses of the crust above the Moho (5 km for ocean basins, 35 km for continents) and the nature of the materials making up the two kinds of crust—on the assumption that over broad segments of the earth weights of equal areas resting on the asthenosphere must be the same? (This assumption, that major blocks of the crust must be in a state of balance because otherwise material of the upper mantle would flow from heavier blocks to lighter blocks, is called the hypothesis of *isostasy* or *isostatic equilibrium.*)

Chapter Fourteen

1. What are two major pieces of evidence for the conclusion that the sea floor on either side of a mid-ocean ridge is moving in opposite directions away from the ridge?
2. Evidence from sea-floor spreading indicates that the mantle is a plastic material capable of flow, while evidence from earthquake waves indicates that mantle material is a rigid solid. How can this apparent contradiction be resolved?
3. Parts of South America were heavily glaciated near the end of the Paleozoic Era, about 250 million years ago. What sort of evidence would you expect to be preserved of such ancient glaciation? Why is this glaciation important in the theory of plate tectonics?
4. Along a transform fault that cuts a mid-ocean rift zone, the direction of displacement of segments of the rift is opposite to the direction of displacement indicated by present-day motion of the two sides of the fault. Explain.
5. Suggest an explanation for the observed fact that epicenters of deep-focus earthquakes are concentrated on the continental side of island arcs, but are rare elsewhere. What other evidence can you cite to support your interpretation?
6. How does the shape of seamounts provide evidence that various parts of the ocean floor have subsided by different amounts?
7. The continental shelf along the east coast of South America is much wider than the shelf on the west coast. Suggest an explanation.
8. Beds of chalk and beds of diatomite were both formed from oozes that accumulated on the sea floor as the remains of tiny organisms, and the two kinds of rock are very similar in appearance. By what simple test could you tell them apart?
9. How is it possible to measure the rate at which different parts of the sea floor are moving away from the mid-ocean ridges? What actual measurements are necessary, and what is the sequence of reasoning from the measurements to rates of movement?
10. The oldest fossils found in sediments of the sea floor date from the Jurassic

Period. In what parts of the oceans would you expect these Jurassic fossils to be found? Do you think there is a good chance that older fossils will someday be found as the sea bottom is more intensively explored?

Chapter Fifteen

1. The San Andreas fault in California is described as a strike-slip fault and also as a transform fault. What do the two terms mean? Is there any contradiction involved in applying both terms to the same fault? What is the evidence that justifies applying each term to the San Andreas fault?
2. Explain, in terms of plate-tectonic theory, why earthquakes are so much less frequent on the east coast of the United States than on the west coast.
3. What is meant by calling the margin of the Pacific Ocean a "circle of fire"? In terms of plate-tectonic theory, why is this term justified?
4. Much of geologic reasoning is embodied in the simple rule "the present is the key to the past"—meaning that geologic processes have acted in much the same way all through earth history, so that past events can be interpreted by careful study of present-day processes. The rule cannot be taken too literally, however, because the geologic record indicates that some events and some conditions of the past were different from those in the modern world. Point out some characteristics of pre-Mesozoic rocks which show that conditions in parts of the Precambrian and Paleozoic were quite different from conditions today.
5. In terms of plate-tectonic theory, why should there be a concentration of active volcanoes along the Mid-Atlantic Ridge? Why a concentration along the west coast of South America? Why are there no active volcanoes in eastern Canada?
6. The idea of drifting continents was generally regarded with skepticism during the first half of this century, chiefly because no force was known large enough to make the continents move. Now continental drift is widely accepted, although we still are not certain of the mechanism. Why did skepticism change to acceptance in a period of just a few years?
7. Recent intensive study of the topography of the moon and Mars has provided no evidence of plate-tectonic movements on these bodies, either present or past. Can you suggest reasons why such movements might be absent on the moon and Mars?

Chapter Sixteen

1. You are looking for gravel deposits along the channel of a meandering stream. Remembering that the size of pebbles carried by running water depends in large part on its speed, in what parts of the channel would you expect to find the coarsest gravel? In what parts would you expect the sediment to be largely fine-grained silt and clay? Sketch a map of the river to show these locations.
2. Weathering of granite ordinarily leads to the formation of sand and clay. In what minerals of the original granite did most of its aluminum occur? Into what mineral or minerals does most of the aluminum go during weathering?

The weathered material generally contains a higher percentage of aluminum than the original granite; why should this be? In places where weathering is especially rapid and complete, the chief ultimate product may be simple hydroxides of aluminum, and these may constitute valuable aluminum ore (*bauxite*). In what parts of the earth would you expect this kind of aluminum ore to be forming? Why?

3. Give two kinds of evidence supporting the hypothesis that some ore deposits are related genetically to igneous rocks or igneous processes. By what mechanism or mechanisms would it be possible for igneous activity to concentrate ore metals?
4. Would you expect any of the motions involved in plate-tectonic theory to play a role in concentration of metallic ores? As an example, could the famous copper deposits of Peru and northern Chile be related to the postulated subduction zone along this coast?
5. A sandstone to be used for making glass must consist of almost pure quartz. By what natural processes can a sandstone of such purity be produced?

Chapter Seventeen

1. A major problem in developing nuclear power is the difficulty of disposing of radioactive wastes. If the waste material is buried underground, its harmful radiation is in large part absorbed by rock and soil. Why, then, can't the problem be solved simply by burial of all such wastes wherever it seems convenient?
2. As reserves of petroleum and natural gas are depleted, other sources of liquid and gaseous fuels will have to be used. Name three other kinds of abundant naturally occurring material from which such fuels can be prepared. Why are these sources not in common use at present?
3. Exploration by seismic methods (earthquake waves produced by artificial explosions and recorded on many small seismographs) has been much more successful in locating oil fields than in finding ore deposits. Suggest a plausible reason.
4. A common structure in which petroleum may be found is a dome or closed arch of sedimentary strata. Suppose that you are making a geologic map of an area in which such a dome is located, and that the area has been eroded to a nearly level surface on which there are numerous rock outcrops. (a) Show by a sketch what your completed map would look like, using symbols to show dips of beds; indicate on the sketch which bed is youngest and which is oldest. (b) Indicate on your map places where a vertical drillhole might strike oil, where it might strike gas, and where you might encounter only salt water.
5. It is a generally accepted hypothesis that petroleum and natural gas originate by partial decay of buried organic matter, but a few geologists prefer the hypothesis that the petroleum hydrocarbons form by inorganic reactions from gases containing carbon which continually escape from the earth's interior. What geologic evidence makes this hypothesis seem improbable?

Index

Index

The Index is supplemented by the Glossary, pages 476–494.